W0051127

PREPARATION OF NUCLEAR TARGETS

IFI DATA BASE LIBRARY

Recent volumes in the series:

BEHAVIORAL DEVELOPMENT OF NONHUMAN PRIMATES
An Abstracted Bibliography
Faren R. Akins, Gillian S. Mace, John W. Hubbard, and Dianna L. Akins

MASTER TABLES FOR ELECTROMAGNETIC DEPTH SOUNDING INTERPRETATION
Rajni K. Verma

SPECTROSCOPIC REFERENCES TO POLYATOMIC MOLECULES
V. N. Verma

RAPIDLY QUENCHED METALS
A Bibliography, 1973–1979
C. Suryanarayana

MOSFET TECHNOLOGIES
A Comprehensive Bibliography
A. H. Agajanian

PARENT–CHILD SEPARATION
Psychosocial Effects on Development
An Abstracted Bibliography
Faren R. Akins, Dianna L. Akins, and Gillian S. Mace

THE BEREAVED CHILD
Analysis, Education, and Treatment
An Abstracted Bibliography
Gillian S. Mace, Faren R. Akins, and Dianna L. Akins

THE HOSPITALIZED CHILD
Psychosocial Issues
An Abstracted Bibliography
Dianna L. Akins, Gillian S. Mace, and Faren R. Akins

PREPARATION OF NUCLEAR TARGETS
A Comprehensive Bibliography
Jozef Jaklovsky

PREPARATION OF NUCLEAR TARGETS

A Comprehensive Bibliography

Jozef Jaklovsky

New England Nuclear Corporation
Boston, Massachusetts

SPRINGER SCIENCE+BUSINESS MEDIA, LLC

Library of Congress Cataloging in Publication Data

Jaklovsky, Jozef.
Preparation of nuclear targets.

(IFI data base library)
Bibliography: p.
Includes index.
1. Targets (Nuclear physics) – Bibliography. I. Title. II. Series.
Z7144.N8J34 [QC787.T35] 016.5397'3 81-12014
ISBN 978-1-4684-9992-6 ISBN 978-1-4684-9990-2 (eBook) AACR2
DOI 10.1007/978-1-4684-9990-2

© 1981 Springer Science+Business Media New York
Originally published by IFI/Plenum Data Company in 1981
Softcover reprint of the hardcover 1st edition 1981

All rights reserved

No part of this book may be reproduced, stored in a retrieval system, or transmitted, in any form or by any means, electronic, mechanical, photocopying, microfilming, recording, or otherwise, without written permission from the Publisher

PREFACE

This comprehensive bibliography (6248 entries) is the first of its kind in the general field of nuclear target preparation for use with particle accelerators. The dates covered are 1936 to June 1980. The bibliography includes thin and thick particle accelerator targets as well as the preparation and use of targets in particle accelerators. The entries in the bibliography are arranged in alphabetical order of authors' names. For ease of use, complete subject, country, and patent indexes have been added.

A special expression of gratitude is owed to my wife, Jolana, and I also wish to thank Bill Evans and Stephen Talutis for helping me to organize this bibliography.

Jozef Jaklovsky

CONTENTS

Abbreviations

AAAC - Australia Atomic Energy Commission
ACNTDS - Chalk River Canada, Nuclear Target Development Soc. Proc.
AECL - Atomic Energy of Canada Ltd., Chalk River
AEET - Atomic Energy Establishment, Trombay, India
AERE - Atomic Energy Research Establishment, Harwell, England
ANL - Argonne National Laboratory, Argonne, Ill., U. S. A.
ANU - Australia National University, Canberra

BMFT-FBK - Federal Republic of Germany
BNL - Brookhaven National Laboratory, Upton, N. Y., U. S. A.
BNWL - Battelle Pacific Northwest Labs, Richland, Wash., U. S. A.
BONN-HE - Bonn University, FRG.

CEA - Cent. Etud. Bruyeres-Le-Chatel, Montrouge, France
CEN (CENBG) - Cent. d'Etudes Nucleaires; Bordeaux - 1 Univ., France
CEA (SC) - Commissariat a L'Energie Atomique, Grenoble, France, Centre d'Etudes Nucleaires
CERN - European Organization for Nuclear Research, Geneva, Switzerland
CMU - Carnegie Mellon University, Pittsburgh, Pa., U. S. A.

DESY - Deutsches Elektronen - Synchrotron, Hamburg, FRG.

GSI - Gesellschaft Fuer Schweriunenforschung m. b. H., Darmstadt, FRG.

HW - Hanford, Richland, Wa., U. S. A.

IA - Israel Atomic Energy Commission, Yavne
IAE - Institut Atomnoi Energy, SSSR, Moscow
IEEE - Institute Electric Electronic Engineering
INP - Institute of Nuclear Physics, Krakow, Poland
INR - Institute of Nuclear Research, Warsaw, Poland
INS-TL - Inst. Nucl. Study, Univ. Tokyo, Tanashi, Japan
IFVE-SEF - Gosudarstvennyi Komitet PO Ispol'zovaniyu Atomnoi Energii SSSR, Serpukhov Institut Fiziki Vysukikh Energii

INTDS - International Nuclear Target Development Society
INTDS - Proceedings 1974, Chalk River, Canada, Oct. 1-3
INTDS - Proceedings 1975, Argonne, U. S. A., Sept. 30 - Oct. 2
INTDS - Proceedings 1976, Los Alamos, U. S. A., Oct. 19-21
INTDS - Proceedings 1977, Berkeley, U. S. A., Oct. 19-20
INTDS - Proceedings 1978, Garching, FRG, Sept. 11-14
INTDS - Proceedings 1979, Boston, U. S. A., Oct. 1-3
ITEF - Institut TeOreticheskoi 1 Eksperimentalno Fiziki, Moscow, USSR

JAERI - Jap. At. Energy Res. Inst., Tokyo
JEN - Junta de Energia Nuclear, Madrid, Spain
JINR - Joint Inst. for Nuclear Research, Dubna, USSR

KAPL - Knolls Atomic Power Lab., Schenectady, NY, U. S. A.

LAL - Laboratoire de l'Accelerateur Lineaire, Univ. de Paris, Orsay, France
LASL(LA) - Los Alamos Scientific Labs., U. S. A.
LBL - Liverpool Univ., England
LNF - Lab. Nazionali di Frascati, Italy

MIT - Mass. Inst. of Tech., Cambridge, U. S. A.
MLM - Mound Lab. Miamisburg, OH, U. S. A.

NASA - National Aeronautics and Space Administration, U. S. A.
NRL - Naval Research Lab., Washington, D. C., U. S. A.

OU(OULNS) - Osaka University, Japan (Lab. of Nuclear Studies)
ORNL - Oak Ridge National Laboratory, Oak Ridge, U. S. A.

PAEC - Philippine Atomic Energy Commission, Manila
PSPSITF - Proceedings of the Seminar on the Preparation and Standardization of Isotopic Targets and Foils held at AERE, Harwell, England on Oct. 20-21, 1965
PTB - Physikalish-Technische Bundesanstalt, Braunschweig, FRG
PURC(PUC) - Princeton University, NJ, U. S. A.

RFP - Rocky Flats Div., Dow Chem., Golden, CO, U. S. A.
RPI - Rensselaer Polytechnic Inst., Troy, NY, U. S. A.
RL(RHEL) - Rutherford Lab., Sci. Res. Counc., Chilton, England

SAND(SC) - Sandia Lab., Albuquerque, NM, U. S. A.
SLAC - Stanford Linear Accelerator Center, Stanford Univ., CA, U. S. A.

TEICOTA - The Eighth International Conference on Cyclotrons and Their Applications, Indiana Univ., U. S. A.

TISRMNM - Third International Symposium on Research Materials for Nuclear Measurements, Gatlinburg, TN, U. S. A., Oct. 5-8, 1971

UCRL - California Univ., Lawrence Livermore Radiation Lab., U. S. A.
UJF - Ustav Jad. Fyz., Czechoslovak Akad. Ved., Rez, Czechoslovakia
UR - Rochester Univ., NY, U. S. A.
USAEC - U. S. At. Energy Comm.
USC - University of Southern California, Los Angeles, U. S. A.
USSR - SSSR

BIBLIOGRAPHY

1. AECL, Chalk River Nuclear Labs., Ontario, Canada, 1980.
2. AECL-4428 (p-18), PR-P-96.
3. AECL-3257 (p-83), PR-P-80, October 1 - December 31, 1968.
4. AECL, (p-79), PR-P-75, Report July 1, 1967 to September 30, 1967.
5. AECL-2201 (p-49), PR-P-64, October 1 to December 31, 1964.
6. ANL, MSD 212 H-118, 9700 S. Cass Avenue, Argonne, IL 60439.
7. ANL-7481, Contract W-31-109-eng-38, p-190, April 1, 1967 - March 31, 1968.
8. ANL-7355, Contract W-31-109-eng-38, p-86, April - June 1967.
9. ANL-7310, pp-377-430, Argonne National Lab., IL.
10. ANL, CAPE-1967.
11. ANL-7210, Contract W-31-109-eng-38, p-446, Dec. 1966.
12. ANL/HEP-7208, Vol. 3, Contract W-31-109-eng-38, p-456, June 3 - August 5, 1971.
13. Arizona Carbon Foil Company, 2239 East Kleindale Road, Tuscon, AZ 85719, U. S. A.
14. Annu. Book ASTM Std. (1975) (No. 45), p-162, ASTM-C-625-72.
15. AERE, Harwell, England, 1968, p-28.
16. AERE, Harwell, England, 1968, p-28, NP-18012.
17. BNL, CAPE-1440.
18. BNL-50042, Contract AT(30-2)-Gen-16, p-78, 1966.
19. BNL-325, 1966, Neutron Cross Sections, Vol. 11C.
20. BMwF-FBK-68-22, Annual Report, 1966, Bonn Univ., May 1968, p-123.
21. Buckbee Mears Co., St. Paul, MN.
22. Calif. Rsch. and Develop. Co., Livermore Rsch. Lab., Livermore, CA, Quarterly Progress Report (FOR) July through September 1953, Apr. 1954, DecI. Mar. 2, 1957, p-68, Contract AT(11-1)-74.
23. Calif. Rsch. and Develop. Co., Livermore Rsch. Lab., Livermore, CA, Contract AT(11-1)-74, April through June 1953.
24. Calif. Rsch. and Develop. Co., Livermore Rsch. Lab., Livermore, CA, Contract AT(11-1)-74, September through November 1952.
25. Carnegie Inst. of Tech., Pittsburgh, PA, Progress Report, 1963-1964, Contract AT(30-1)-844, p-51, July 31, 1964.
26. CEA-N-1232, pp-81-159.
27. CERN Cour., July - Aug. 1979, Vol. 19(5), pp-193-195.

28. 10 CFR 20 Proposed Rule Making 44 FR 10388, 2/20/79, Notices, Instructions and Reports to Workers.
29. Colorado Univ., Boulder, Progress Report, November 1, 1967 - November 1, 1968, Contract AT(11-1)-535, p-101.
30. Colorado Univ., Boulder, Nov. 1, 1967, Contract AT(11-1)-535, p-93.
31. Chromium Corp. of America, Waterburg 20, CT, U. S. A.
32. Columbia Univ., NY, Progress Report, January 1 - December 31, 1970, Contract AT(30-1)-73, p-61.
33. Com. Naz. Energ. Nucl., Notiz.; 16: No. 4, 73-4, Apr. 1970, in Italian.
34. Daresbury Lab., Science Research Council, Daresbury (UK), DL/NSF/TM-35, Apr. 1978, p-29.
35. Duke Univ., Durham, NC, Nuclear Structure Lab., Dec. 31, 1967, Contract AT(40-1)-1067, p-55.
36. EUR-2641.d,f,e, Liege Univ., Belgium, 1966, p-372.
37. EUR-1815.e (Rev.), p-248, CONF-640206.
38. EUR-1815.e (Rev.), pp-186-8, Liege Univ.
39. EUR-1815.d,f,e, Liege Univ., 1964, p-188.
40. Ferrofluidies Corp., 144 Middlesex Tpke., Burlington, MA 01803.
41. Fulmer Research Inst., Ldt., Stoke Poges, England, Composite Report No. E.4680, (1966) p-32.
42. GSI-Jahresbericht 1977, J - 1 - 78, Mai 1978, GSI-Darmstadt, Germany.
43. GSI-73-7, CONF-730368-, Aug. 1973, p-244.
44. Handbook of Chemistry & Physics, 52nd Edition, The Chemical Rubber Company.
45. HEDL TME-73-31, Dec. 1972, Jan. - Feb. 1973, 7th Quarterly Progress Report.
46. HUX-2752-1, Harvard Univ., Cambridge, MA, Dec. 30, 1964, Contract AT(30-1)-2752, p-23.
47. IA-1082, pp-289-94.
48. IA-920, July - December 1963, p-196.
49. ICRP 2, Report of Committee II on Permissible Dose for Internal Radiation, Pergamon Press, London, 1959.
50. Istitute for Nuclear Study, University of Tokyo, Midoricho 3-2-1, Tanashi-City, Japan.
51. Instituut voor Kernphysisch Onderzoek, Amsterdam, 1976, pp-32-39.
52. Institute of Physical and Chemical Research, Tokyo, Japan, Progress Report 1969, Vol. 3, Dec. 1969, p-125.
53. Internal Report ISN (June 1976); Annual Reports, ISN (1973, 1974, 1975, 1976), France.
54. Kansas Univ., Contract AT(11-1)-1120, p-60, 1968-1969.
55. Kindermann & Co. GmbH Ochsenfurt, Germany.
56. Kobenhauns Universitet, Niels Bohr Institute, Riso 4000, Roskilde, Denmark.
57. LAL-1136, Octobre 15, 1964 - Avril 15, 1965, p-89.
58. McMaster Univ., Hamilton (Ontario), Jan. 1965, p-75.

59. Micro Matter Company, 197 34th Ave., East, Seattle, WA 98112.
60. Micro Matter Company, 197 34th Ave., East, Seattle, WA 98112.
61. Minnesota Univ., Minneapolis, John H. Williams Lab. of Nuclear Physics, Aug. 1968, Contract AT(11-1)-1265, p-224.
62. Minnesota Univ., Minneapolis, John H. Williams Lab. of Nuclear Physics, Aug. 1967, Contract AT(11-1)-1265, p-168.
63. MIT-2098-142, Nov. 1, 1964, Contract AT(30-1)-905, p-151.
64. MIT, May 1, 1959, Contracts AT(30-1)-2098, p-167.
65. MLM-1611, Feb. 1970, Contract AT-33-1-GEN-53, p-44.
66. MLM-1608, Nov. 15, 1969, Contract AT-33-1-GEN-53, p-48.
67. NEC Equipment Corp., Newton Highlands, MA.
68. NBS Circular 585, January 1958, U. S. Department of Commerce, NY.
69. NMI-1170, Nuclear Metals, Inc., Cambridge, MA, June 30, 1956, February 18, 1957, Contract AT(30-1)-1565, p-145.
70. Nuclear Structure Research Lab., Univ. of Rochester, Annual Report, 1973.
71. ORNL Isotope Sales, Materials, P. O. Box X, Bldg. 3037, Oak Ridge, TN 37830, U. S. A.
72. ORNL-5328, Oct. 1977, pp-168-175.
73. ORNL-5297, Sept. 1977, pp-84-112.
74. ORNL-4770, June 30, 1971, pp-176-81.
75. ORNL-4570, June 30, 1970, pp-27-30.
76. ORNL-4570, June 30, 1970, pp-201-4.
77. ORNL-4545, December 31, 1969, pp-124-37.
78. ORNL-4306, pp-22-45.
79. ORNL-3830, pp-140-56.
80. ORNL-3782, pp-144-52.
81. ORNL-5378, pp-46-59.
82. ORNL, CAPE-1188.
83. ORNL Catalog, Isotopes Target Center for all AECL Laboratories and Contractors.
84. ORNL, CAPE-1188.
85. ORNL, Isotopes Development Center 05 Program Review, May 26, 1970, pp-59-72.
86. OU-LNS-69-2, December 1968, p-51.
87. PAEC(A)AR651, 1964 - 1965, p-188.
88. Phys. Lett., B, Mar. 29, 1976, Vol. 61(3), pp-234-236.
89. Phys. Lett., B, Mar. 29, 1976, Vol. 61(3), pp-234-236.
90. PUC-937-307, April 1967 - April 1968, Contract AT(30-1)-937, p-42.
91. PURC(PUC), April 1965 - April 1966, Contract AT(30-1)-937, p-47.
92. Proceedings of the International Conference on Low and Intermediate Energy Electromagnetic Interactions, Dubna, USSR, February 7-15, 1967, Vol. 4. Moscow; Academy of Sciences of the USSR (1967), p-344, CONF-670203-(Vol. 4).

93. Proceedings of the American Nuclear Society National Topical Meeting, April 19-21, 1971, Augusta, GA, DuPont de Nemours (E. I.) and Co., Aiken, SC, Savannah River Lab., April 1971, Contract AT(07-2)-1, p-173.
94. Proceedings of Annual Symposia which have appeared regularly since 1961, Vacuum Microbalance Techniques (New York: Plenum).
95. Proceedings of the International Conference, Saclay, France, December 5-9, 1966.
96. Proceedings of the Conference on the Use of Small Accelerators for Teaching and Research held at the Oak Ridge Associated Universities, Oak Ridge, TN, April 8-10, 1968, CONF-860411, pp-426-437.
97. Purdue Univ., Lafayette, IN, March 29, 1968, Contract AT(11-1)-1420, p-90.
98. RPI-328-142, pp-68-116.
99. Rice Univ., Houston, TX, 1968, Contract AT(40-1)-1316, p-21.
100. Rockefeller Univ., NY, Feb. 14, 1978, Contract EY-76-C-2232, p-12.
101. RL(RHEL)/R-136, pp-131-51.
102. RL(RHEL)/R-144, pp-41-59.
103. RL(RHEL)/R-129, pp-45-77.
104. RPI, October 1966 - December 1966, Contract AT(30-3)-328, p-125.
105. RPI-328-97, pp-35-95.
106. SAND-79-0002, Feb. 1979, pp-97-110, Progress Report, Oct. 1977 - March 1978.
107. CEA-CONF-3180, 1975, 11, p-3, National Soviet Conference on Neutron Physics, Kiev, USSR, June 9, 1975.
108. Specialty Metals Division, Grove City, OH 43123, U. S. A.
109. Stanford Univ., CA, Dept. of Physics, March 15 - October 14, 1965, Contract Nonr-225(86), 1967, p-2.
110. Texas Univ., Austin, TX, U. S. A., Progress Report, April 1, 1977 - March 31, 1978, ORO/5224-2, Dec. 31, 1977, pp-3-16.
111. Texas Univ., Austin, TX, Center for Nuclear Studies, Progress Report No. 7, ORO-2972-80, Dec. 1968, Contract AT(40-1)-2972, p-54.
112. UCRL-18667, pp-343-80.
113. UCLA-34P106-1, May 10, 1966 - May 24, 1967, Contract AT(11-1)-34, p-29.
114. UCRL-2200, March 15, 1953 to April 15, 1953. May 1, 1953. Decl. Mar. 6, 1957, Contract W-7405-eng-48, p-18.
115. UCRL-2043(Del.), June, July, August, 1952. Dec. 8, 1952. Decl. with deletions Mar. 5, 1957, Contract W-7405-eng-48, p-148.
116. UCRL-1903(Del.), Mar. 5, 1957, Contract W07405-eng-48, p-183.
117. UCRL-1680(Del.), September, October, November 1951. Feb. 29, 1952. Decl. with deletions Mar. 5, 1957, Contract W-7405-eng-48, p-127.

118. UCRL, CAPE-1582.
119. UCRL, CAPE-1151.
120. UCRL, CAPE-983.
121. USC-136-146, Oct. 1968, Contract AT(04-3)-136, p-59.
122. USC-136-132, Oct. 1967, Contract AT(04-3)-136, p-59.
123. Universitat Munchen, 8046 Garching-AM, Coulombwall 1, FRG.
124. University of Arizona, Dept. of Physics, Tucson, AZ 25721, U. S. A.
125. Varian/NRC 3117 Vacuum Coating System manufactured by Varian, Palo Alto, Vacuum Division, 611 Hansen Way, Palo Alto, CA 94303, U. S. A.
126. Veeco Instruments, Inc., Kronos Thickness Monitor, Model QM321. Distributed by Veeco Instruments, Inc., Terminal Drive, Plainview, NY 11803.
127. Weizmann Institute of Science, Rehovot, Israel.
128. Washington Univ., Seattle, Nuclear Physics Lab., June 1967, Contract AT(45-1)-1388, p-114.
129. Aarts, H. J. M., Engelbertink, G. A. P., et al., INKA-Conf-79-004-050, Verh. Dtsch. Phys. Ges. (1979) (No. 3), pp-717-718.
130. Abdivaliev, A., Angelov, N., et al., JINR-R-1-12125, 1979, p-14.
131. Abe, K., DeLillo, T., et al., AIP Conf. Proc., 73, No. 12,85-98.
132. Abe, Y., AIP Conf. Proc. ISSN 0094-243X, (1978) (No. 47), pp-132-143.
133. Abe, T., Kasuya, K., et al., Inst. Plasma Phys., 79, (IPPJ-
134. Abegg, R., Falk, W. R., et al., AIP Conf. Proc. ISSN 0094-243X, (1978) (No. 47), pp-692-693.
135. Abegg, R., Birchall, J., et al., AIP Conf. Proc. ISSN 0094-243X, (1978) (No. 47), pp-614-615.
136. Abele, H. K., Glaessel, P., et al., Nucl. Instr. and Meth. 137(1), Aug. 15, 1976, pp-157-167.
137. Abele, H. K., Glaessel, P., et al., Proceedings of the Fourth Annual International Conference of the Nuclear Target Development Society, ANL/PHY/MSD-76-1, 1975, pp-117-128.
138. Abele, H. K., Glaessel, P., et al., Internal Report, JINR-1975.
139. Aberle, J. A., MacDonald, A. R., et al., AEC Access. Nos., 66, (AD 636185), p-19.
140. Abgrall, Y., Labarsouque, J., et al., Nucl. Phys., A, Nov. 2, 1976, 271(2), pp-477-494.
141. Abragam, A., and Borghini, M., Prog. Low Temp. Phys., 4: pp-384-449, 1964.
142. Abragam, A., Kirsch, J., et al., New York and London, Academic Press, 1967.
143. Abragam, A., from Conference on Polarized Targets and Sources, Saclay, France, Dec. 5-9, 1966.

144. Abragam, A., Bacchella, G. L., et al., Proceedings of the Conference on Nuetron Scattering 1103-9, 1976.
145. Abragam, A., AIP Conf. Proc., 79, 51, pp-1-14.
146. Abragam, A., Bouffard, V., et al., Seances Acad. Sci., Ser. B 1980, 290(10), pp-203-5.
147. Abramov, B. M., Dukhovskoy, I. A., et al., Nucl. Phys. B, ISSN 0029-5582, Sept. 24, 1979, 157(2), pp-189-196.
148. Abramov, B. M., and Smorodinskaya, N. Ya, Seminar on Nuclear Theory in ITEP, Sept. - Dec. 1978, ITEF-68(1979), 1979.
149. Abul-Magd, A. Y., and El-Abed, K. I., Prog. Theor. Phys. (Kyoto), Feb. 1975, 53(2), pp-480-488.
150. Abul-Magd, A. Y., and Simbel, M. H., Phys. Lett., B, ISSN 0370-2693, April 23, 1979, 83(1), pp-27-30.
151. Acerbi, E., Castiglioni, M., et al., Nucl. Instr. and Meth. 85, Aug. 1, 1970, pp-45-8.
152. Acerbi, E., Birattari, C., et al., The Seventh International Conf. on Cyclotrons and their Application, Zurich, August 19-22, 1975.
153. Acerbi, E., Bellomo, G., IEEE Trans. Nucl. Sci. V. NS-24, June 3, 1977; Proc. of the Part Accel. Conf., 7th: Accel. Eng. and Tech., Chicago, IL, March 16-18, 1977, pp-1112-1114.
154. Ackermann, A., Anders, B., et al., Proceedings on Nuclear Cross Section and Technology. Washington, District of Columbia, U. S. A., March 3, 1975.
155. Adair, H. L., Isotopes Development Center - 05 Program Review, Oak Ridge National Laboratory, 1970, pp-15-24.
156. Adair, H. L., J. Inorg. Nucl. Chem. 32, 1970, pp-1173-1181.
157. Adair, H. L., and Kobisk, E. H., 1962 Trans. 9th Nat. Vacuum Symp., Chicago: American Vacuum Society, pp-125-8.
158. Adair, H. L., Kobisk, E. H., et al., Third International Conf. on Electron and Ion Beam Science and Tech., ed. by R. A. Bakish, Electrochemical Society, Inc., NY, 1968, p-530.
159. Adair, H. L., ORNL-4339, 1969.
160. Adair, H. L., CONF-711002, Oct. 1971, pp-307-20
161. Adair, H. L., Nucl. Instr. and Meth., 72, 102(3), pp-599-610.
162. Adair, H. L., and Kuehn, P. R., Nucl. Instr. and Meth. 114, 1974, pp-327-332.
163. Adair, H. L., and Kobisk, E. H., 1973 Winter meeting of the American Nuclear Society, San Francisco, CA.
164. Adair, H. L., Nucl. Instr. and Meth., 73, 113(4), pp-545-8.
165. Adair, H. L., and Kobisk, E. H., ACNTDS; CRNL, Oct. 1-3, 1974.
166. Adair, H. L., and Kobisk, E. H., ACNTDS Proceedings 1974, CRNL, Oct. 1-3.
167. Adair, H. L., INTDS Newsletter, July 1979.
168. Adair, H. L., and Kobisk, E. H., INTDS Proceeding 1975, Argonne, U. S. A., Sept. 30 - Oct. 2.

169. Adair, H. L., Gibson, J. R., et al., INTDS Proceedings 1976, Los Alamos, U. S. A., Oct. 19-21.
170. Adair, H. L., INTDS Proceedings 1978, Garching, FRG., Sept. 11-14.
171. Adair, H. L., INTDS Proceedings 1977, Berkeley, U. S. A., Oct. 19-20.
172. Adair, H. L., INTDS Newsletter, Jan. 1979.
173. Adam, J., Kuklik, A., et al., Czech. J. Phys. ISSN 0011-4626, 1978, 28(8), pp-857-864.
174. Adam, I., Venos, D., et al., Conf. on Nucl. Spectroscopy and Atomic Nucl. Structure, 1978, p-118.
175. Adam, R. W., Z. Naturforsch. 23a, 1968, p-1526.
176. Adamov, V. M., Alexandrov, B. M., et al., Natl. Bur. Stand. (U. S.), Spec. Publ., Oct. 1977, No. 493, pp-313-318.
177. Adeev, G. D., Kazantseva, A. M., et al., Yad. Fiz., 1977, 26(4), pp-683-690.
178. Adel-Fawzy, Foertsch, H., et al., Annual Report 1978, Gemeinsamer Jahresbericht 1978, ZfK-385, Jun 1979, pp-17-18.
179. Adel-Fawzy, Foertsch, H., et al., Annual Report 1978, Gemeinsamer Jahresbericht 1978, ZfK-385, Jun 1979, pp-15-16.
180. Adelberger, E. G., McDonald, A. B., et al., Annual Reoprt 1977, RLO/1388-362, Jun 1977, pp-17-20.
181. Adler, D. B., and Adler, F. T., Trans. Am. Nucl. Soc. (Nov. 1976), 24, pp-450-452.
182. Adomssent, S., Isotopenpraxis 6, Jul. 1970, pp-232-4.
183. Adzhei, A. K., Galakhmatova, B. S., et al., Vestn. Mosk. Univ., Ser. III. Fiz. Astron. 12, No. 4, Jul-Aug. 1971, pp-393-7.
184. Aeystoe, J., and Valli, K., Nucl. Instr. and Meth. 111, No. 3, 1973, pp-531-537.
185. Afonin, I. P., Babykin, M. V., et al., Proc. of the Symp. on Eng. Probl. of Fusion Res., 7th, Knoxville, TN, Oct. 25-28, 1977, Publ. by IEEE, NY, NY, 1977, 1, pp-269-272.
186. Afanas'ev, N. G., Savitskij, G. A., et al., Conf. on Nucl. Spectroscopy and Atomic Nucl. Structure, Leningrad, Nauka. 1978, p-256.
187. Agababyan, N. M., Atayan, M. P., et al., JINR-R-1-11158, 1977, p-8.
188. Agarwal, N. K., and Haubold, A. D., 1977 Thin Solid Film pp-299-308.
189. Agarwal, S., Galin, J., et al., Nucl. Phys., A, Dec. 12-19, 1977, 293(1-2), pp-230-247.
190. Agranovich, V. L., Antuf'ev, Yu. P., et al., KFTI-77-9, 1977, Voprosy Atomnoj Nauki I Tekhniki No. 2(19).
191. Ah-Hot, L., Baron, E., et al., Int. Cyclotron Conf., St. Catherine's Coll., Oxford, Sept. 17-20, 1969.

192. Ah-Hot, L., Baron, E., et al., Int. Cyclotron Conf., St. Catherine's Coll., Oxford, Sept. 17-20, 1979.
193. Ahmad, I., and Khan, Z. A., Nucl. Phys., A, Dec. 21, 1976, 274(3-4), pp-519-524.
194. Ahmad, I., International Centre for Theoretical Physics, Trieste (Italy), IC-77-24, Mar. 1977, p-12.
195. Ahmad, I., International Centre for Theoretical Physics, Trieste (Italy), IC-77-27, Apr. 1977, p-22.
196. Ahmad, I., and Khan, Z. A., Proc. of the Nucl. Phys. and Solid State Phys. Symposium (held at) Pune, Dec. 26-30, 1977, 20B.
197. Ahmad, I., Sjoblom, R. K., et al., Phys. Rev., C, June 1978, 17(6), pp-2163-2175.
198. Ahmad, I., and Khan, Z. A., Phys. Scr. ISSN 0031-8949, July 1979, 20(1), pp-26-28.
199. Ahmad, S. S., and Robson, B. A., Sixth AINSE Nucl. Phys. Conf. Feb. 9-11, 1976.
200. Ahmad, S. S., Barrett, R. F., et al., Nucl. Phys., A, Feb. 9, 1976, 257(3), pp-378-388.
201. Ahmad, Shoaib, Farmery, B. W., et al., Nucl. Instr. and Meth. 1980, 170(1-3), pp-327-30.
202. Ahmed, Motaheruddin, AECD/EP-22, Jun 1971, p-5.
203. Aiguabella, R., Strasbourg Univ., Faculte des Sciences, 1969, p-59.
204. Aitken, D., Hofstadter, R., et al., 1962 International Conf. on High-Energy Phys. at CERN, pp-185-93.
205. Aitken, J. H., Univ. of Toronto, Private Communication.
206. Ajzenberg, F., 1952 Rev. Sci. Instr. 40, p-683.
207. Ajzenberg-Selove, F., Flynn, E. R., et al., Phys. Rev., C, Apr 1978, 17(4), p-1283-1293.
208. Akhavan-Rezayat, A., and Quinn, C. M., INTDS Newsletter, July 1979.
209. Akimov, Yu. K., and Selivanov, G. I., JINR-E13-5621, 1971, p-5.
210. Akimtsev, A. I., Aseev, A. A., et al., Instr. Exp. Tech. 19,
211. Albanese, J. P, Arvieux, J., et al., Int. Conf. on High-Energy Phys. and Nucl. Structure, Zurich, Switzerland, Aug. 29 - Sept. 2, 1977.
212. Albanese, J.P., Boschitz, E., et al., Int. Conf. on High-Energy Phys. and Nucl. Structure, Zurich, Switzerland, Aug. 29 - Sept. 2, 1977.
213. Albert, D. J., Nagl, A., Phys. Rev., C, Aug. 1977, 16(2), pp-503-512.
214. Albert, Thomas E., Univ. of Florida, Gainesville, FL, 69, p-128, Coden: DABSAQ.
215. Albert, T. E., and Carroll, E. E., Nucl. Instr. and Meth. 94, No. 1, 1971, pp-173-7.
216. Alberigi Quaranta, A., Canali, C., et al., Rev. Sci. Instr. 41, Aug. 1970, pp-1205-13.

217. Albrecht, R., Damjantschitsch, H., et al., Max-Planck-Inst., Annual Report 1976.
218. Albrecht, R., Damjantschitsch, H., Max-Planck-Inst., Annual Report 1975.
219. Alburger, D. E., Phys. Rev., C, Aug. 1977, 16(2), pp-889-890.
220. Alburger, D. E., Phys. Rev., C, Dec. 1977, 16(6), pp-2394-2400.
221. Alburger, D. E., Balamuth, D. P., et al., Phys. Rev., C, May 1978, 17(5), pp-1525-1530.
222. Alburger, D. E., Phys. Rev., C, July 1978, 18(1), pp-576-578.
223. Al'bui, Zh., Ozhe, Zh., et al., Izv. Akad. Nauk SSSR, Ser. Fiz. ISSN 0367-6765, Jan. 1979, 43(1), pp-32-36.
224. Alder, J. C., Gabioud, B., et al., AIP Conf. Proc. 1976, No. 33, pp-624-625.
225. Alder, J. C., Gabioud, B., et al., AIP Conf. Proc. 1976, No. 33, pp-626-627.
226. Alder, J. C., Gabioud, B., et al., AIP Conf. Proc. 1976, No. 33, pp-628-629.
227. Aleksandrov, B. M., Krivokhatskij, A. S., et al., Yad. Fiz. ISSN 0044-0027, 1978, 28(5), pp-1165-1169.
228. Aleksandrov, Yu. M., Zinevich, A. N., et al., Tr. Fiz. Inst., Akad. Nauk SSSR, 34, 1966, pp-206-12 (in Russian).
229. Aleonard, M. M., Deleplanque, M. A., et al., Int. Workshop on Gross Properties of Nuclei and Nucl. Excitations, Hirschegg, Austria, Jan. 16-21, 1978.
230. Alevra, A., Duma, M., et al., Rev. Roum. Phys. ISSN 0035-4090, 1978, 23(3), pp-289-297.
231. Alexander, T. K., and Bell, A., Nucl. Instr. and Meth. 81, 1970, pp-22-6.
232. Alexander, Y., and Rinrat, A. S., Nucl. Phys., A, Mar. 14, 1977, 278(3), pp-525-532.
233. Alexander, Y., and Moffa, P. J., ORO-5126-26, Aug. 1977, p-16.
234. Alexander, Y., and Moffa, P. J., Phys. Rev., C, Feb. 1978, 17(2), pp-676-680.
235. Alfassi, Z. B., 7th Int. Conf. on Cyclotrons and their Application, Zurich, Switzerland, Aug. 19-22, 1975.
236. Alford, W. P., Anderson, R. E., et al., Nucl. Phys., A, Dec. 12-19, 1977, 293(1-2), pp-83-91.
237. Alfimenkov, P., Akopyan, G. G., et al., Yad. Fiz., 1977, 25(6), pp-1145-1149.
238. Alfimenkov, V. P., Akopyan, G. G., et al., Yad. Fiz., May 1977, 25(5), pp-930-937.
239. Alfimenkov, V. P., Borzakov, S. B., et al., JINR-E-15-12380, 1979, p-16.
240. Alger, D. L., and Steinberg, R., NASA-TN-D-5598; E-5073, 1970, p-24.
241. Alger, D. L., and Steinberg, R., IEEE, Trans. Nucl. Sci., NS-20, No. 1, Feb. 1973, pp-420-425.

242. Al-Janabi, M. A. A., and Al-Hashimi, H. N., J. Radioanal. Chem. ISSN 0134-0719, 1979, 53-(1-2), pp-321-326.
243. Alkhazov, G. D., Belostotsky, S. L., et al., Phys. Lett., B, Sept. 12, 1977, 70(1), pp-20-22.
244. Alkhazov, G. D., Belostotskij, S. L., Yad. Fiz., 1977, 26(4), pp-673-682.
245. Alkhazov, G. D., Bertej, R., et al., Izv. Akad, Nauk SSSR, Ser. Fiz. ISSN 0367-6765, Sept. 1978, 42(9), pp-1952-1956.
246. Alkhazov, G. D., Belostotsky, S. L., et al., Phys. Lett., B, ISSN 0370-2693, Jul 30, 1979, 85(1), pp-43-46.
247. Allardyce, B. W., Fiebig, A., et al., Status Report on CERN SC; TEICOCTA; IU, U. S. A., Sept. 18-21, 1978.
248. Allen, B. J., Boldeman, J. W., et al., NBS (U. S.) Spec. Publ., 1 Oct. 1975, pp-360-362. See NBS-SPEC. PUBL.-425, 1.
249. Allen, B. J., Musgrove, A. R., et al., Nucl. Phys., A, Oct. 5, 1976, 269(2), pp-408-428.
250. Allen, B. J., Musgrove, A. R., et al., Nucl. Phys., A, Jun 6, 1977, 283(1), pp-37-44.
251. Allen, K. W., Dolan, S. P., et al., Nucl. Instr. and Meth. 134(1), Apr. 1, 1976, pp-1-8.
252. Allen, K. W., Dolan, S. P., et al., INTDS Proc. 1974, Chalk River, Canada, Oct. 1-3.
253. Allen, K. W., Dolan, S. P., et al., Nucl. Phys. Lab., Oxford, UK, Internal Report, 1975.
254. Allen, K. W., Dolan, S. P., ACNTDS, CRNL, Oct. 1-3, 1974.
255. Allen, L. D. F., LA-2769, Contract W-7405-eng-36, Sept. 1962, p-37.
256. Allen, P. D., 7th AINSE Nucl. Phys. Conf., 1978.
257. Allard, C., Proc. of the 3rd Conf. on Accel. Targets Designed for the Prod. of Neutrons, Liege, Belgium, Sept. 18-19, 1967, pp-75-102.
258. Allard, C., Euratom Rev., 7, June 1968, pp-34-41.
259. Allard, C., EUR-1815.e (Rev.), pp-271-82.
260. Alm, A., and Lindgren, L. J., Nucl. Phys., A., Oct. 26, 1976, 271(1), pp-1-14.
261a. Almen, O., and Bruce, G., Nucl. Instr. and Meth. 11, 1961, pp-257-279.
261b. Almond, R. P., Otte, A. V., et al., IEEE Trans. on Nucl. Sci., NS-26(1), Feb. 1979.
262. Alsmiller, R. G., Jr., and Barish, J., ORNL, Oak Ridge, TN, Particle Accel., 5(3), 1973, pp-155-9.
263. Alstad, J., Bergersen, B., et al., Dept. Chem., Univ. of Oslo, Blindern, Norway, 1970, pp-109-22.
264. Alt, E. O., Sandhas, W., et al., Phys. Rev. Lett., Dec. 6, 1976, 37(23), pp-1537-1540.
265. Alter, H., BNL-NCS-50446, Apr. 1975, pp-126-142.
266. Althoff, K. H., Herr, H., et al., From 2nd Int. Conf. on Polarized Targets, Berkeley, CA, Aug. 30, 1971.

267. Alton, G. D., and Love. L. O., Nucl. Instr. and Meth. 102(3), 1972, pp-379-88.
268. Alton, G. D., and Love, L. O., CONF-711002, ORNL, Oct. 1971, pp-10-20.
269. Amakawa, H., and Kubo, K. I., Nucl. Phys., A, Aug. 9, 1976, 266(2), pp-521-532.
270. Amann, J. F., Barnes, P. D., et al., Phys. Rev. Lett., Mar. 20, 1978, 40(12), pp-758-761.
271. Ambard, G., Nucl. Instr. and Meth., 1969, 73, pp-223-4.
272. Ambrosius-Olesen, P., Soerensen, O. W., et al., Nucl. Instr. and Meth., May 1965, 34, pp-255-7.
273. Amos, K., Nesci, P., et al., Aust. J. Phys., Aug. 1976, 29(4), pp-233-243.
274. Amphlett, B. C., Proc. of Conf. held at St. Catherine's Coll., Oxford, Sept. 22-23, 1969.
275. Amsel, G., Ann. Phys. (Paris) 9, 1964, p-309.
276. Amsel, G., and d'Artemare, E., IEEE Trans. on Nucl. Sci., NS-14(1), Feb. 1967, pp-189-203.
277. Amsel, G., Nadai, J. P., et al., Nucl. Instr. and Meth. 149 (1-3), 1978.
278. Amsel, G., Nadai, J. P., et al., Proc. of 3rd Int. Conf. on Ion Beam Anal., Washington, DC, June 27-July 2, 1977.
279. Amsel, G., Nadai, J. P., et al., Univ. Paris, France, Internal Report, 1976
280. Amsden, A. A., Ginocchio, J. N., et al., Phys. Rev. Lett., May 9, 1977, 38(19), pp-1055-1058.
281. Amsden, A. A., Harlow, F. H., et al., Phys. Rev., C, June 1977, 15(6), pp-2059-2071.
282. Amten, L., Goenczi, L., et al., Uppsala, TLU-26/74, 1976.
283. Amus'ya, M. Ya., Talov, M. B., et al., Conf. on Nucl. Spectro. and Nucl. Structure, Baku, Feb. 3-6, 1976.
284. Amus'ya, M. Ya., Zhalov, M. B., et al., Conf. on Nucl. Spectro. and Atomic Nucl. Structure, Leningrad, Nauka, 1978, p-169.
285. Anan'ev, V. D., and Matora, I. M. This paper was summarized in At. Energ. (USSR), 29, No. 4, Oct. 1970, pp-285-6.
286. Anan'in, P. S., Glavanakov, L. V., et al., Conf. on Nucl. Spectro. and Atomic Nucl. Structure, 1978, p-249.
287. Anan'in, P. S., Glavanakov, I. V., et al., Izv. Vyssh. Uchebn. Zaved., Fiz. ISSN 0021-3411, No. 5, 1979, p-121.
288. Anantaraman, N., Gove, H. E., et al., Phys. Lett., B, Jan. 5, 1976, 60(2), pp-149-152.
289. Anantaraman, N., Gove, H. E., et al., Phys. Lett., B, Apr 10, 1978, 74(3), pp-199-201.
290. Anderl, R. A., Gursky, J. C., et al., 5th Annual Conf. of the INTDS, LA-6850-C, Jun 1977, pp-48-58.
291. Anderl, R. A., Harker, Y. D., et al., Trans. Am. Nucl. Soc., 28, June 1978, p-745.
292. Andersen, V., Christensen, C. J., et al., Danish Atomic Energy Comm. 269(2), 1976, pp-338-348.

293. Anderson, V., and Christensen, C. J., Danish Atomic Energy Commission, Report 1975.
294. Anderson, A. N., Cameron, J. M., et al., Univ. of Alberta, Edmonton, Canada, Report 1977.
295. Anderson, A. N., Cameron, J. M., et al., Phys. Rev. Lett. 40(24), Jun 12, 1978, pp-1553-1555.
296. Andersson, C., Grapengiesser, B., et al., Proc. Int. EMIS Conf. Low Energy Ion Accel. Mass Sep., 8th, 1973, pp-463-8.
297. Andersson, C., Grapengiesser, B., et al., Swed. Res. Counc. Lab., Studsvik, Nykoping, Report 1973.
298. Andersen, C. A., Roden, H. J., et al., J. Appl. Phys., 40, July 1969, pp-3419-20.
299. Andersen, C. A., Roden, H. J., et al., Appl. Res. Labs., Goleta, CA, Report 1968.
300. Anderson, D. W., and Swift, S. C., Phys. Med. Biol., 15, Apr 1970, pp-349-54.
301. Anderson, D. W., Univ. of Okla., Oklahoma City, Report 1969.
302. Anderson, D. W., Univ. of Okla., Oklahoma City, Report 1968.
303. Anderson, D. W., Univ. of Okla., Oklahoma City, Report 1969.
304. Anderson, D. W., Petry, R. F., et al., Nucl. Phys., A, 262(1), May 10, 1976, pp-91-95.
305. Andersson, G., Areskoug, M., et al., Phys. Lett., B, 64(4), Oct. 11, 1976, pp-421-23.
306. Andersson, G., Nucl. Instr. and Meth. 139(1), 1976, pp-169-73.
307. Andersen, L. M., Andersen, A. S., et al., Nucl. Phys., A, 153, 1970, p-17.
308. Anderson, L. W., Nucl. Instr. and Meth., 167(3), 1979, pp-363-70.
309. Andersen, O., Garrett, J. D., et al., Phys. Rev. Lett. ISSN 0031-9007, 43(10), Sep 3, 1979, pp-687-690.
310. Anderson, R. E., Kraushaar, J. J., et al., Colorado Univ., Boulder, U. S. A., Nov. 1, 1975, pp-10-13, COO-535-733.
311. Anderson, R. E., Kraushaar, J. J., et al., Colorado Univ., Boulder, U. S. A., Nov. 1, 1975, pp-100-103, COO-535-733.
312. Anderson, R. E., Kraushaar, J. J., et al., Nucl. Phys. A, 287(2), Sept. 5, 1977, pp-265-279.
313. Anderson, R. E., Batay-Csorba, P. A., et al., Phys. Rev. Lett., 39(16), Oct. 17, 1977, pp-987-990.
314. Anderson, R. E., Batay-Csorba, P., et al., TID-27909, Colorado Univ., Boulder, U. S. A., Nov. 1, 1977, pp-56-59.
315. Anderson, R. E., Batay-Csorba, P. A., et al., TID-27909, Colorado Univ., Boulder, U. S. A., Nov. 1, 1977, pp-60-61.
316. Anderson, R. E., Batay-Csorba, P. A., et al., TID-27909, Colorado Univ., Boulder, U. S. A., Nov. 1, 1977, pp-62-64.

317. Anderson, R. E., Batay-Csorba, P. A., et al., TID-27909, Colorado Univ., Boulder, U. S. A., Nov. 1, 1977, pp-67-70.
318. Anderson, R. E., Batay-Csorba, P. A., et al., TID-27909, Colorado Univ., Boulder, U. S. A., Nov. 1, 1977, pp-54-56.
319. Anderson, R. D., and Oscarson, E. E., from 16th Ann. Health Phys. Soc. Meeting, NY, NY, July 12, 1971.
320. Anderson, R. L., Nucl. Instr. and Meth., 70, Apr. 1, 1969, pp-87-9.
321. Anderson, R. L., Ash, W. W., et al., Phys. Rev. Lett., 38(6), Feb. 7, 1977, pp-263-266.
322a. Anderson, H. L., Nucl. Instr. and Meth. 12, 111, 1961.
322b. Ando, T., Ikeda, K., et al., AIP Conf. Proc. ISSN 0094-243X, 47, 1978, pp-516-517.
323. Andrade, J., Silva, E., et al., Nucl. Instr. and Meth. 26, Feb. 1964, pp-22-8.
324. Andre, C., Gauvin, H., et al., J. Phys. (Paris), 37(1), Jan. 1976, pp-5-15.
325. Andre, S., Genevey-Rivier, J., et al., Nucl. Phys., A, ISSN 0375-9474, 325(2-3), Aug 20-27, 1979, pp-445-462.
326. Andreev, E. A., Sit'ko, S. P., et al., Prib. Tekh. Eksp. 14, 2, Mar-Apr 1971, pp-40-3.
327. Andreev, E. A., Blazhenkov, V. V., et al., Prib. Tekh. Eksp. 15, 1, 1972, pp-73-5.
328. Andrews, H. R., et al., Proc. of the 4th Int. Conf. on Hyperfine Interactions, Madison, NJ, 1977.
329. Anikeev, E. F., Andryushchenko, V. I., et al., Izv. Vyssh. Uchebn. Zaved., Chern. Metall. 5, 1977, pp-145-147.
330. Angeleschu, T., Mihul, A., et al., Stud. and Cercet. Fiz., 31(1), 1979, pp-103-12.
331. Angelov, N., Vishnevskaya, K. P., et al., JINR-R-1-9238, 1975, p-11.
332. Angelov, N., Ivanovskaya, I. A., et al., Yad. Fiz. 27(1), 1978, pp-190-193.
333. Angelov, N., Anoshin, A. I., et al., JINR-R-1-11506, 1978, p-9, 19th Int. Conf. on High Energy Phys., Tokyo, Japan.
334. Anholt, R., Z. Phys., A, 292(2), Sept. 1979, pp-123-130.
335. Anikin, L. T., Kravetskij, G. A., et al., Svar. Proizvod, 1, Jan 1977, pp-33-35.
336. Anisimov, S. I., Ivanov, M. F., et al., Pis'ma Zh. Ehksp. Teor. Fiz. 22(6), Sep 20, 1975, pp-343-346.
337. Anisimov, S. I., Bespalov, V. E., et al., Pis'ma Zh. Eksp. Teor. Fiz. 31(1), 1980, pp-67-70 (Russ.)
338. Anne, R., Delpierre, P., et al., J. Phys. (Paris), Colloq. 5, 1975, p-C5.129-C5.131.
339. Annegarn, H. J., Univ. of the Witwatersrand, Johannesburg, Thesis (Ph.D.) 1976, p-366.
340. Ansaldo, E. J., Garvy, G. T., et al., Nucl. Inst. and Meth. 125(3), 1975, pp-429-33.

341. Anthony, J. M.,INTDS Proceedings 1979, Boston, U. S. A., Oct. 1-3.
342. Antoinette Saettel-Marie, INTDS Newsletter, July 1979.
343. Antolvovicj, B., Fizika (Zagreb), 8(2/3), 1976, pp-163-171.
344. Antonenko, V. G., Vinogradov, A. A., et al., Pis'ma Zh. Ehksp. Teor. Fiz. ISSN 0370-274X, 30(8), 20 Oct. 1979, pp-536-541.
345. Antonini, M., and Nicotera, E., J. Phys., E (London), 2, Oct. 1969, pp-890-1.
346. Antropov, A. E., Baiyumi, T., et al., Yad. Fiz. ISSN 0044-0027, 29(6), 1979, pp-1432-1442.
347. Antropov, A. E., Afonin, O. F., et al., Sov. Tech. Phys. Lett. (Engl. Transl.), 4(5), May 1978, pp-224-225.
348. Antsipov, G. V., Konshin, V. A., INDC(CCP)-77/U, 1976, p-7.
349. Antsipov, G. V., Benderskij, A. R., et al., INDC(CCP)-76/U, 1976, p-4.
350. Anttila, A., Bister, M., et al., Nucl. Instr. and Meth. 96(1), 1971, pp-141-3.
351. Antuf'ev, Yu.P., Agranovich, V. L., et al., Yad. Fiz. 21(6), Jun 1975, pp-1206-1214.
352. Anufriev, V. A., Kolesov, A. G., et al., TsNIIAtominform. 1977, pp-263-266.
353. Aoki, Y., Kato, S., et al., Phys. Lett., B, 61(5), 26 Apr 1976, pp-437-440.
354. Aoki, Y., Sanada, J., et al., UTTAC-11, 1978, p-32.
355. Apokin, V. D., Vasil'ev, A. N., et al., Yad. Fiz. 25(3), Mar 1977, pp-555-560.
356. Appel, C., Eberhard, K. A., et al., Verh. Dtsch. Phys. Ges. 4, 1978, pp-912-913.
357. Arakawa, R., Saito, T., et al., Nucl. Instr. and Meth. 13(2), 1976, p-369.
358. Arakelyan, S. G., and Shakhbazyan, B. A., JINR-1-12347, 1979, p-16.
359. Aref'ev, A. V., and Suchkov, D. A., Cryogenics 16(9), 1976, pp-548-9.
360. von Ardenne, M., Schiller, S., et al., Nachrichtentech, 15, 1965, p-306.
361. von Ardenne, M., Tabellen der Elektronenphysik, Ionenphysik und Ubermikroscopie, (VEB Deutscher Verlag de Wissenschaten, 1968).
362. Ardisson, G., Radiochem. Radioanal. Lett. 18(6), 1974, pp-365-372.
363. Aref'ev, A. V., and Suchkov, D. A., Prib. Tekh. Ehksp. 1, Jan 1976, pp-22-23.
364. Arends, J., Eyink, J., et al., Verh. Dtsch. Phys. Ges. 4, 1978, pp-814-815.
365. Arends, J., Eyink, J., et al., Bonn Univ., GFR., AED-CONF-78-084-115.
366. Arenhoevel, H., and Fabian, W., Nucl. Phys., A, 292(3), 5 Dec 1977, pp-429-436.

367. Argan, P., Audit, G., et al., CEA-N-1959 (nd), 1 Sept 1975 - 30 Sept 1976, pp-158-159.
368. Argan, P., Audit, G., et al., CEA-N-1959 (nd), 1 Sept 1975 - 30 Sept 1976, pp-161-163.
369. Argan, P., Audit, G., et al., CEA-CONF-4082, 1977, p-1.
370. Argan, P. E., Audit, G., et al., CEA-CONF-4097, 1977, p-1.
371. Argan, P., Audit, G., et al., CEA-CONF-4572, 1979, p-2.
372. Argan, P. E., Audit, G., et al., Nucl. Phys., A, 296(3), 27 Feb 1978, pp-373-337.
373. Argan, P., Audit, G., et al., AIP Conf. Proc. ISSN 0094-243X, 54(1), 15 Jul 1979, pp-428-429.
374. Arima, A., Seki, R., et al., AIP Conf. Proc. ISSN 0094-243X, 54(1), 15 Jul 1979, pp-592-593.
375. Arminen, E., Dept. Phys., Univ. Helsinki, Finland, Rep. 1969.
376. Arminen, E., Nucl. Instr. and Meth. 85(1), 1970, pp-109-14.
377. Armitage, B. H., Hughes, J. D. M., et al., 1978 Daresbury Laboratory Report DL/NSF/p-76.
378. Armitage, B. H., Hughes, J. D. H., et al., Nucl. Instr. and Meth. 155, 1978, p-565.
379. Armitage, B. H., Hughes, J. D. H., et al, INTDS Proceedings 1978, Garching, FRG, Sept. 11-14.
380. Armitage, F. G., Melbourne Univ., Parkville (Australia), Thesis (Ph. D), 1974.
381a. Arnell, S. E., Nucl. Phys. 24, 1961, p-500.
381b. Arnesen, A., Noreland, T., et al., Nucl. Sci. Abstr. 31(5), 1975.
382. Arnison, G. T. J., AERE-R-5097, Paper 4, p-6.
383. Arnison, G. T. J., Nucl. Instr. and Meth. 40, Mar. 1966, p-359.
384. Arnison, G. T. J., and Gilmour, R. J., Nucl. Instr. and Meth. 45, Nov. - Dec. 1966, p-178.
385. Arnison, G. T. J., 1967a AWRE Report No. 0-32/67.
386. Arnison, G. T. J., Nucl. Instr. and Meth. 53, 1967, pp-357-8.
387. Arnison, G. T. J., AWRE Report 0-32/67, May 1967, p-42.
388. Arnold, L. G., Clark, B. C., et al., Phys. Rev., C, 14(5), Nov. 1976, pp-1878-1884.
389. Arnold, R. G., Chertok, B. T., et al., Phys. Rev. Lett. 40(22), 29 May 1978, pp-1429-1432.
390. Arnold, L. G., and Seyler, R. G., Phys. Rev., C, ISSN 0556-2813, 20(5), Nov. 1979, pp-1917-1925.
391. Arnold, R. G., SLAC-PUB-2373, Aug. 1979, p-18.
392. Arnold, R. T., Vanderbilt Univ., Nashville, (Thesis), June 1956, p-43.
393. Arnold, W., Berg, H., et al., Verh. Dtsch. Phys. Ges. 4, 1978, pp-832-833.
394. Aron, J., Benaroya, R., et al., Proc. of the 1979 Linear Accel. Conf.; and references therein..
395. Arrott, A. S., Templeton, T. L., et al., TRI-77-1, Aug. 1977, Simon Fraser Univ., Burnaby, Canada, p-43.

396. Arruda Neto, J. D. T., Herdade, S. B., et al., Phys. Rev., C, 14(4), Oct. 1976, pp-1499-1505.
397. Artamonov, S. A., Bunakov, V. E., et al., Izv. Akad. Nauk SSSR, Ser. Fiz. ISSN 0367-6765, 43(1), Jan 1979, pp-165-166.
398. Artemov, K. P., Gol'dberg, V. Z., et al., Yad. Fiz. 24(1), 1976, pp-3-8.
399. Artemov, K. P., Gol'dberg, V. Z., et al., Yad. Fiz. 27(1), 1978, pp-3-5.
400. Artemov, S. V., Zaparov, Eh.A., et al., Izv. Akad. Nauk SSSR, Ser. Fiz. ISSN 0367-6765, 42(11), Nov 1978, pp-2409-2412.
401. Arthur, E. D., Drake, D. M., et al., Natl. Bur. Stds. (U. S.), Spec. Publ., 425, Oct. 1975, pp-770-773.
402. Arthur, E. D., Chiang, D., et al., Wash. Univ., Seattle (U. S. A.), Nuclear Physics Lab., Annual Report, 1977, RLO/1388-362, June 1977, pp-127-128.
403. Artukh, A. G., Gierlik, E., et al., JINR-E-7-10464, 1977, p-19.
404. Artukh, A. G., Gridnev, G. F., et al., Nucl. Phys., A, 283(2), 13 Jun 1977, pp-350-364.
405. Artyukh, A. G., Volkov, V. V., et al., Yad. Fiz. 27(1), 1978, pp-29-36.
406. Artyukh, A. G., Volkov, V. V., et al., Yad. Fiz. ISSN 0044-0027, 28(5), 1978, pp-1154-1164.
407. Arutyunyan, V. N., Badalyan, G. V., et al., Izv. Akad. Nauk Arm. SSR, Fiz 14(3), 1979, pp-172-9.
408. Albanese, J. P., Arvieux, J., et al., Phys. Lett., B, 73(2), 13 Feb 1978, pp-119-122.
409. Artz, J. L., Greenfield, M. B., et al., Phys. Rev., C, 13(1), Jan. 1976, pp-156-163.
410. Asai, J., and Slobodrian, R. J., Phys. Lett., B, 64(3), 27 Sep 1976, pp-257-258.
411. Aschenbach, J., Fiedler, G., Nucl. Phys., A, 260(2), 5 Apr 1976, pp-287-291.
412. Asghar, M., D'hondt, P., et al., Nucl. Phys., A, 292(1-2), 21-28 Nov 1977, pp-225-236.
413. Ash, W., Nucl. Instr. and Meth. 134(1), 1976, pp-9-10.
414a. Ash, W. W., Symp. on High Energy Phys. with Polarized Beams and Targets, 23-27 Aug. 1976, Argonne, U. S. A.
414b. Ash, W., (ed.), BNL-20415, Dec. 1975, pp-309-314.
415. Ashery, D., Zisman, M. S., et al., Calif. Univ., Berkeley (U. S. A.), Lawrence Berkeley Lab. Nucl. Sci. Ann. Report, 1975, p-97.
416. Ashery, D., Barbarino, S., et al., CEA-N-1959, 1 Sep 1975 - 30 Sep. 1976, pp-186-187.
417. Ashibe, K., and Taketani, H., Nucl. Phys., A, 255(2), 22 Dec 1975, pp-360-386.
418. Aslanides, E, Bauer, T., et al., CEA-CONF-4083, 1977 Sept. 2, p-1.

419. Aspelund, O., Hrehuss, G., et al., Nucl. Phys., A, 253(2), 24 Nov 1975, pp-263-273.
420. Assmann, W., Ehrenberg, J., et al., Verh. Dtsch. Phys. Ges. 4, 1978, p-902.
421. Assmus, K. H., Jäger, K, et al., T.E.I.C.C.T.A. I.U., U. S. A., Sept. 18-21, 1978.
422. Atkinson, H. H., Belcher, B. E., et al., V Inter. Conf. on High Energy Accel., Rome, Comitato Nazionale per l'Energia Nucleare, 1966, pp-559-64.
423. Atkinson, H. H., From Conf. on Polarized Targets and Sources, Saclay, France, Dec. 5-9, 1966.
424. Atkinson, H. H., Advanced Cryogenics, Bailey, C. A. (ed.), London, Plenum Press, 1971, pp-345-413.
425. van Atta, C. M., Lee, J. D., et al., UCRL-52144, 11 Oct. 1976, p-49.
426. Auble, R. L., Nucl. Data Sheets 16(3), Nov. 1975, pp-417-444.
427. Auble, R. L., Galbraith, D. M., et al., INTDS Proceedings 1979, Boston, U. S. A., Oct. 1-3.
428. Auchampaugh, G. F., Ragan, C. E., III, et al., LASL-6761, June 1977, p-18.
429. Auchampaugh, G. F., Plattard, S., et al., CEA-CONF-3722, 1976, p-1.
430. van Audenhove, J., Eschbach, H. L., et al., Nucl. Instr. and Meth. 24, 1963, p-465.
431. van Audenhove, J., P.S.P.S.I.T.F., A.E.R.E., Oct. 20-21, 1965.
432. van Audenhove, J., and Brulmans, J., Geel, Belgium, Report, 1965.
433. van Audenhove, J., Rev. Sci. Instr. 36, 1965, p-383.
434. van Audenhove, J., and Joyeux, J., J. Nucl. Mater. 19, 1966, p-97.
435. van Audenhove, J., and Joyeux, J., Nucl. Instr. and Meth. 57, 1967, p-157.
436. van Audenhove, J., Nucl. Instr. and Meth. 65, 1968, p-115.
437. van Audenhove, J., Joyeux, J., et al., Supplement to No. 136, LeVide, 1969, pp-69-74.
438. van Audenhove, J., and Joyeux, J., T.I.S.R.M.N.M., ORNL, Oct. 5-8, 1971.
439. van Audenhove, J., and Joyeux, J., Nucl. Instr. and Meth. 102, 1972, pp-409-415.
440. van Audenhove, J., and Joyeux, J., Geel, Belgium, Report, 1972.
441. van Audenhove, J., Verdingh, V., et al., A.C.N.T.D.S., CRNL, Oct. 1-3, 1974, p-1974.
442. van Audenhove, J., DeBievre, P., et al., INTDS Proceedings 1978, Garching, FRG, Sept. 11-14.
443. van Audenhove, J., DeBievre, P., et al., Nucl. Instr. and Meth. 167(1), 1979, pp-61-3.
444. van Audenhove, J., INTDS Proceedings 1979, Boston, U. S. A., Oct. 1-3.

445. van Audenhove, J., European Atomic Energy Community, Geel (Belgium), AERE-R-5097, Paper 23, p-8.
446. Audit, G., Schuhl, C., et al., CEA-CONF-3682, 1976, p-7.
447. Auer, I. P., and Caplan, H. S., Nucl. Instr. and Meth. 135(1), 15 May 1976, pp-27-28.
448. Auerbach, N., and Warszawski, J., AIP Conf. Proc. 33, 1976, pp-322-323.
449. Auerbach, N., Paris-11 Univ., 91 - Orsay (France). Inst. de Physique Nucleaire. IPNO-TH-7702, 1977, p-4.
450. Auerbach, N., LYCEN-7702 (pt. 1), 28 Fev - 4 Mar 1977, pp-S2.1-S2.4.
451. Auerbach, N., IPNO-TH-7651, Dec 1976, p-9.
452a. Auerbach, J. M., Manes, K. R., et al., UCRL-83057 CONF-791135-2, 10 Nov. 1979, p-18.
452b. Auger, F., Berthier, B., et al., CEA-N-1861 (nd), 1 Oct 1974 - 1 Sep 1975, p-21.
453. Auger, F., Badawy, I., et al., CEA-N-1959 (nd), 1 Sep 1975 - 30 Sep 1976, pp-27-30.
454. Auger, F., Basrak, Z., et al., CEA-N-1959 (nd), 1 Sep 1975 - 30 Sep 1976, pp-30-31.
455. Auger, F., Basrak, Z., et al., CEA-N-1959 (nd), 1 Sep 1975 - 30 Sep 1976, pp-57-58.
456. Auger, F., Basrak, Z., et al., CEA-N-1959 (nd), 1 Sep 1975 - 30 Sep 1976, pp-23-24.
457. Augustynyak, V., Kherman, M., et al., Proc. of the 4th All-Union Conf. on Neutron Physics, 1977, pp-221-227.
458. Aumann, D. C., and Muellen, G., Nucl. Instr. and Meth. 115(1), 1974, pp-75-81.
459. Aumann, D. C., Faleschini, H., et al., Nucl. Instr. and Meth. 158(2), 1 Apr 1978, pp-233-9.
460. Aune, R. G., Hartwig, A., et al., Int. Conf. Accel. Dosim. Exper., (Proc.), 2nd, 1970, pp-860-8.
461. Aune, R. G., Contract W-7405-Eng-48, from 2nd Int. Conf. on Accel. Dosim. and Exp., Stanford, CA, p-11.
462. Auriol, E., Bernheim, M., et al., CEA-N-1959 (nd), 1 Sep 1975 - 30 Sep 1976, pp-174-175.
463. Auriol, E., Bernheim, M., et al., CEA-N-1959 (nd), 1 Sep 1975 - 30 Sep 1976, pp-173-174.
464. Auriol, E., Bernheim, M., et al., CEA-N-1959 (nd), 1 Sep 1975 - 30 Sep 1976, pp-172-173.
465. Avan, L., Avan, M., et al., Clermont-Ferrand, France, Univ. Laboratoire de Physique Nucleaire et Corpuschulaire, NP-11592, Vol. I, pp-307-25.
466. Avasthi, D. K., Mittal, V. K., et al., Proc. Nucl. Phys. Solid State Phys. Symp. 21B, 1978, p-354.
467. Avchukhov, V. D., Akhmed, M. R., et al., Izv. Akad. Nauk SSSR, Ser. Fiz. ISSN 0367-6765, 42(9), Sep 1978, pp-1937-1941.

468. Avchukhov, V. D., Al'-Amili, M. A., et al., Izv. Akad. Nauk SSSR, Ser. Fiz. ISSN 0367-6765, 42(9), Sep 1978, pp-1942-1947.
469. Averill, R. J., Cambridge Electron Accel., MA, Contract AT (30-1)-2076, Apr. 1965, p-16.
470. Avida, R., and Mustachi, A., Nucl. Instr. and Meth. 50, 1967, pp-351-2.
471. Avishai, Y., Nucl. Phys., A, ISSN 0375-9474, 326(2-3), 10 Sep 1979, pp-352-384.
472. Avril, M., Cunsolo, A., et al., CEA-N-1861 (nd), 1 Oct 1974 - 1 Sep 1975, pp-9-12.
473. Ayer, J. E., Soppet, F. E., et al., Contract W-31-109-Eng-38, (ANL-7075), Feb. 1966, p-17.
474. Awschalom, M., Larsen, F. L., et al., U.S.A.E.C., 67 (PPAD-627E), p-19.
475. Awschalom, M., Larsen, F. L., et al., Health Phys. 14, Apr. 1968, pp-345-51.
476. Awschalom, M., Larsen, F. L., et al., Nucl. Instr. and Meth. 75, 1969, pp-93-102.
477. Awschalom, M., Borak, T. B., et al., Trans. Nucl. Sci. NS-18(3), June 1971, pp-739-40.
478. Awschalom, M., Gollon, P. J., et al., Nucl. Instr. and Meth. 131(2), 1976, pp-235-41.
479. Ayoub, E. E., Ascuitto, R. J., et al., Comput. Phys. Commun. 10(4), Oct. 1975, pp-203-222.
480. Axel, P., Contract Nonr-1834(05), Mar. 1961, p-30.
481. Axen, D., Duesdieker, G., et al., Nucl. Phys., A, 256(3), 19 Jan 1976, pp-387-413.
482. Azevedo, L. J., and Azevedo, L. J., Rev. Sci. Instr. 50(2), Feb. 1979, pp-231-232.
483. Azhgirey, L. S., Ignatenko, M. A., et al., AIP Conf. Proc. ISSN 0094-243X, 47, 1978, pp-670-671.
484. Azhgirey, L. S., Ignatenko, M. A., et al., AIP Conf. Proc. ISSN 0094-243X, 47, 1978, pp-632-633.
485. Azhgirey, L. S., Ignatenko, M. A., et al., JINR-E-1-12296, 1979, p-22.
486. Azimov, S. A., Gulamov, K. G., et al., Nucl. Phys., B, 107(1), 26 Apr 1976, pp-45-64.
487. Azimov, S. A., Arushanov, G. G., et al., Yad. Fiz. 27(2), Feb 1978, pp-388-392.
488. Baba, C. V. K., Fossan, D. B., et al., Nucl. Phys., A, 257(1), 26 Jan 1976, pp-135-143.
489. Baba, K., Endo, I., et al., AIP Conf. Proc. 33, 1976, pp-614-615.
490. Baba, S., JAERI-M-5567, pp-231-235, in Proc. of Meeting on Atomic Energy Rsch. with Heavy Ions.
491. Babin, A., Contract W-7405-eng-48, LBL-755, Mar. 1972, p-21.
492. Babinet, R., Moretto, L. G., et al., Nucl. Phys., A, 258(1), 16 Feb. 1976, pp-172-188.

493. Bachelier, D., Boyard, J. L., et al., CEA-N-1861 (nd), 1 Octobre 1974 - 1 Septembre 1975, pp-161-162.
494. Bachelier, D., Boyard, J. L., et al., J. Phys. (Paris), Colloq. 5, 1975, pp-C5.81-C5.82.
495. Bachelier, D., Boyard, J. L., et al., J. Phys. (Paris), Colloq. 5, 1975, pp-C5.127-C5.128.
496. Bachelier, D., Bernas, M., et al., Nucl. Phys., A, 251(3), 27 Oct. 1975, pp-433-445.
497. Bachelier, D., Boyard, J. L., et al., Paris-11 Univ., 91-Orsay, Internal Report, 1974.
498. Bachelier, D., Boyard, J. L., et al., AIP Conf. Proc. 33, 1976, pp-262-263.
499. Bachelier, D., Boyard, J. L., et al., CEA-N-1959 (nd), 1 Septembre 1975 - 30 Septembre 1976, p-191.
500. Bacher, A. D., Conzett, H. E., et al., Helv. Phys. Acta. ISSN 0018-0238, 51(5-6), 31 Jul 1979, pp-680-684.
501. Bachmann, L., Sawyer, D. L., et al., J. Appl. Phys. 36. 1965, p-304.
502. Bachmann, L., and Hilbrand, H., Basic Problems in Thin Film Physics, Vandenhoeck and Ruprecht, Gottingen, 1966.
503. Bachrach, R. Z., and Bianconi, A., Nucl. Instr. and Meth. ISSN 0029-554X, 152(1), 1 Jun 1978, pp-53-56.
504. Back, B. B., Betts, R. R., et al., Phys. Rev., C, 13(2), Feb. 1976, pp-875-878.
505. Back, N. L., Cramer, J. G., et al., Wash. Univ., Seattle, U. S. A., Nuclear Phys. Lab. Ann. Report RLO/1388-362, June 1977, pp-61-65.
506. Back, N. L., Baker, M. P., et al., Phys. Rev., C, 17(6), June 1978, pp-2053-2060.
507. Backenstoss, G., Brandao, A., et al., Helv. Phys. Acta. 50(5), 30 Nov 1977, pp-567-568.
508. Baertschi, P., Gruetter, A., et al., Nucl. Phys., A, 294(3), 16 Jan 1978, pp-369-375.
509. Bacon, F. M., and Reidel, A. A., IEEE Trans. Nucl. Sci. U. S. A., NS-26(1), pt. 2, Feb. 1979, pp-1505-8.
510. Bacon, F. M., and Reidel, A. A., Report 78, (SAND-78-2122C, CONF-781113-22), p-4.
511. Badovskii, V. P., Bobkov, V. G., et al., At. Energ. 1977, 43(3), pp-209-10.
512. Baggerly, L. L., Davis, D. V., et al., Proceedings of the 2nd Oak Ridge Conf. on the Use of Small Accel. for Teaching and Rsch., March 23-25, 1970, CONF-700322, pp-379-92.
513. Baghuis, L. C. J., Duys, A. M. W., et al., Proceedings of the 6th Int. Cyclotron Conf., Vancouver, Canada, July 18-21, 1972.
514. Bagieu, G., Cole, A. J., et al., Inst. des Sciences Nucleaires, 1976, p-2.
515. Bailar, J. C., ed., Comprehensive Inorganic Chem. 5, Pergamon Press 1973.

516. Baillon, Y., Bertin, P., et al., CEA-N-1861 (nd), 1 Septembre 1975, pp-157-158.
517. Bair, J. K., Nucl. Sci. Eng. ISSN 0029-5639, 71(1), July 1979, pp-18-28.
518. Baisden, P. A., and Seaborg, G. T., LBL-8151, 1978, pp-35-37.
519. Bakanov, L. V., Bunakov, V. E., et al., Pis'ma Zh. Ehksp. Teor. Fiz. 25(7), 5 May 1977, pp-337-341.
520. Bakalov, T., Ilchev, G., et al., JINR-R-3-12404, 1979, p-12.
521. Baker, F. T., Glashausser, C., et al., Nucl. Phys., A, 253(2), 24 Nov. 1975, p-461-468.
522. Baker, F. T., Kruse, T. H., et al., Nucl. Phys., A, 258(1), 16 Feb. 1976, pp-43-60.
523. Baker, F. T., Scott, A., et al., Nucl. Phys., A, 266(2), 9 Aug. 1976, pp-337-345.
524. Baker, F. T., Scott, A., et al., Phys. Lett., B, 70(2), 26 Sept. 1977, pp-167-169.
525. Baker, F. T., Scott, A., et al., Vanderbilt Univ., Nashville, TN, 1977.
526. Baker, F. T., Scott, A., et al., Nucl. Phys., A, ISSN 0375-9474, 325(2-3), 20-27 Aug. 1979, pp-525-532.
527. Baker, M. P., Wash. Univ., Seattle, U. S. A., Thesis (Ph.D).
528. Baker, M. P., Blair, J. S., et al., RLO-1388-308, CONF-750829-8, 1975, p-2.
529. Baker, M. P., Blair, J. S., et al., RLO-1388-307, CONF-750829-12, 1975, p-2.
530. Baker, M. P., Trainor, T. A., et al., RLO-1388-309, CONF-750829-9, 1975, p-2.
531. Baker, P. S., Duncan, F. R., et al., Nucl. Sci. and Eng. 7, 1960, pp-325-6.
532. Baker, S. D., from 2nd Int. Conf. on Polarized Targets, Berkeley, CA, 30 Aug. 1971, LBL-500, pp-85-8.
533. Baker, S. D., Polarization Phenomena Nucl. React., Proc. Int. Symp., 3rd, 1971, pp-899-902.
534. Balabanov, N. P., Gledenov, Yu.M., et al., JINR-R-3-8653, 1975, p-16.
535. Balabanov, N. P., Gledenov, Yu.M., et al., Nejtronnayafizika, Chast' 4, 4, 1976, pp-60-64.
536. Balabanov, N., Gledenov, Yu.M., et al., Yad. Fiz. ISSN 0044-0027, 28(5), 1978, pp-1148-1153.
537. Balakrishnan, M., Kailas, S., et al., Proc. of the Nucl. Phys. and Solid State Phys. Symp., Bombay, Dec. 27-31, 1974, Vol. 17B.
538. Balakrishnan, M., Kailas, S., et al., Proc. of the Nucl. Phys. and Solid State Phys. Symp., Bombay, Dec. 27-31, 1974, Vol. 17B.
539. Balandiko, N. I., Belushkin, V. A., et al., Cryogenics 6, June 1966, pp-158-64.
540. Baldridge, W. J., Penn. State Univ., University Park, U. S. A., Thesis (Ph.D), 1975.

541. Balestra, F., Bollini, E., et al., Proc., Delhi, India, 29 Dec. 1975-3 Jan. 1976, pp-315-316.

542. Balestra, F., Bussa, M. P., et al., AIP Conf. Proc. ISSN 0094-243X, 47, 1978, pp-604-605.

543. Balestri, B., Fournier, G., et al., CEA-CONF-4574, 1979, p-1, 5. Biennial Session on Nucl. Phys., Aussois, France, 5-6 Mar. 1979.

544. Balestri, B., Fournier, G., et al., CEA-CONF-4591, 1979, p-1, 2. Inter. Conf. on Meson-Nuclear Phys., Houston, TX, 5-6 Mar. 1979.

545. Balestrini, S. J., and Forman, L., from 8th Int. Electromagnetic Isotope Separator Conf., Skovde, Sweden, 12 June 1973, CONF-730618-4.

546. Baldin, A. M., Viryasov, N. M., et al., JINR-P-1-6212, 1972, p-40.

547. Ball, A. E., Int. Conf. on Instrumentation for High Energy Physics, 1973, pp-726-30.

548. Ball, G. C., Davies, W. G., et al., Phys. Lett., B, 60(3), 19 Jan. 1976, pp-265-268.

549. Ball, J. B., and Fulmer, C. B., ORNL-3554, Contract W-7405-eng-26, March 1964, p-13.

550. Ball, J. B., Fulmer, C. B., et al., Nucl. Phys., A, 252(1), 3 Nov. 1975, pp-208-236.

551. Balodis, M. K., Plate, M. N., et al., Conf. on Nucl. Spectroscopy and Nucl. Structure, Baku, Russia, 3-6 Feb. 1976.

552. Baltz, A. J., ANL/PHY-76-2(1), May 1976, pp-65-93.

553. Baltz, A. J., Phys. Rev. Lett. 38(21), 23 May 1977, pp-1197-1200.

554. Balzer, D., and Bonani, G., INTDS Proc., 1978, Garching, FRG, Sept. 11-14.

555. Balzer, D., and Bonani, G., Nucl. Instr. and Meth. 167(1), 1979, pp-129-33.

556. Balzer, D., INTDS Proc., 1979, Boston, U. S. A., Oct. 1-3.

557. Balzer, D., INTDS Proc., 1979, Boston, U. S. A., Oct. 1-3.

558. Balzer, R., Hugi, M., et al., Nucl. Phys., A, 293(3), 26 Dec 1977, pp-518-530.

559. Bance, D. A., Proc. of the 3rd Conf. on Accel. Targets Designed for the Prod. of Neutrons, Liege (Belgium), Sept. 18-19, 1967, pp-181-9.

560. Bander, N., Rutherford High Energy Lab., Chilton, Berks., England, 1972.

561. Banerjee, S. N., Proc. of the 7th Int. Conf. on Few Body Problems in Nuclear and Particle Physics, Delhi, India, 29 Dec 1975 - 3 Jan. 1976.

562. Bangerter, R. O., BNL 50769, Proc. Heavy Ion Fusion Workshop, 1977, pp-78-9.

563. Bangerter, R. O., Lee, E. P., et al., Law. Liv. Lab., 1978.

564. Bangerter, R. O., Report, 1978, UCRL-82026, CONF-7806114-2, p-49.

565. Bangester, R. O., Lee, E. P., et al., UCRL-82120, 29 Dec. 1978, p-5.
566. Banks, P. H. T., Cragg, D. A., et al., Rutherford High Energy Lab., (Rep.), 70, (RPP/A-81), p-16.
567. Banks, P. H. T., Cragg, D. A., et al., CONF-700811-2, 27 Aug. 1970, from 3rd Int. Symp. on Polarization Phenomena in Nuclear Reactions, Madison, WI, p-16.
568. Bannerman, D. E., Proc. of the 7th Int. Vacuum Congress and the 3rd Int. Conf. on Solid Surfaces, Vol. 3, Vienna, Austria.
569. Bano, N., and Ahmad, I., Proc. of the Nucl. Phys. and Solid State Phys. Symp., Calcutta, Dec. 22-26, 1975.
570. Barakat, N., Appl. Phys. 14(3), Nov. 1977, pp-319-323.
571. Baranov, I. B., Mashgiz-Leningrad, 1959, translated by E. Bishop.
572. Bar-Avraham, E., and Lee, L. C., Nucl. Instr. and Meth. 64, Sept. 15, 1968, pp-141-7.
573. Bar-Nir, I., and Shuster, M. D., Nucl. Phys., B, 103(1), 26 Jan. 1976, pp-103-108.
574. Bar-Noy, T., and Moreh, R., Reactor Centrum Nederland, Mar. 1975, pp-485-488.
575. Bar-Noy, T., and Moreh, R., Nucl. Phys., A, 288(2), 26 Sep. 1977, pp-192-200.
576. Barbarino, J., Doron, A., et al., CEA-N-1861 (nd), 1 Oct. 1974 - 1 Sept. 1975, pp-160-161.
577. Barbarino, S., Cassagnou, Y., et al., CEA-N-1959 (nd), 1 Sept. 1975 - 30 Sept. 1976, pp-181-185.
578. Barbarino, S., Cassagnou, Y., et al., CEA-N-1959 (nd), 1 Sept. 1975 - 30 Sept. 1976, pp-62-64.
579. Barbier, M., and Fulmer, C. B., CERN-63-19, pp-171-4.
580. Bardin, B. M., and Rickey, M. E., Rev. Sci. Instr. 35, July 1964, pp-902-3.
581. Bardin, C., Duclos, J., et al., CEA-CONF-4575, 5. Biennial Session on Nucl. Phys., Aussois, France, 5-6 Mar. 1979, p-6.
582a. Bardin, T. T., Pronko, J. G., et al., Phys. Rev., C, 14(5), Nov. 1976, pp-1782-1788.
582b. Barfoot, K. M., Mitchell, J. V., et al., Nucl. Instr. and Meth. ISSN 0029-554X, 168(1-3), Jan. 1980, pp-131-138.
583. Bargholtz, Chr., Fransson, K., et al., Nucl. Instr. and Meth. 143(2), 1 Jun 1977, pp-273-275.
584. Bargmann, H., Nucl. Eng. Des. 30(2), Sept. 1974, pp-234-241.
585. Barish, B. C., Bingham, G., et al., U. S. A. E. C. 69, (TID-25473) Vol. 1, pp-33-9.
586. Barit, I. Ya., Balashko, Yu. G., et al., Prib. Tekh. Eksp. 14(1), Jan-Feb. 1971, pp-57-60.
587. Barkan, S., D'Auria, M. J., et al., Proc. of Conf. held at St. Catherine's Coll., Oxford, 22-23 Sept. 1969.
588. Barkov, B. P., Katinov, Yu. V., et al., ITEF-23, 1975, p-24.

589. Barkov, L. M., Zolotorev, M. S., et al., Instr. Exp. Tech. 18(4), pt. 1, July-Aug. 1975, pp-1032-1033.
590. Barna, A., Barna, P. B., et al., Nucl. Instr. and Meth. 102, 1972, pp-549-52.
591. Barna, A., Barna, P. B., et al., T.I.S.R.M.N.M., ORNL, Oct. 5-8, 1971.
592. Barnes, P. D., AIP Conf. Proc. 33, 1976, pp-281-296.
593. Barnett, G. A., Crosby, J., et al., P.S.P.S.I.T.F., AERE, Oct. 20-21, 1965.
594. Barr, S. M., and DelVecchio, R. M., Phys. Rev., C, 15(1), Jan. 1977, pp-114-122.
595. Barreau, P., Leger, P., et al., J. Phys. Rad. 15, Suppl. No. 14A-7A, 1954.
596. Barreto, J., Inst. de Physique Nucleaire, 1976, p-47.
597. Barreto, J., Detraz, C., et al., European Physical Soc., Geneva, Switz., Conf. on Nucl. Phys. with Heavy Ions, Sept. 1976.
598. Barrett, R. F., Allen, B. J., et al., Phys. Lett., B, 61(5), 26 Apr. 1976, pp-441-443.
599. Barrette, M., Barrette, J., et al., Bull. Am. Phys. Soc., Ser. 2, 17, 1972, p-551.
600. Barrette, J., Braun-Munzinger, P., et al., Nucl. Phys., A, 261(3), 3 May 1976, pp-491-497.
601. Barrette, J., Gamp, A., et al., European Physical Soc., Conf. on Nucl. Phys. with Heavy Ions, Sept. 1976.
602. Barrette, J., Gamp, A., et al., Internal Report, Max Planck Institut (FRG.), 1976.
603. Barrette, J., Braun-Munzinger, P., et al., European Physical Soc., Geneva, Switz., Conf. on Nucl. Phys. with Heavy Ions, Sept., 1976.
604. Barrette, J., and Braun-Munzinger, P., Nucl. Phys., A, 287(1), 29 Aug. 1977, pp-195-204.
605. Barrette, J., LeVine, M. J., et al., Phys. Rev. Lett., 40(7), 13 Feb. 1978, pp-445-448.
606. Barrette, J., LeVine, M. J., et al., Phys. Rev., C, ISSN 0556-2813, 20(5), Nov. 1979, p-1759-1767.
607. Barrow, C., Guest, G. H., et al., J. Sci. Instr. 40, May 1963, pp-260-1.
608. Barrus, D. M., and Blake, R. L., 5th Ann. Conf. of the Int. Nucl. Target Develop. Soc., LASL, Oct. 19-21, 1976.
609. Barrus, D. M., and Blake, R. L., Rev. Sci. Instr. 48, 1977, p-116.
610. Barshay, S., Dover, C. B., et al., Brookhaven National Lab.
611. Barshay, S., and Troost, W., Phys. Lett., B, 73B(4-5), 13 March 1978, pp-437-9.
612. Bartenev, V. D., Belushkina, A. A., et al., Proc. of the Int. Conf. on Instr. for High Energy Physics, Dubna, USSR, 8 Sept. 1970, pp-16-25.

613. Bartenev, V. D., Valevich, A. L., et al., JINR-P13-6058, 1971, p-20.
614. Bartenev, V. D., Belushkina, A. A., et al., JINR-P13-6324, 1972, p-13.
615. Bartenev, V. D., Kuznetsov, A., et al., Advan. Cryog. Eng. 18, 1973, pp-460-6
616. Bartenev, V., Klen, J., et al., Appl. of Cryogenic Tech., Vol. V, Whitestone, NY, Scholium International, Inc., 1973, pp-263-268.
617. Bartenev, V. D., Valevich, A. I., et al., Cryogenics 13(4), 1973, pp-239-42.
618. Bartenev, V. D., Beluchkina, A. A., et al., Instr. Exp. Tech. 16(1), 1973, pp-28-31.
619. Bartle, C. M., Nucl. Instr. and Meth. 144(3), 1 Aug. 1977, p-599.
620. Bartle, C. M., and Meyer, H. O., Nucl. Instr. and Meth. 112, 1973, p-615.
621. Bartle, C. M., Chapman, N. G., et al., Nucl. Instr. and Meth. 95(2), 1971, pp-221-8.
622. Bartsch, H., Huber, K., et al., Nucl. Phys., A, 256(2), 12 Jan. 1976, pp-243-252.
623. Bartsch, H., Huber, K., et al., Nucl. Phys., A, 252(1), 3 Nov. 1975, pp-1-7.
624. Baryshevskii, V. G., and Lyuboshits, V. L., JINR-P-1999, 1965, p-6.
625. Basak, A. K., Griffith, J. A. R., et al., Nucl. Phys., A, 275(2), 10 Jan. 1977, p-381-394.
626. Basak, A. K., Griffith, J. A. R., et al., Nucl. Phys., A, 286(3), 22 Aug. 1977, p-420-430.
627. Basak, A. K., Griffith, J. A. R., et al., Nucl. Phys., A, 295(1), 23 Jan. 1978, p-111-124.
628. Basrak, Z., Auger, F., et al., J. Phys. (Paris), Lett., 37(6), June 1976, p-L.131-L.134.
629. Basrak, Z., Cindro, N., et al., European Physical Soc., Geneva, Switz., Conf. on Nucl. Phys. with Heavy Ions, 6-10 Sept. 1976.
630. Basrak, Z., Auger, F., et al., Phys. Lett., B, 65(2), 8 Nov. 1976, pp-119-121.
631. Basov, N. G., Krokhin, O., et al., P. N. Lebedev Phys. Inst., Moscow, 4A, 1977, pp-15-42.
632. Bassalleck, B., Engelhardt, H. D., et al., Phys. Lett., B, 65(2), 8 Nov. 1976, pp-128-130.
633. Bassalleck, B., Klotz, W., et al., Phys. Rev., C, 16(4), Oct. 1977, pp-1526-1539.
634. Bassett, G. A., and Pashley, D. W., J. Instr. Meth. 87, 1959, pp-449.
635. Batist, L. H., Vitman, V. D., et al., Nucl. Phys., A, 254(2), 8 Dec. 1975, pp-480-484.
636. Batty, C. J., RHEL/R-170, Progress Report, Nov. 1968, p-166.

637. Batty, C. J., Czech. J. Phys. 25(3), 1975, pp-286-287.
638. Baturin, P. I., Popov, S. G., et al., Zh. Tekh. Fiz. 46(3), Mar 1971, pp-637-640.
639. Bauer, E., Green, A. K., et al., Basic Problems in Thin Film Physics, Vandenhoeck and Ruprecht, Göttingen, 1966.
640. Bauer, E, Chapter 16 in Techniques for the Direct Observation of Structure and Imperfections, Interscience, NY, 1969.
641a. Bauer, G. S., Ber. Kernforschungsanlage Juelich (Conf.)1980, Jul-Conf-34.
641b. Bauer, G. S., Meeting on Targets for Neutron Beam Spallation Sources, Juelich, June 11-12, 1979.
642. Bauer, T., Boudard, A., et al., CEA-N-1861 (nd), 1 Oct. 1974-1 Sept. 1975, pp-205-207.
643. Bauer, T., Beurtey, R., et al., CEA-N-1861 (nd), 1 Oct. 1974-1 Sept. 1975, pp-204-205.
644. Bauer, T., Beurtey, R., et al., Verh. Dtsch. Phys. Ges. 10(7), 1975, pp-770-771.
645. Bauer, T., Bertini, R., et al., CEA-N-1861 (nd), 1 Oct. 1974-1 Sept. 1975, pp-191-194.
646. Bauer, T., Verh. Dtsch. Phys. Ges. 10(7), 1975, pp-721-722.
647. Bauer, T., Beurtey, R., et al., CEA-CONF-3186, 1975, p-1.
648. Bauer, T., Boudard, A., et al., CEA-CONF-3188, 1975, p-1.
649. Bauer, T., Beurtey, R., et al., CEA-N-1959 (nd), 1 Sept. 1975-30 Sept. 1976, pp-244-247.
650. Bauer, T., Beurtey, R., et al., Phys. Lett., B, 69(4), 29 Aug. 1977, pp-433-436.
651. Bauer, T., Beurtey, R., et al., CEA-CONF-4709, 1979, p-1.
652. Baukloh, and Henke, Z. Anorg.-Chemie 234, 1937, p-307
653. Baumann, H., Thesis, Heidelberg, 1973.
654. Baumann, H., and Bethge, K., Nucl. Instr. and Meth. 122, 1974, p-517.
655. Baumann, H., Bethge, K., et al., Trans. Nucl. Sci. NS 23(2), 1976, p-1081.
656. Baumann, H., and Wirth, L. H., INTDS Proc., 1977, LBL, U. S. A., Oct. 19-20.
657. Baumann, H., Schuehrer, B., et al., 7th INTDS, Univ. und Tech. Univ. München, Sept. 11-14, 1978.
658. Baumann, H., and Wirth, H. L., Nucl. Instr. and Meth. 167(1), 1979, pp-71-2.
659. Baur, G., Roesel, F., et al., Nucl. Phys., A, 252(1), 3 Nov. 1975, pp-77-89.
660. Baur, G., Phys. Lett., B, 60(2), 5 Jan. 1976, pp-137-140.
661. Baur, G., Roesel, F., et al., Verh. Dtsch. Phys. Ges. 6, 1977, pp-835-836.
662. Baur, G., Nucl. Phys., A, 283(3), 20 Jun. 1977, pp-521-525.
663. Baur, G., Roesel, F., et al., Nucl. Phys., A, 288(1), 19 Sept. 1977, pp-113-131.
664. Baxter, A. M., Ikossi, P. G., et al., Nucl. Phys., A, 291(2), 14 Nov. 1977, pp-282-292.

665. Baxter, A. M., McDonald, A. M., et al., Australian Inst. of Nucl. Sci and Eng., Lucas Heights, 7th AINSE Nucl. Phys. Conf., 1978.
666. Baybarz, R. D., and Adair, H. L., J. Inorg. Nucl. Chem. 34, 1972, pp-3127-3130.
667. Baye, D., and Reidemeister, G., Nucl. Phys., A, 258(1), 16 Feb. 1976, pp-157-171.
668. Baye, D., and Heeneen, P. H., European Physical Soc., Geneva, (Switz.), European Conf. on Nucl. Phys. with Heavy Ions, 6-10 Sept. 1976.
669. Bayfield, J. E., Rev. Sci. Instr., 40(7), 1969, pp-869-74.
670. Bayukov, Yu. D., Fedorov, V. B., et al., Nucl. Phys., A, 282(3), 30 May 1977, pp-389-396.
671. Baz', A. I., Vydrug-Vlasenko, S. M., et al., Nauka. 1978. p-150.
672. Baz', A. I., Kiselev, S. M., et al., Nauka. 1978, p-212.
673. Bazell, C. G., and Davies, R. V., AEEW-R 391, UKAEA, 1964.
674. Beale, D. J., and Poletti, A. R., Nucl. Phys., A, 261(2), 26 Apr. 1976, pp-238-252.
675. Beart, G., and Reidemeister, G., AIP Conf. Proc. ISSN 0094-243X, 47, 1978, pp-502-503.
676. Beaudet, G., and Shaviv, G., Astrophys. Space Sci. 51(2), Oct. 1977, pp-395-400.
677. Beaver, J. E., and Hupf, H. B., Private Communication, 1974..
678. Becchetti, F. D., Janecke, J., et al., AIP Conf. Proc. ISSN 0094-243X, 47, 1978, pp-752-753.
679. Becchetti, F. D., Jaenecke, J., et al., AIP Conf. Proc. ISSN 0094-243X, 47, 1978, pp-728-729.
680. Bechtold, V., Friedrich, L., et al., Verh. Dtsch. Phys. Ges. 10(7), 1975, p-866.
681. Bechtold, V., Friedrich, L., et al., Nucl. Phys., A, 288(2), 26 Sep. 1977, pp-189-191.
682. Bechtold, V., Bialy, J., et al., Verh. Dtsch. Phys. Ges. 4, 1978, p-767.
683. Beck, F. A., Byrski, T., et al., Phys. Rev., C, 13(5), May 1976, pp-1792-1800.
684. Beck, F., Mueller, K., et al., Phys. Rev. Lett. 40(13), 27 Mar. 1978, pp-837-840.
685. Beck, S. M., NASA-TN-D-7925, L-9941, Oct. 1975, p-64.
686. Beck, S. M., and Powell, C. A., NASA-TN-D-8119, L-10597, Apr. 1976, p-143.
687. Becker, F., Strykowski, I., et al., CEA, 75 - Paris (France).
688. Becker, F., and Strykowski, I., Nucl. Phys., A, 289(2), 17 Oct. 1977, pp-446-460.
689. Beckert, K., Herrmann, F., et al., Gemeinsamer Jahresbericht 1976, ZfK-315, Aug. 1976, pp-9-10.
690. Becret, C., Champlong, P., et al., Etablissement Central de l'Armement ARCUEIL, France, Note no. 76-R-121.
691. Beeman, W. T., Wessman, R. A., et al., CONF-710809, 30 Aug. 1971, from tritium symposium, Las Vegas, NE, U. S. A.

692. Beene, J. R., and DeVries, R. M., Phys. Rev. Lett. 37(15), 11 Oct. 1976, pp-1027-1030.
693. Beer, H., and Rohr, G., Z. Phys., A, 277(2), May 1976, pp-181-188.
694. Beer, H., Spencer, R. R., et al., NBS Spec. Publ. 425, Oct. 1975, pp-816-818.
695. Beer, H., and Kaeppeler, F., Inst. fuer Angewandte Kernphysik, Annual Report KFK-2868, Oct. 1979, pp-12-13.
696. Beer, H., Wisshak, K., et al., Inst. fuer Angewandte Kernphysik, Annual Report KFK-2868, Oct. 1979, pp-14-15.
697. Beer, H., and Spencer, R. R., Inst. fuer Angewandte Kernphysik, Annual Report KFK-2868, Oct. 1979, p-21.
698. Beer, J., Pulfer, P., et al., Helv. Phys. Acta. 49(5), 29 Oct. 1976, p-754.
699. Beer, M., Kalos, M. H., et al., Chalk River Exp. on Cross Sections of Fissile Nuclides, Final Report PB-253100, Dec. 1975, p-80.
700. Beer, M., Kalos, M. H., et al., Trans. Am. Nucl Soc., 23, Jun 1976, p-509.
701. Beetz, R., Boer, F. W. N., de, et al., Spring Meeting of the Dutch Physical Society, Mar 1978, p-23.
702. Bederka, S., Nucl. Instr. and Meth. 163(1), 1979, pp-271-3.
703. Bedford, L. A. W., Harwell, Eng., Publ. 66, Series: AERE-5217, p-17.
704. Behar, M., Filevich, A., et al., Nucl. Phys., A, 261(2), 26 Apr 1976, pp-317-327.
705. Behar, M. Filevich, A., et al., Nucl. Phys., A, 287(2), 5 Sep 1977, pp-255-264.
706. Behbehani, A. H., et al., Phys. Lett. 74B, 1978, p-219.
707. Behrndt, K. H., Z. Angew. Phys. 8, 1956, p-453.
708. Behrndt, K., and Love, R. W., Vac. Symp. Trans. 7, 1960, p-87.
709. Behrndt, K. H., and Love, R. W., Vacuum 12, 1962, pp-1-9.
710. Behrndt, K. H., ASM Seminar 1963.
711. Behrens, J. W., Magana, J. W., et al., UCID-17442, Apr 1977, p-8.
712. Behrens, J. W., Newbury, R. S., et al., Nucl. Sci. and Eng. 66(3), June 1978, pp-433-441.
713. Beil, H., Bergere, R., et al., CEA-N-1861 (nd), 1 Oct. 1974-1 Sept. 1975, pp-85-86.
714. Beil, H., Bergere, R., et al., CEA-N-1959 (nd), 1 Sept. 1975 - 30 Sept. 1976, pp-85-89.
715. Beitins, M. R., Kramer, N. D., et al., Nucl. Phys., A, 262(2), 17 May 1976, pp-273-300.
716. Belanova, T. S., Kolesov, A. G., et al., At. Ehnerg. ISSN 0004-7163 47(3), Sep 1979, pp-206-207.
717. Bell, R., Clay, H., et al., Trans. Nucl. Sci. NS-16, June 1969, pp-631-2.
718. Bellamy, E. H., Hogg, W. R., et al., Nucl. Instr. and Meth. 7, June 1960, pp-293-6.

719. Bellicard, J., Frois, B., et al., CEA-N-1861 (nd), 1 Oct. 1974 - 1 Sept. 1975, pp-146-147.
720. Bellicard, J. B., Frois, B., et al., CEA-N-1861 (nd), 1 Oct. 1974 - 1 Sept. 1975, pp-148-149.
721. Bellicard, J. B., Cavedon, J. M., et al., CEA-N-1959 (nd), 1 Sept. 1975 - 30 Sept. 1976, pp-164-166.
722. Bellicard, J. B., Cavedon, J. M., et al., CEA-N-1959 (nd), 1 Sept. 1975 - 30 Sept. 1976, pp-170-172.
723. Bellotti, E., Fiorini, E., et al., Proc. of the Int. Symp. on Interaction Studies in Nuclei, 17-20 Feb. 1975.
724. Belovitskij, G. E., Presnyak, O. S., et al., Nejtronnaya fizika, Chast' 4, (4), 1976, pp-209-214.
725. Belyakov, A. V., Kurtseva, N. N., et al., Tr. Mosk. Khim.-Tekhnol. Inst. 76, 1973, pp-91-93.
726. Belyaev, V. B., Wrzecionko, J., et al., Phys. Lett., B, ISSN 0370-2693 83(1), 23 Apr 1979, pp-19-21.
727. Bem, P., Burjan, V., et al., Ustav Jaderne Fyziky, Feb. 1977, p-44.
728. Bemis, C. E., Jr., Oliver, J. H., et al., Nucl. Sci and Eng. 63(4), Aug. 1977, pp-413-417.
729. Benayoun, J. J., Chauvin, J., et al., Phys. Rev. Lett. 36(24), 14 Jun 1976, pp-1438-1440.
730. Bendt, P. J., Jackson, J. A., et al., Nucl. Instr. and Meth. 83(2), June 15, 1970, pp-201-7.
731. Bendt, P. J., 3rd Int. Symp. on Polarization Phenomena in Nuclear Reactions, Madison, WI, August 31 - Sept. 4, 1970.
732. Benetskij, B. A., Klyachko, A. V., et al., TsNIIAtominform. 1977, pp-44-46.
733. Benetskij, B. A., Klyachko, A. V., et al., TsNIIAtominform. 1977, pp-47-51.
734. Benjamin, R. W., McCrosson, F. J., et al., Trans. Am. Nucl. Soc. 23, June 1976, pp-545-546.
735. Benjamin, R. W., McCrosson, F. J., et al., DuPont de Nemours (E. I.) and Co., Aiken, SC, Savannah River Lab., Jan. 1977, p-31.
736. Benjamin, R. F., and White, L., Appl. Opt. 17(8), 15 April 1978, p-1160.
737. Benjamin, R. F., Schappert, G. T., et al., J. Appl. Phys. 50(1), 1979, pp-7-10.
738. Benka, O., Geretschlager, M., et al., J. Appl. Phys. 47(11), Nov. 1976, pp-5090-5093.
739. Bennett, C. L., Fulbright, H. W., et al., Verh. Dtsch. Phys. Ges. 4, 1978, pp-920-921.
740. Bennett, G. W., Trans. Nucl. Sci. NS-16, June 1969, pp-637-9.
741. Bennett, M. J., Feldl, E., et al., Nucl. Instr. and Meth. 68, Feb. 15, 1969, pp-229-234.
742. Bennink, A. H., Nucl. Instr. and Meth. 146(3), 1 Nov. 1977, p-591.

743. Bennink, A. H., Nucl. Instr. and Meth. 146(3), 1 Nov. 1977, p-591.
744. Bennink, A. H., and Tuintjer, T. W., INTDS Proc., 1979, Boston, U. S. A., Oct. 1-3.
745. Bennink, A. H., INTDS Newsletter, Jan. 1979.
746. Bennink, A. H., and Tuintjer, T. W., INTDS Newsletter, July 1979.
747. Bennink, A. H., and Tuintjer, T. W., INTDS Newsletter, July 1979.
748. Bennink. A. H., and Tuintjer, T. W., INTDS Newsletter, July 1979.
749. Bensussan, A., and Azam, G., French patent document 2386109/A.
750. Bent, R. D., Debevec, P. T., et al., Phys. Rev. Lett. 40(8), 20 Feb. 1978, pp-495-498.
751. Bentley, G. F., Barnes, J. W., et al., Trans. Am. Nucl. Soc. 30, 1978, p-808, from 1978 winter meeting of American Nuclear Society, Washington, DC, U. S. A., 12 Nov.
752. Bentley, G. E., Barnes, J. W., et al., Proc. Conf. Remote Syst. Tech. 26, 1979, pp-378-81.
753. Berceanu, I., Borcea, C., et al., Nucl. Instr. and Meth. 146(1), 1 Oct. 1977, p-319.
754. Berdnikov, Ya. A., Gismatullin, Yu. R., et al., Yad. Fiz. 25(5), May 1977, pp-938-944.
755. Berecz, I., Karolyi, J., et al., Nucl. Instr. and Meth. 33, March 1965, p-364.
756. Berezhnoj, Yu. A., and Shlyakhov, N. A., Izv. Akad. Nauk SSSR, Ser. Fiz. ISSN 0367-6765 42(4), Apr 1978, pp-743-746.
757. Berg, E. R., 7th Int. Conf. on Cyclotron and their Application, Zurich (Switz.), Aug. 19-22, 1975.
758. Berg, H. L., Hietzke, W., et al., Nucl Phys., A, 276(1), 17 Jan. 1977, pp-168-188.
759. Berg, G., Kuehn, W., et al., Nucl. Phys., A, 254(1), 1 Dec. 1975, pp-169-182.
760. Berg, G. P. A., Das, R., et al., Nucl. Phys., A, 289(1), 10 Oct. 1977, pp-15-35.
761. Berge, J. P., Fermi National Accelerator Lab., Batavia, IL, 1975, p-vp.
762. Berger, J., Duflo, J., et al., Phys. Rev. Lett. 37(18), 1 Nov. 1976, pp-1195-1198.
763. Berger, J., Bauer, T., et al., CEA-CONF-4132, 1977, p-1.
764. Berger, M. J., Seltzer, S. M., et al., Trans. Am. Nucl. Soc. 14(2), Oct. 1971, pp-887-8.
765. Berger, R., Boucher, R., et al., Info. Bull. Isotope Generators 11, 1971, pp-10-36.
766. Bergere, R., Beil, H., et al., Verh. Dtsch. Phys. Ges. 4, 1978, p-816.
767. Berghaus, U., Brueckmann, H., et al., Verh. Dtsch. Phys. Ges. 4, 1978, p-788.

768. Berghe, G. V., Nucl. Phys., A, 265(3), 26 July 1976, pp-479-492.
769. Berghofer, D., Hasinoff, M. D., et al., Nucl. Phys., A, 263(1), 31 May 1976, pp-109-130.
770. Bergstroem, I., Fransson, K., et al., Voorjaarsvergadering, 26 en 27 April 1976, Amsterdam, p-55.
771. Bergstrom, I., Brown, F., et al., Nucl. Instr. and Meth. 21, 1963, p-249.
772. Bergstrom, J. C., and Tomusiak, E. L., Nucl. Phys., A, 262(2), 17 May 1976, pp-196-204.
773. Bergstrom, J. C., Auer, I. P., et al., Nucl. Phys., A, 251(3), 27 Oct. 1975, pp-401-417.
774. Bergstrom, J. C., Deutschmann, U., et al., Nucl. Phys., A, ISSN 0375-9474 327(2), 24 Sept. 1979, pp-439-457.
775. Beringer, R., and Rall, W., Rev. Sci. Instr. 28(2), Feb. 1957, pp-77-9.
776. Berlanger, M., Deleplanque, M. A., et al., CEA-N-1959 (nd), 1 Sept. 1975 - 30 Sept. 1976, pp-120-121.
777. Berlanger, M., Hanappe, F., et al., European Physical Society, Geneva (Switz.), Sept. 1976.
778. Berlanger, M., Hanappe, F., et al., European Physical Society, Geneva (Switz.), Sept. 1976.
779. Berlanger, M., Hanappe, F., et al., Nucl. Phys., A, 276(2), 24 Jan. 1977, pp-347-353.
780. Berley, D., U. S. Atomic Energy Comm. TID-25473, (1), 1969, pp-251-7.
781. Berman, P., Harry Diamond Labs., Washington, D. C., June 25, 1963, p-17.
782. Berndt, D., Harney, H. L., et al., Verh. Dtsch. Phys. Ges. 4, 1978, p-914.
783. Bernhardt, K. G., Eberhard, K. A., et al., Univ. of Wash., Seattle, U. S. A., 1975.
784a. Bernhardt, K. G., Bohn, H., et al., Wash. Univ., Seattle (U. S. A.), Nuclear Physics Lab., Annual Report, June 1977, pp-100-105.
784b. Bernhardt, K. G., Eberhardt, K. A., et al., Wash. Univ., Seattle (U. S. A.), Nuclear Physics Lab., Annual Report, June 1977, pp-79-80.
785. Bernhardt, K. G., Bohn, H., et al., Wash. Univ., Seattle (U. S. A.), Nuclear Physics Lab., Annual Report, June 1977, pp-118-120.
786. Bernhardt, K. G., Bohn, H., et al., Verh. Dtsch. Phys. Ges. 4, 1978, pp-872-873.
787. Bernhardt, K. G., Bohn, H., et al., Verh. Dtsch. Phys. Ges. 4, 1978, p-873.
788. Bernhardt, K. G., Bohn, H., et al., Verh. Dtsch. Phys. Ges. 4, 1978, p-902.
789. Bernheim, M., Bussiere, A., et al., CEA-N-1861 (nd), 1 Oct. 1974 - 1 Sept. 1975, pp-151-152.

790. Bernheim, M., Bussiere, A., et al., CEA-CONF-4081, 1977, p-1.
791. Bernstein, A. M., Mass. Inst. of Tech., Cambridge (U. S. A.), June 13-24, 1977.
792. Bernstein, A. M., Paras, N., et al., LA-7892C, July 1979.
793. Bernstein, M. E., and Ferguson, M. S., IEEE Trans. on Nucl. Sci., NS-26(1), Feb. 1979.
794. Berovic, N., Clews, C. J., et al., Nucl. Instr. and Meth. 113(3), 1973, pp-417-22.
795. Berrier, G., Vergnes, M., et al., J. Phys. (Paris), 37(4), Apr. 1976, pp-311-328.
796. Berrier-Ronsin, G., Vergnes, M., et al., Inst. de Physique Nucleaire, 1977, p-37.
797. Berrier-Ronsin, G., Vergnes, M., et al., Phys. Rev., C, 17(2), Feb. 1978, pp-529-543.
798. Berry, H. G., INTDS Proc. , 1975, Argonne, U. S. A., Sept. 29 - Oct. 2.
799. Berry, H. G., Rep. Prog. Phys. 40, 1977, p-155.
800. Berry, R. W., Hall, P. M., et al., Thin Film Tech. (NY, Van Nostrand Reinhold), 1968.
801. Bersohn, R., and Lin, S. H., Advan. Chem. Phys. 16, 1969, pp-67-100.
802. Bertault, D., Quebert, J. L., et al., CENBG-7611, 1976, p-18.
803. Bertault, D., Quebert, J. L., et al., Nucl. Phys., A, 276(2), 24 Jan. 1977, pp-229-236.
804. Berthet, B., Inst. des Sciences Nucleaires, 1976, p-61.
805. Berthier, B., Bianchi, L., et al., European Physical Soc., Geneva (Switz.), 6-10 Sept. 1976.
806. Berthier, B., Bianchi, L., et al., CEA-N-1959 (nd), 1 Sept. 1975 - 30 Sept. 1976, pp-37-40.
807. Berthot, J., Fain, J., et al., Univ., Clermont, France
808. Bertin, A., Bois, R., et al., CEA-N-1875, Apr. 1976, pp-71-76.
809. Bertin, A., Bois, R., et al., Trans. Am. Nucl. Soc. 22, Nov. 1975, pp-667-668.
810. Bertin, P., Coupat, B., et al., AIP Conf. Proc. 33, 1976, pp-34-35.
811. Bertini, R., Beurtey, R., et al., CEA-N-1861 (nd), 1 Oct. 1974 - 1 Sept. 1975, pp-203-204.
812. Bertocchi, L., and Troncon, C., Int. Centre for Theoretical Phys., Trieste (Italy), Dec. 1977, p-22.
813. Bertolini, G., Cappellani, F., et al., Nucl. Instr. and Meth. 27, 1964, p-281.
814. Bertolini, G., de Pasquali, G., et al., Nucl. Instr. and Meth. 32, 1965, pp-355-6.
815. Bertolucci, E., Mannelli, I., et al., Nucl. Instr. and Meth. 69, Mar. 1969, pp-21-4.
816. Bertozzi, W., Cambridge Univ. (UK), Dept. of Physics, 1976.

817. Bertrand, F. E., and Kocher, D. C., Phys. Rev., C, 13(6), June 1976, pp-2241-2246.
818. Bertrand, F. E., Oak Ridge National Lab., TN (U. S. A.), DOE/TIC-10198, 1979, p-70.
819. Besch, H. J., Eisermann, H. W., et al., Z. Phys., A, 292(2), Sept. 1979, pp-197-203.
820. Betak, E., Acta Phys. Slovaca 25(4), 1975, pp-264-276.
821. Betak, E., Wiss. Z. Tech. Univ. Dres. 27(2), 1978, pp-305-306.
822. Bethge, K., and Günther, G., Z. Angew. Phys. 17, 1964, p-548.
823. Betts, R. R., Back, B. B., et al., Niels Bohr Inst., Copenhagan, 1975.
824. Betts, R. R., Fortune, H. T., et al., Nucl. Phys., A, 292(1-2), 21-28 Nov. 1977, pp-281-287.
825. Betts, R. R., DiCenzo, S. B., et al., Phys. Rev. Lett. 39(19), 7 Nov. 1977, pp-1183-1186.
826. Betz, H. D., Trans. Nucl. Sci., NS-18(3), June 1971, pp-1110-14.
827. Betz, P., Freiburg Univ., 14 Nov. 1974, p-67.
828. Beurtey, R., and Borghini, M., J. Phys. (Paris), 30, Colloq. C2, May-June 1969, pp-56-63.
829. Beurtey, R., Boudard, A., et al., CEA-N-1959 (nd), 1 Sept. 1975 - 30 Sept. 1976, pp-220-221.
830. Beurtey, R., Bimbot, L., et al., Service de Physique Nucleaire a Moyenne Energie, Chapter 5, Pt. 1.5.
831. Beuscher, H., Davidson, W. F., et al., Verh. Dtsch. Phys. Ges. 10(7), 1975, p-811.
832. Beyer, G. J., Knotek, O., et al., Zentralinst Kernforsch., Rossendorf Dresden (Ber.), 1977, ZfK-332, p-16.
833. Bhang, H., Halpern, I., et al., Nucl. Phys. Lab. Annual Report, 1977, Washington Univ., RLO/1388-362, June 1977, pp-66-69.
834. Bhasin, V. S., Delhi University, India, 1976.
835. Bhattacharyya, P., Chattopadhyay, R. K., et al., Nuovo Cim., A, ISSN 0369-3546, 51(3), 1 June 1979, pp-419-430.
836. Bhowmik, R. K., Chang, C. C., et al., Phys. Rev., C, 13(6), June 1976, pp-2105-2115.
837. Bhowmik, R., Doering, R. R., et al., Z. Phys., A, 280(3), Feb. 1977, pp-267-269.
838. Bhowmik, R. K., Pollacco, E. C., et al., Phys. Rev. Lett. ISSN 0031-9007, 43(9), 27 Aug. 1979, pp-619-623.
839. Bialas, A., Report 1978, FERMILAB-CONF-78/75-THY, CONF-780788-5, p-20.
840. Boridy, E., and Del Bianco, W., Nucl. Instr. and Meth. 92, 1971, pp-111-23.
841. Del Bianco, W., Kundu, S., et al., Can. J. Phys. 55(4), 15 Feb. 1977, pp-302-304.
842. Lo Bianco, G., Molho, N., et al., Phys. Rev., C, 15(6), June 1977, pp-2245-2247.

843. Van Bibber, K., Ledoux, R. J., et al., ANL/PHY-76-2(2), May 1976, pp-811-818.

844. Van Bibber, K., Hendrie, D. L., et al., Lawrence Berkeley Lab., Nuclear Science Annual Report, July 1, 1977 - June 30, 1978, LBL-8151, pp-118-119.

845. Van Bibber, K., Hendrie, D. L., et al., Phys. Rev. Lett. ISSN 0031-9007, 43(12), 17 Sept. 1979, pp-840-844.

846. Bibby, D. M., Oldham, G., et al., Nucl. Energy 11, May-June 1970, pp-68-71.

847. Bibby, D. M., and Oldham, G., Radiochem. Radioanal. Lett. 6(2), 1971, pp-109-13.

848. Bibby, D. M., Oldham, G., et al., Nucl. Instr. and Meth. 94(3), 1971, pp-397-400.

849. Bice, A. N., de Meijer, R. J., et al., Lawrence Berkeley Lab., Nuclear Science Annual Report, July 1, 1977 - June 30, 1978, LBL-8151, pp-3-4.

850. Bice, A. N., Stahel, D. P., et al., Kernfysisch Versneller Inst. KVI Annual Report 1978, pp-25-28.

851. Bichwiller, C., and Meens, A., INTDS Proceedings, 1979, Boston, U. S. A., Oct. 1-3.

852a. Bickes, R. W., Jr., and Bernstein, R. B., Rev. Sci. Instr. 41, May 1970, pp-759-68.

852b. Bieg, K. W., and Chang, J., Tropical meeting on Inertial Confinement Fusion, San Diego, CA, U. S. A., 26-28 Feb. 1980.

853. Biel, J., Bromberg, C., et al., Univ. of Rochester, Rochester, NY, Report 1975, CONF-750654-6, p-14.

854. Biel, J., Bleser, E., et al., Phys. Rev. Lett. 36(17), 26 Apr. 1976, pp-1004-1007.

855. Bieszk, J. A., Notre Dame Univ., Ind. (U. S. A.), Thesis (Ph.D.), 1977, p-177.

856. De Bièvre, P., Lauer, K. F., et al., British Nuclear Energy Society Meeting, Canterbury, Session 4, Paper 28, Sept. 1971.

857. De Bièvre, P., Lauer, K. F., et al., EANDC Symposium on Neutron Standards and Flux Normalization, Argonne National Lab., U. S. A., October 1970.

858. De Bièvre, P., and Debus, G. H., J. Mass Spectro. and Ion Physics 2, 1969, pp-15-23.

859. De Bièvre, P., and Del Bino, G., Anal. Chim. 50, 1970, pp-526-530.

860. Bigler, R. E., Zaidi, S. A. A., et al., Phys. Rev., C, 13(2), Feb. 1976, pp-528-535.

861. Bignami, A., Coceva, C., et al., Comitato Nazionale per l'Energia Nucleare, Bologna (Italy), Sept. 1974, p-40.

862. Bihoreau, B., Coupat, B., et al., CEA-N-1861 (nd), 1 Oct. 1974 - 1 Sept. 1975, pp-158-160.

863. Bilaniuk, O. M., Crozier, D. J., et al., Phys. Rev., C, 16(1), July 1977, pp-55-60.

864. Bilen'kii, S. M., Lapidus, L. I., et al., Tom III, Dubna, Joint Inst. for Nuclear Rsch., 1964.
865. Bilenkii, S. M., Lapidus, L. I., et al., Fortschr. Physik 13, 1965, pp-1-70.
866. Bilenky, S. M., Lapidus, L. I., et al., JINR-P-1634, 1964, p-80.
867. Billen, J. H., Phys. Rev. C, 20(5), Nov. 1979, pp-1648-1672.
868. Billerey, R., Cerruti, C., et al., AIP Conf. Proc. 47, 1978, pp-666-667.
869. Billquist, P. J., Symp. of Northeastern Accelerator Personnel, Oak Ridge, TN, U. S. A., 23-25 Oct. 1978, p-8.
870a. Bimbot, R., Gardes, D., et al., Inst. Phys. Nucl., Univ. Paris-11, Orsay, Fr., Report 1978, IPNO-RC-78-09, p-19.
870b. Bimbot, R., Girard, J., et al., CEA-N-1861 (nd), 1 Oct. 1974 - 1 Sept. 1975, pp-102-103.
871. Bimbot, R., Girard, J., et al., CEA-N-1861 (nd), 1 Oct. 1974 - 1 Sept. 1975, pp-103-104.
872. Bimbot, R., Girard, J., et al., CEA-N-1959 (nd), 1 Sept. 1975 - 30 Sept. 1976, pp-108-113.
873. Binder, I., Fowler, M. M., et al., Lawrence Berkeley Lab., Nuclear Science Annual Report, 1975, LBL-5075, pp-158-9.
874. Bindewald, H., and Knaelmann, M., Atomkernenergie 14(6), 1969, pp-431-4.
875. Binkley, M., Gaines, I., et al., Phys. Rev. Lett. 37(10), 1976, pp-571-4.
876. Bin'kovskii, Yu. A., Nemets, O. F., et al., Pribory i Tekh. Ekspt. 6(5), Sept. - Oct. 1961, p-190.
877. Bini, M., Bizzeti-Sona, A. M., et al., Phys. Rev., C, 18(1), July 1978, pp-108-113.
878. Del Bino, G., Lauer, K. F., et al., Nucl. Instr. and Meth. 93(2), 1971, pp-205-9.
879. Del Bino, G., Lauer, K. F., et al., Int. Rep., Bur. Cent. Mes. Nucl.; C.C.R., Geel, Belgium, 1970.
880. Birchall, J., Baumgartner, E., et al., Helv. Phys. Acta 50(4), 10 Nov. 1977, pp-509-512.
881. Birchall, J., Svenne, J. P., et al., Phys. Rev., C, ISSN 0556-2813, 20(4), Oct. 1979, pp-1585-1588.
882. Birmingham, W. B., Chelton, B. D., et al., Am. Soc. Testing Matls. -Bul. N240, Sept. 1959, pp-34-9.
883. Biro, J., Central Rsch. Inst. for Phys., Budapest, Akademizi Kiado, 1973, pp-625-629.
884. Biron, R. D., MIT-2098-259, 1966, p-14.
885. Biryukov, N. S., Zhuravlev, B. V., et al., Fiziko-Ehnerge-ticheskij Inst., FEI-687, 1976, p-21.
886. Biryukov, N. S., Zhuravlev, B. V., et al., TsNII Atominform. 1977, pp-22-26.
887. Biryukov, N. S., Zhuravlev, B. V., et al., TsNII Atominform. 1977, pp-27-31.

888. Biryukov, N. S., Zhuravlev, B. V., et al., Yad. Fiz. ISSN 0044-0027 29(6), 1979, pp-1443-1448.
889. Bishop, H. E., Brit. J. Appl. Phys. 18, June 1967, pp-703-15.
890. Bishop, R. G., Preston, G., et al., J. Sci. Instr. 31(2), Feb. 1954, pp-64-6.
891. Bister, M., Anttila, A., et al., Phys. Rev., C, 16(4), Oct. 1977, pp-1303-1308.
892. Bister, M., and Anttila, A., Nucl. Instr. and Meth. 77, 1970, pp-315-19.
893. Bistirlich, J. A., Cooper, S., et al., LBL-4268, Jan. 1976, p-12.
894. Bistirlich, J. A., Cooper, S., et al., Phys. Rev. Lett. 36(16), 19 Apr. 1976, pp-942-945.
895. Biswas, S., and Porile, N. T., Phys. Rev., C, 20(4), Oct. 1979, pp-1467-1478.
896. Bittner, G., Kretschmer, W., et al., Verh. Dtsch. Phys. Ges. 4, 1978, pp-808-809.
897. Bittner, G., Kretschmer, W., et al., INTDS Proc., 1978, Garching, FRG, Sept. 11-14.
898. Bittner, G., Kretschmer, W., et al., Nucl. Instr. and Meth. 167(1), 1979, pp-1-8.
899. Bizard, G., Bonthonneau, F., et al., Nucl. Instr. and Meth. 111(3), 1973, pp-445-9.
900. Bizard, G., Bonthonneau, F., et al., Nucl. Instr. and Meth. 129(2), 15 Nov. 1975, pp-569-573.
901. Bizard, G., AIP Conf. Proc., 47, 1978, pp-620-621.
902. Bizzeti, P. C., and Perego, A., Phys. Lett., B, 64(3), 27 Sept. 1976, pp-298-300.
903. Bjerregaard, H. J., Knudsen, P., et al., INTDS Proc., 1979, Boston, U. S. A., Oct. 1-3.
904. Bjørnholm, S., Dam, PH., et al., Nucl. Instr. and Meth. 5, Sept. 1959, pp-196-8.
905. Blair, J. S., Conjeaud, M., et al., CEA-N-1861 (nd), 1 Oct. 1974 - 1 Sept. 1975, pp-15-17.
906. Blair, J. S., Cuengco, B., et al., Wash. Univ., Nuclear Phys. Lab., Annual Report, 1977, RLO/1388-362, June, 1977, pp-60-61.
907. Blanchard, R. L., Kahn, B., et al., Health Physics 2, 1960, pp-246-55.
908. Blankleider, B., and Afnan, I. R., Australian Inst. of Nucl. Sci. and Eng., Lucas Heights, 7th AINSE Nuclear Phys. Conf., 1978.
909. Blann, M., Doering, R. R., et al., Nucl. Phys., A, 257(1), 26 Jan. 1976, pp-15-28.
910. Blann, M., and Komoto, T. T., UCRL-83247, Aug. 1979, p-59.
911. Blanpied, G., Liljestrand, R., et al., Phys. Rev., C, 12(6), Dec. 1975, pp-1726-1729.
912. Blanpied, G. S., Thesis (Ph.D.), 1977, p-175.

913. Blanpied, G. S., Coker, W. R., et al., Phys. Rev. Lett 39(23), 5 Dec. 1977, pp-1447-1450.

914. Blanpied, G. S., LA-7262-T, May 1978, p-137.

915. Blanpied, G., Hintz, N. M., et al., Phys. Rev., C, ISSN 0556-2813, 20(4), Oct. 1979, pp-1490-1497.

916. Blatt, L. S., Nichols, B. D., et al., Nucl. Instr. and Meth. 61(2), May 1, 1968, pp-232-4.

917. Blecher, M., Gotow, K., et al., Phys. Rev., C, ISSN 0556-2813, 20(5), Nov. 1979, pp-1884-1890.

918. Bleszynski, M., and Jaroszewicz, T., Nucl. Phys., A, 256(3), 19 Jan. 1976, pp-429-443.

919. Block, R. C., Harris, D. R., et al., Trans. Am. Nucl. Soc. 27, 1977, pp-868-869.

920. Block, R. C., Nakagome, Y., et al., Trans. Am. Nucl. Soc. 28, June 1978, pp-719-721.

921. Block, R. C., Rensselaer Polytechnic Inst., Troy, NY (U. S. A.), COO-2479-16, June 1979, p-8.

922. Blokhin, A. I., Ignatyuk, A. V., et al., Phys. Rev., C, 17(1), Jan. 1978, pp-227-236.

923. Blommestijn, G. J. F., Haitsma, Y., et al., Delhi, India, ISBN 0720404819, 29 Dec. 1975 - 3 Jan. 1976, pp-212-14.

924. Blons, J., Mazur, C., et al., CEA-CONF-3181, 1975, p-10.

925. Blons, J., Mazur, C., et al., Natl. Bur. Stands. Spec. Publ. 425, Oct. 1975, pp-642-645.

926. Blons, J., Paya, D., et al., CEA-N-1959 (nd), 1 Sept. 1975 - 30 Sept. 1976, pp-90-93.

927. Blons, J., and Derrien, H., J. Phys. (Paris), 37(6), June 1976, pp-659-669.

928. Bloomster, C. H., Battelle-Northwest, Richland, WA, BNWL-519, Dec. 1967, Contract AT(45-1)-1830, p-18.

929. Bloser, M., JUL-834-RG, March 1972, Thesis, p-101.

930. Blosser, H, Gordon, M. M., et al., 5th Int. Cyc. Conf., St. Catherine's Coll., Oxford, 17-20 Sept. 1969.

931. Blue, J. W., and Sodd, V. J., from Int. Conf. on the Use of Cyc. in Chemistry, Metallurgy, and Biology, Oxford, Eng., see CONF-690924.

932. Blue, J. W., and Benjamin, P. P., J. Nucl. Med. 12, 1971, p-416.

933. Blue, J. W., Leonard, R., et al., Report No. NASA-TM-X-73409, E-8740, Conf. presented at Am. Phys. Soc. Meeting, Washington, DC, 26-29 Apr. 1976.

934. Blum, D., Boucrot, J., et al., Nucl. Instr. and Meth. 104(3), 1 Nov., 1972, pp-509-16.

935. Blum, D., Boucrot, J., et al., Laboratoire de l'Accelerateur Lineaire, May 1972, p-20.

936. Blumberg, L. N., Stein, P., et al., Contract W-7405-eng-36, May 1962, LA-2711, p-12.

937. Blumberg, L. N., Gross, E. E., et al., Nucl. Instr. and Meth. 39, Jan. 1966, pp-125-32.

938. Blumberg, L. N. Barton, M. Q., et al., CONF-690811-8, from 7th Int. Conf. on High Energy Acc., Moscow, USSR, 1969.
939a. Blumberg, L. N., and Webster, A. E., IEEE Trans. Nucl. Sci. NS-24(3), June 1977.
939b. Boal, D. H., and Woloshyn, R. M., Phys. Rev., C, ISSN 0556-2813, 20(5), Nov. 1979, pp-1878-1883.
940. Bochin, V. P., IAE-973, 1965, p-20.
941. Bochin, V. P., Isotopenpraxis 3, 1967, pp-317-21.
942. Bochkarev, V. V., and Levin, V. I., CONF-700204-15, from IAEA meeting on Rsch. Reactor Utilization, Rome, Italy, 2 Feb. 1970, pp-243-86.
943. Bockisch, A., Braun, G., et al., Nucl. Instr. and Meth. 161(3), 1979, pp-361-4.
944. Bodansky, D., Jacobs, W. W., et al., Wash. Univ., RLO-1388-297, 10 April 1975, p-38.
945. Bodansky, D., Chiang, D., et al., Wash. Univ., Nucl. Phys. Lab., Annual Report, RLO/1388-362, June 1977, p-1-3.
946. Bodart, F., and Deconninck, G., Proc. of 3rd Conf. on Accel. Targets Designed for the Prod. of Neutrons, Liege (Belgium), Sept. 18-19, 1967, pp-225-37.
947. Bodek, K., Jarczyk, L., et al., INP-939/PL, 1977, p-54.
948. Boehm, R., Christiansen, J., et al., Verh. Dtsch. Phys. Ges. 1, 1977, p-307.
949. de Boer, J. H., and Fast, J. D., Z. Anorg. Chem. 187, 1930, pp-177-198.
950. de Boer, J., Petten, the Netherlands, Reactor Centrum Nederland, March 1975, pp-609-613.
951. de Boer, J., and Seyfarth, H., Autumn meeting, Najaarsvergadering, 15 Oct. 1976, p-8.
952. de Boer, J., Robert van de Graaff Lab., Annual Report, 1977, March 1978, pp-18-19.
953. de Boer, J., Spring meeting of the Dutch Physical Society, March 1978, p-16.
954. Boerma, D. O., Grueebler, W., et al., Nucl. Phys., A, 270(1), 12 Oct. 1976, pp-15-28.
955. Boettger, W., Casel, A., et al., BONN-HE-79-18, Sept. 1979, p-10.
956. Boehin, P. V., Gavrilov, E. B., et al., Isotopenpraxis 7(6), 1971, pp-232-5.
957. Boettinger, J., Davies, J. A., et al., Nucl. Instr. and Meth. 109(3), 1973, pp-579-583.
958. Bogdanovic, M., Koicki, S., et al., Fizika (Zagreb) 9(1), 1977, pp-22-24.
959. Bogomolov, A. V., Vovrhenko, et al., JINR-P-396, 1959, p-37.
960. Borghini, M., Tr. Mezhdunar. Konf. Elektromagn. Vzaimodeistviyam Nizkikh Srednikh Energ., Dubna, USSR, 67, 1967, pp-4, 277-99.
961. Bogoyavlenskii, I. V., and Bereznyak, N. G., Ukr. Fiz. Zh. 10, Dec. 1965, pp-1376-7.

962. Bohlen, H. G., Bohne, W., et al., Phys. Rev. Lett 37(4), 26 July 1976, pp-195-198.
963. Bohlen, H. G., Oertzen, W., von, et al., Z. Phys., A, 292(1), Aug. 1979, pp-105-106.
964. Bokacheva, L. P., Kiselev-Fedorov, V. P., et al., Inzh. Fiz. Zh., Akad. Nauk Belorussk. SSR, 6(9), Sept. 1963, pp-47-51.
965. Bokharee, S. A., State Univ. of NY, Albany (U. S. A.), Thesis (Ph.D.), 1976, p-296.
966. Bokharee, S. A., Emmett, R. W., et al., Contract ET-78-S-02-4933, from ANS annual meeting, Atlanta, GA, U. S. A., 3 June 1979, p-7.
967. Bokov, N. N., Bunatyan, A. A., et al., Pis'ma Zh. Ehksp. Teor. Fiz. 26(9), 5 Nov. 1977, pp-630-634.
968. Boland, B. C., Carne, A., et al., Report, 1978, RL-78-070, p-18.
969. Boldeman, J. W., Musgrove, A. R., de L., et al., Nucl. Phys., A, 269(1), 28 Sept. 1976, pp-31-45.
970. Boldeman, J. W., Musgrove, A. R., de L., et al., Nucl. Sci. and Eng. 64(3), Nov. 1977, pp-744-748.
971. Bolger, J. E., Braithwaite, W. J., et al., Phys. Rev. Lett. 37(18), 1 Nov. 1976, pp-1206-1208.
972. Bolger, J. E., Texas Univ., Austin (U. S. A.), Thesis (Ph.D), 1977, p-163.
973. Bol'shov, V. I., Dubinin, A. A., et al., At. Energ. 28, May 1970, pp-388-92.
974. Bolton, C., Schier, W. A., et al., Phys. Rev., C, 18(1), July 1978, pp-293-300.
975. Bommer, J., Ekpo, M., et al., Nucl. Phys., A, 251(2), 20 Oct. 1975, pp-246-256.
976. Bommer, J., Ekpo, M., et al., Nucl. Phys., A, 251(2), 20 Oct. 1975, pp-257-268.
977. Bommer, J., Ekpo, M., et al., Phys. Rev., C, 12(3), Sept. 1975, pp-1069-1071.
978. Bommer, J., Hahn-Meitner-Institut fuer Kernforschung, Berlin G. m. b. H, p-35.
979a. Bonardi, M., Radiochem. Radioanal. Lett. ISSN 0079-9483, 42(1), 3 Jan. 1980, pp-35-44.
979b. Bonazzola, G. C., Bressani, T., et al., AIP Conf. Proc. 33, 1976, pp-536-537.
980. Bonbright, D. I., Roberts, D. J., et al., University of Manitoba, Winnipeg, Canada, 1975.
981. Bonbright, D. I., Elbakr, S. A., et al., University of Manitoba, Winnipeg, Canada, 1975.
982. Bond, C. D., McGarry, W. J., et al., Nucl. Instr. and Meth. 85, 1 Aug. 1970, pp-85-91.
983. Bond, P. D., Korner, H. J., et al., Phys. Rev., C, 16(1), July 1977, pp-177-182.

984. Bondar, A. D., Emlyaninov, A. S., et al., Pribory i Tekhnika Experimenta 3, 1960, p-134.
985. Bondar, A. D., Emlyaninov, A. S., et al., Instr. Exp. Tech. 3, 1960, pp-493-6.
986. Bondar, A. D., Karev, V. N., et al., Pribory i Tekh. Ekspt. 6(4), July-Aug. 1961, pp-136-9.
987. Bondarenko, V. A., and Prokof'ev, P. T., Nauka, 1978, p-72.
988. Bondelid, R. O., Trans. Nucl. Sci., NS-12(3), June 1965, pp-508-11.
989. Bondouk, I. I., and Saad, S., Atomkernenergie 29(4), 1977, pp-270-271.
990. Boneh, Y., Blocki, J. P., et al., Lawrence Berkeley Lab., LBL-4374, Feb. 1976, p-33.
991. Bonet-Maury, P., Radioprotection 6(2), 1970, pp-143-7.
992. Bonetti, C., CEA-R-4141, Jan. 1971, p-14.
993. Bonetti, C. P., Guay, P. M., et al., INTDS Proc., 1975, Argonne, U. S. A., Sept. 30 - Oct. 2.
994. Bonetti, C. P., Internal Report CEA-CONF-3289, 1975.
995. Bonetti, C. P., Guay, P. M., et al., INTDS Proc., 1975, Argonne, U. S. A., Sept. 30 - Oct. 2
996. Bonetti, C. P., INTDS Proc., 1975, Argonne, U. S. A., Sept. 30 - Oct. 2.
997. Bonetti, C. P., Guay, P. M., et al., Report, 1975, CEA-CONF-3290, p-15.
998. Bonetti, C. P., INTDS Proc., 1978, Garching, FRG, Sept. 11-14.
999. Bonetti, C., Nucl. Instr. and Meth. 167(1), 1979, pp-13-16.
1000. Boos, A. H., Brandt, R., et al., from Int. Topical Conf. on Nucl. Track Registration in Insulating Solids and Applications, Clermont-Ferrand, France, 6 May 1969.
1001. Booth, E. C., Chasan, B., et al., Phys. Rev., C. ISSN 0556-2813, 20(4), Oct. 1979, pp-1217-1220.
1002a. Booth, P. S. L., Boyes, E., et al., LBL-500, 1971, pp-213-15.
1002b. Booth, L., Hill, R. S., et al., J. Sci. Instr. 35, 1958, p-24-6.
1003. Booth, R., UCRL-70183, Feb. 27, 1967, p-13.
1004. Booth, R., IEEE Trans. Nucl. Sci, NS-14, 1967, p-943.
1005. Booth, R., Nucl. Instr. and Meth. 59, Feb. 1968, UCRL-70661, pp-131-5.
1006. Booth, R., and Barschall, H. H., Nucl. Instr. and Meth. 99(1), 1972, pp-1-4.
1007. Booth, R., Barschall, H. H., et al., UCRL-74285, 21 Feb. 1973, p-5.
1008. Booth, R., Nucl. Instr. and Meth. 120(2), 1974, p-353.
1009. Booth, R., et al., Nucl. Instr. and Meth. 145, 1977, p-25.
1010. Booth, R. S., ORNL-TM-2925, 31 March 1970, p-57.
1011. Boothe, T. E., and Ache, H. J., J. Nucl. Mater. ISSN 0022-3115, 84(1-2), Oct. 1979, pp-85-92.

1012. Borie, E., Phys. Lett., B, 68(5), 4 July 1977, pp-433-435.
1013. Borchers, R. R., and Wood, R. M., Nucl. Instr. and Meth. 35, 1965, pp-138-40.
1014. Borg, K., van der, Meyer, R. J., de, et al., Nederlandse Natuurkundige Vereniging, Amsterdam, Spring meeting, 26-27 April 1976, Amsterdam, Voorjaarsvergadering, 26-27 April 1976, Amsterdam, p-56.
1015. Borghini, M., "Progress in Fast Neutron Physics", Chicago, Univ. of Chicago Press, 1963, pp-341-7.
1016. Borghini, M., Bull. Inform. Sci. Tech. (Paris), 106, July-Aug. 1966, pp-17-34.
1017. Borghini, M., CERN-66-3, (Switz.), Jan. 31, 1966, p-29.
1018. Borghini, M., CONF-661205, from Conf. on Polar. Targets and Sources, Saclay, France, Dec. 5-9, 1966.
1019. Borghini, M., Roubeau, P., et al., Nucl. Instr. and Meth. 49, 1967, pp-248-58.
1020. Borghini, M., Proc. of High Energy Phys. Meeting, Pisa, Italy, Oct. 9-11, 1967, pp-33-41.
1021. Borghini, M., Chamberlain, O., et al., Nucl. Instr. and Meth. 84, 1970, pp-168-172.
1022. Borghini, M., from Second Int. Conf. on Polar. Targets, Berkeley, CA, 30 Aug. 1971.
1023. Borghini, M., from Conf. of Moriond Meeting on the Phenomenology of Weak and Electromagnetic Interactions, Meribel-les-Allues, France, 7 March 1971.
1024. Borghini, M., Masaike, A., et al., Nucl. Instr. and Meth. 97(3), 1971, pp-577-9.
1025. Borghini, M., and Scheffler, K., Nucl. Instr. and Meth. 95(1), 1971, pp-93-8.
1026. Borghini, M., Niinikoski, T. O., et al., Nucl. Instr. and Meth. 105(2), Dec. 1972, pp-215-220.
1027. Borghini, M., BNL-20415, June 3-8, 1974, Dec. 1975, pp-315-23.
1028. Borghini, M., BNL-20415, June 3-8, 1974, Dec. 1975, pp-325-30.
1029. Born, M., and Wolf, E., Principles of Optics, Pergamon Press, 1970.
1030. Bornand, M., Plattner, G. R., et al., Helv. Phys. Acta 49(5), 29 Oct. 1976, p-788.
1031. Le Bornec, Y., Tatischeff, B., et al., Phys. Lett., B, 61(1), 1 March 1976, pp-47-49.
1032. Borovikova, N. V., Vesna, V. A., et al., Pis'ma Zh. Ehksp. Teor. Fiz. ISSN 0370-274X, 30(8), 20 Oct. 1979, pp-527-532.
1033. Borovlev, V. I., and Lopatko, I. D., Izv. Akad. Nauk SSSR, Ser. Fiz. 39(8), Aug. 1975, pp-1758-1760.
1034. Borsay, F. L., Nucl. Instr. and Meth. 52, 1967, pp-338-40.
1035. Borukhovich, G. Z., Zvezdkina, T. K., et al., Nauka, 1978, p-366.

1036. Borzunov, Yu. T., Golovanov, L. B., et al., JINR-P-8-5212, 1970, p-17.
1037. Borzunov, Yu. T., Golovanov, L. B., et al., Prib. Tekh. Eksp. 14(5), Sept.-Oct. 1971, pp-48-50.
1038. Borzunov, Yu. T., Golovanov, L. B., et al., Prib. Tekh. Eksp. 14(3), May-June 1971, pp-52-4.
1039. Borzunov, Yu. T., Golvanov, L. B., et al., Cryogenics 12(3), June 1972, pp-235-6.
1040. Borzunov, Yu. T., Golovanov, L. B., et al., Cryogenics 12(6), 1972, pp-462-3.
1041. Borzunov, Yu. T., Bolovanov, L. B., et al., JINR-8-7413, 1973, p-11.
1042. Borzunov, Yu. T., Golovanov, L. B., et al., Instr. Exp. Tech. 17(4), 1974, pp-963-66.
1043. Borzunov, Yu. T., Golovanov, L. B., et al., JINR-8-8991, 1975, p-14.
1044. Bosman, M., Jongen, Y., et al., Nucl. Instr. and Meth. 126(3), 1975, pp-465-6.
1045. Bosman, J. J., Delhey, P. P. J., et al., Reactor Centrum Nederland, March 1975, pp-638-642.
1046. Bosman, M., Leleux, P., et al., Nucl. Instr. and Meth. 148(2), Jan. 15, 1978, pp-363-367.
1047. Bosted, P. E., Blomqvist, K. I., et al., MIT (USA), 1979.
1048. Bothur, E. H., Clausnitzer, G., et al., Nucl. Instr. and Meth. 121(3), 1 Nov. 1974, pp-533-535.
1049. Bouchard, C. A., INTDS Proc., 1974, Chalk River, Canada, Oct. 1-3.
1050. Boudrie, R. L., Brissaud, I., et al., Colo. Univ., Boulder (U. S. A.), 1 Nov. 1977, pp-106-107.
1051. Bougon, M., Marquet, M., et al., Proc. Congr. Refrigeration, 10th Congr., Copenhagen, 1, 1959, pp-207-10.
1052. Bounin, P., Lab. Naz. Frascati-65/22, 19 Jul. 1965, p-28.
1053. Bourgoignie, R. R., Nucl. Instr. and Meth. 72, 1969, pp-277-84.
1054. Boussard, D., No. 1316, Paris, Presses Universitaires de France, 1968, p-136.
1055. Bowen, T., DeLise, D. A., et al., Phys. Rev., C, 14(1), July 1976, pp-195-203.
1056. Bowles, T., Geesaman, D. F., et al., Phys. Rev. Lett. 40(2), 9 Jan. 1978, pp-97-99.
1057. Bowles, T. J., Robertson, R. G. H., et al., ANL-78-66, 1 April 1977 - 31 March 1978, pp-80-81.
1058. Bowles, T. J., Holt, R. J., et al., CONF-790610-5, 1979, ANL, p-25.
1059. Bowman, C. D., Auchampaugh, G. F., et al., Phys. Rev. 166, 1968, p-1219.
1060. Bowman, C. D., IEEE Trans. Nucl. Sci., NS-26(1), 2 Feb. 1979.
1061. Bowman, J. D., Proc. of the Int. Symp. on Interaction Studies in Nuclei held in Mainz, F. R. Germany, 17-20 Feb. 1975.

1062. Bowman, D., Baer, H., et al., LA-7892C, July 1979.
1063. Bowman, J. D., LA-UR-79-2377, CONF-790847-3, 1979, p-10.
1064. Bowman, W. W., Sugihara, T. T., et al., Nucl. Instr. and Meth. 103(1), 1972, pp-61-7.
1065. Box, W. D., TISRMNM, ORNL, Oct. 5-8, 1971.
1066. Boyard, J. L., Inst. de Physique Nucleaire, Orsay, France, These (D. es S.), 1975, p-66.
1067. Boyd, R. N., Elmore, D., et al., Phys. Rev., C, 14(1), July 1976, pp-4-7.
1068. Boyer, A. L., Med. Phys. ISSN 0094-2405, 6(5), Sept. 1979, pp-454-456.
1069. Boyer, K., Goves, E. H., et al., Rev. Sci. Instr. 22(5), 1951, pp-310-20.
1070. Boyes, E., Court, G. R., et al., LBL-500, 1971, pp-403-6.
1071. Bozek, E., Gehringer, C., et al., Centre de Recherches Nucleaires, 1975, p-6.
1072. Bozek, E., Hrynkiewicz, A. Z., et al., Phys. Rev., C, 12(6), Dec. 1975, pp-1873-1877.
1073. Bozler, H. M., and Graf, E. H., LBL-500, 1971, pp-103-6.
1074. Bradley, D. E., Brit. J. Appl. Phys. 5, 1954, p-55.
1075. Bradley, D. E., Technique for Electron Microscopy, ed. D. Kay, Oxford: Blackwell, 1965, pp-63-67.
1076. Brady, F. P., King, N. S. P., et al., Phys. Rev., C, 16(1), July 1977, pp-31-41.
1077. Brady, F. P., Viggars, D. A., et al., Phys. Rev. Lett. 39(14), 3 Oct. 1977, pp-870-873.
1078. Braendle, H., Meyer, V., et al., Nucl. Phys., A, 256(1), 5 Jan. 1976, pp-141-151.
1079. Braithwaite, W. J., and Obst, A. W., Phys. Rev., C, 16(1), July 1977, pp-224-226.
1080. Bramlitt, E. T., Ark. Univ., Fayetteville, 1962, p-227.
1081. Brandan, M. E., and Haeberli, W., Nucl. Phys., A, 287(2), 5 Sept. 1977, pp-213-219.
1082. Brandan, M. E., Plattner, G. R., et al., Nucl. Phys., A, 263(2), 7 June 1976, pp-189-192.
1083. Brandan, M. E., and Haeberli, W., Nucl. Phys., A, 287(2), 5 Sept. 1977, pp-205-212.
1084. Brandolini, F., Signorini, C., et al., Nucl. Instr. and Meth. 91, 1971, pp-341-4.
1085. Branford, D., Nagorcka, B. N., et al., Australian National Univ., Sept. 1977, p-22.
1086. Branson, J. G., Sanders, G. H., et al., Phys. Rev. Lett. 38(23), 6 June 1977, pp-1334-1337.
1087. Branson, J. G., Sanders, G. H., et al., Phys. Rev. Lett. 38(23), 6 June 1977, pp-1331-1334.
1088. Braski, D. N., TISRMNM, ORNL, Oct. 5-8, 1971.
1089. Braski, D. N., Nucl. Instr. and Meth. 102, 1972, p-553.
1090. Braumandl, F., Egidy, T. von, et al., Z. Phys., A, 292(4), Nov. 1979, pp-397-398.

1091. Braun-Munzinger, P., Gelbke, C. K., et al., Phys. Rev. Lett. 36(15), 12 April 1976, pp-849-852.
1092. Bray, K. H., Frawley, A. D., et al., Nucl. Phys., A, 288(2), 26 Sept. 1977, pp-334-350.
1093. Bremen, W., Naoumidis, A., et al., Inst. fuer Reaktorwerkstoffe und Heisse Zellen, June 1977, p-67.
1094. Bressers, J., Cassanelli, G., et al., Comm. Eur. Communities, (Rep.) EUR, 1978, (EUR 5908), p-35.
1095. Breynat, G., Fiat, G., et al., EUR-1815-E (Rev.), 1964, p-263.
1096. Breynat, G., Dubus, M., et al., Bull. Inform. Sci. Tech. (Paris), 98, 1965, p-81.
1097. Breynat, G., Fiat, G., et al., EUR-1815-E (Rev.), p-263-70.
1098. Breynat, G., and Manin, A., Bull. Inform. Sci. Tech. (Paris), 178, Feb. 1973, pp-59-64.
1099. Brice, D. K., Appl. Phys. Lett. 29(1), 1976, pp-10-12.
1100. Bridwell, L., Beyer, L. M., et al., Nucl. Instr. and Meth. 90, 1970, pp-187-96.
1101. Briggs, A. B., KAPL-M-S3G-RE-507, SAR Project, Contract W-31-109-eng-52, Sept. 5, 1956, p-12.
1102. Brink, B. O., ten, Nes, P., van, et al., Vrije Universiteit, Amsterdam, Natuurkundig Lab.), Oct. 1977, p-13.
1103. Brinkmann, U., Hartig, W., et al., VIII Int. Conf. on Quantum Elect., San Francisco, 1974.
1104. Brissaud, I., and Brussel, M. K., Phys. Rev., C, 15(1), Jan. 1977, pp-452-455.
1105. Brissaud, I., Berrier-Ronsin, G., et al., Z. Phys., A, 293(1), 1979, pp-1-4.
1106. Britt, Gavron, et al., BNL-NCS-22500, March 1977, pp-126-127.
1107. Britt, Gavron, et al., BNL-NCS-22500, March 1977, pp-108-10.
1108. Brochard, F., Chevallier, P., et al., Centre de Recherches Nucleaires, Strasbourg, 1975, pp-524-525.
1109. Brochard, F., Chevallier, P., et al., Centre de Recherches Nucleaires, Strasbourg, 1976, p-27.
1110. Broda, R., Kulessa, R., et al., CEA, 1976, p-93.
1111. Broda, R., Kulessa, R., et al., CEA, 1976, p-92.
1112. Broerse, J. J., Engels, A. C., et al., Eur. J. Cancer 7(2-3), May 1971, pp-105-14.
1113. Broerse, J. J., Engels, A. C., et al., Int. J. Appl. Radiat. Isotop. 22(8), Aug. 1971, pp-486-9.
1114. Brogden, T. W. P., AERE-R-5328, Dec. 1966, p-26.
1115. Brolley, E. J., 1973 Part. Accel. Conf., Accel. Eng. and Tech., San Francisco, CA, March 5-7.
1116. Bromberg, C., Fox, G., et al., Phys. Rev. Lett. 38(25), 20 June 1977, pp-1447-1450.
1117. Bron, J., Hesselink, W. H. A., et al., Dutch Physical Soc., March 1978, p-17.
1118. Brooks, J. M., and Otavka, M. A., Rev. Sci. Instr. 39, Sept. 1968, pp-1348-50.

1119. Brotschi, U., Menetrey, A., et al., Nucl. Instr. and Meth. 129(1), 1 Nov. 1975, pp-39-41.
1120. Brown, Haglund, et al., BNL-NCS-22500, March 1977, pp-106-107, 115.
1121. Brown, B. A., Fossan, D. B., et al., Phys. Rev., C, 14(3), Sept. 1976, pp-1016-1022.
1122. Brown, B. A. Jachcinski, C., et al., Phys. Rev., C, 16(1), July 1977, pp-480-481.
1123. Brown, D., (John Wiley & Sons, Ltd., London), 1971.
1124. Brown, D. R., Wash. Univ., Seattle (U. S. A.), Thesis (Ph.D), 1975, p-282.
1125. Brown, D. R., Halpern, I., et al., Phys. Rev., C, 14(3), Sept. 1976, pp-896-907.
1126. Brown, G. E., BNL-50445, 1974, pp-18-33.
1127. Brown, L. C., Callahan, A. P., et al., Int. J. Appl. Radiat. Isotop. 24(11), 1973, pp-651-5.
1128. Brown, R. E., John H. Williams Lab. of Nuclear Phys., Univ. of Minn., Annual Report, 1975, Sept., pp-44-55.
1129. Brown, S. C., Irvine, J. W., Jr., et al., J. Chem. Phys. 12, 1944, pp-132-4.
1130. Brown, W. K., and Watkins, D. W., LA-4443, Contract W-7405-eng-36, May 1970, p-7.
1131. Browne, C. P., and Michael, I. Phys. Rev. B, 134, 1964, p-133.
1132. Browne, J. C., Lamaze, G. P., et al., Phys. Rev., C, 14(3), Sept. 1976, pp-1287-1288.
1133. Brownell, R. B., McLennan, W. D., et al., Rev. Sci. Instr. 35, 1964, p-1147.
1134. Bruandet, J. F., Agard, M., et al., Phys. Rev., C, 12(6), Dec. 1975, pp-1739-1744.
1135. Bruandet, J. F., Berthet, B., et al., Inst. des Sciences Nucleaires, 1976, p-21.
1136. Bruandet, J. F., Berthet, B., et al., Phys. Rev., C, 14(1), July 1976, pp-103-108.
1137. Brueckmann, H., Lueck, W., et al., from 3rd Int. Symp. on Polarization Phenomena in Nuclear Reactions, Madison, WI, Aug. - Sept. 1970.
1138. Brueckner, W., Granz, B., et al., Phys. Lett., B, 62(4), 21 June 1976, pp-481-484.
1139. Bruggeman, A., Maenhaut, W., et al., J. Radioanal. Chem. 23(1-2), 1974, pp-131-146.
1140. Brunet, J. C., and Hauviller, C., Nucl. Instr. and Meth. ISSN 0029-554X, 153(1), 1 July 1978, pp-59-60.
1141. Bruneton, C., Bystritski, I., et al., Prib. Tekh. Eksp. 19(5), Pt. 1, Sept. - Oct. 1976, pp-46-51.
1142. Bruneton, C., Instrum. Exp. Tech. 19(5), Pt. 1, Sept. - Oct. 1976, pp-1300-1305.
1143. Bruno, M., D'Agostino, M., et al., Nuovo Cimento Soc. Ital. Fis. 22(14), 1978, pp-556-8.

1144. Bruninx, E., and Rudstam, G., Nucl. Instr. and Meth. 13, 1961, p-131.
1145. Bruninx, E., and Crombeen, J., Proc. of the 3rd Conf. on Accel. Targets Designed for the Prod. of Neutrons, Liege (Belgium), Sept. 18-19, 1967, pp-247-60.
1146. Bruninx, E., Proc. of a Conf. held at St. Catherine's Coll., Oxford, 22-23 Sept. 1969.
1147. Bryant, F. J., and Radford, C. J., Cryogenics 10, Aug. 1970, pp-329-31.
1148. Bubb, E., Altemus, R., et al., ORO-4043-37, Dec. 1, 1976 - Nov. 30, 1977, pp-26-29, Virg. Univ., U. S. A.
1149. Bubb, I. F., Nucl. Instr. and Meth. 92, 1971, pp-29-32.
1150. Buch, J., Balalikin, N. I., et al., Nucl. Instr. and Meth. 137(2), 1 Sept. 1976, pp-381-384.
1151. Buchanan, J. A., and Clement, J. M., IEEE Trans. Nucl. Sci., NS-26(4), 1979, pp-4614-17.
1152. Bucher, W., Hollandsworth, C. E., et al., Natl. Bur. of Stds. (U. S.), Spec. Publ. 425, Oct. 1975, pp-946-949.
1153. Buchnea, A., Johnson, R. G., et al., Can. J. Phys. 55(4), 15 Feb. 1977, pp-364-369.
1154. Buchnea, A., Johnson, R. G., et al., Can. J. Phys. 56(1), 1 Jan. 1978, pp-47-51.
1155. Buck, W., Schenk, K., et al., Univ. Tuebingen (Ger.), 1975.
1156. Buck, W., Hoyler, F., et al., AIP Conf. Proc. ISSN 0094-243X, 47, 1978, pp-694-695.
1157. Bucurescu, D., Dragulescu, E., et al., Rev. Roum. Phys. 21(1), 1976, pp-97-104.
1158. Budal, K., IEEE Trans. Nucl. Sci., NS-14, June 1967, pp-1132-7.
1159. Budal, K., CERN-67-17, July 11, 1967, p-104.
1160. Budker, G. I., Onuchin, A. P., et al., Yadern. Fiz. 6, Oct. 1967, pp-775-9.
1161. Budker, G. I., Dikanskij, N. S., et al., 10th Int. Conf. on High Energy Accel., 1977, p-78-79.
1162. Budzanowski, A., Jarczyk, L., et al., Nucl. Phys., A, 265(3), 26 July 1976, pp-461-478.
1163. Budzanowski, A., Alderliesten, C., et al., Inst. fuer Angewandte Kernphysik 1, July 1979.
1164. Bueker, H., Kerntechnik 9, Sept. 1967, pp-409-13.
1165. Buelow, B., Johnson, B., et al., Z. Phys., A, 278(1), July 1976, pp-89-95.
1166. Buelow, B., Johnsson, B., et al., Z. Phys., A, 282(3), July 1977, pp-261-265.
1167. Buenerd, M., Gelbke, C. K., et al., Lawrence Berkeley Lab., Annual Report, LBL-5075, 1975, pp-115-117.
1168. Buenerd, M., Inst. des Sciences Nucleaires, These (D. es S.), 1975, p-146, ISN-75-24.
1169. Buenerd, M., Phys. Rev., C, 13(1), Jan. 1976, pp-444-446.

1170. Buenerd, M., Gelbke, C. K., et al., Phys. Rev. Lett 37(18), 1 Nov. 1976, pp-1191-1194.
1171. Buenerd, M., Gelbke, C. K., et al., CA Univ., Berkeley, 1976.
1172. Buenerd, M., Martin, P., et al., Inst. des Sciences Nucleaires, March 1977, p-46.
1173. Buenerd, M., Martin, P., et al., Nucl. Phys., A, 286(3), 22 Aug. 1977, pp-377-402.
1174. Buford, F. A., Freeman, J. H., et al., AERE-R-5097, Paper 13, 1965, p-12.
1175. Buford, F. A., Freeman, J. H., et al., Proc. Semin. on Prep. and Stand. of Isotopic Targets and Foils, Harwell, Oxon, AERE-R5097, 1965, pp-66-7.
1176. Bugg, D. V., Carter, A. A., et al., Birmingham Univ. (England), Report No. RPP/H-28, 1967, p-23.
1177. Buhler, S., from 3rd Int. Cryo. Eng. Conf., Berlin, Germany, 25 May 1970.
1178. Buhler, S., Proc. of the 5th Int. Cryo. Eng. Conf., Kyoto, 1974, pp-553-555.
1179. Bulaev, O. F., Gorbunov, V. I., et al., Defektoskopiya 2, 1968, pp-68-70.
1180. Bulgakov, M. I., Gul'ko, A. D., et al., ITEF-875, 1971, p-29.
1181. Bulkin, V. S., Gorpinich, O. K., et al., Nauka, 1978, p-197.
1182. Bunakov, V. E., Ermakov, K. N., et al., Izv. Akad. Nauk SSSR, Ser. Fiz. ISSN 0367-6765, 42(9), Sept. 1978, pp-1906-1910.
1183. Burch, D. R., Cramer, J. S., et al., Annual Report, Univ. of Wash. at Seattle, 1975, pp-18-20.
1184. Burch, W. D., ORNL-3482, Oct. 8, 1963, Contract W-7405-eng-26, p-68.
1185. Burch, W. D., ORNL-3597, May 1964, Contract W-7405-eng-26, p-78.
1186. Burch, W. D., ORNL-3651, Oct. 1964, Contract W-7405-eng-26, p-75.
1187. Burch, W. D., ORNL-3739, Contract W-7405-eng-26, May 1965, p-79.
1188. Burch, W. D., ORNL-3847, Contract W-7405-eng-26, Sept. 1965, p-60.
1189. Burch, W. D., ORNL-3880, Contract W-7405-eng-26, Jan. 1966, p-60.
1190. Burcham, W. E., Squier, G. T. A., et al., Nucl. Instr. and Meth. ISSN 0029-554X, 164(3), 1 Sept. 1979, pp-533-537.
1191. Burford, F. A., Freeman, J. H., et al., PSPSITF, AERE, Oct. 20-21, 1965.
1192. Burge, E. J., Nucl. Instr. and Meth. 50, 1967, pp-209-12.
1193. Burgess, C. A., BNWL-CC-368, Contract E(45-1)-1830, 22 Nov. 1965, p-22.
1194. Burgerjon, J. J., Arrott, A. S., et al., IEEE Trans. Nucl. Sci., NS-26(3), Pt. 1, June 1979.

1195. Burget, J., Odehna, M., et al., Czech. Report, 1972, UJF-2899-F, p-67.
1196. Burget, J., Petricek, V., et al., Czech. J. Phys. B17, 1967, pp-1041-7.
1197. Burget, J., Odehnal, M., et al., Nucl. Instr. and Meth. 85, 1 Aug. 1970, pp-81-4.
1198. Burke, D. G., and Tippett, J. C., Nucl. Instr. and Meth. 63, Aug. 15, 1968, pp-353-4.
1199. Burleson, G., Hicks, G., et al., Phys. Rev., D, 12(9), 1 Nov. 1975, pp-2557-2560.
1200. Burman, C., and Bakhru, H., Nucl. Instr. and Meth. ISSN 0029-554X, 165(2), 1 Oct. 1979, p-355.
1201. Burnereau, N., Centre d'Etudes Nucleaires, 1975, p-87.
1202. Burns, G. W., and Reed, V. R. C., Proc. of Conf. held at St. Catherine's Coll., Oxford, 22-23 Sept. 1969.
1203. Burn, N., and Bender, L. B., ACNTDS, CRNL, Oct. 1-3, 1974, p-1974.
1204. Burns, E. J. T., Johnson, D. J., et al., Appl. Phys. Lett. 35(2), 15 July 1979, pp-140-142.
1205. Burrill, E. A., and Hirschfield, J., ANL-6515, p-9-27.
1206. Burson, Z. G., Trans. Amer. Nucl. Soc. 13, June-July 1970, p-265.
1207. Burson, Z. G., U. S. A. E. C., CEX-65.04, 1970, p-82.
1208. Burson, Z. G., U. S. A. E. C., CEX-65.04, March 1971, pp-18-23.
1209. Burt, R. J., Meyer, S. F., et al., J. Vac. Sci. Technol. 17(1), 1980, pp-407-10.
1210. Burtsev, V. A., Grad, G. A., et al., Summaries of reports of the all-union conf. on Thermonuclear Reactor Tech., Leningrad, 28-30 June 1977.
1211. Buschner, R., Junge, J., et al., Verh. Dtsch. Phys. Ges. 4, 1978, p-912.
1212. Buschmann, J., Gils, H. J., et al., Verh. Dtsch. Phys. Ges. 4, 1978, p-945.
1213. Busharov, N. P., Gorbatov, E. A., et al., J. Nucl. Mater. 63, Dec. 1976, pp-230-234.
1214. Busharov, N. P., Gorbatov, E. A., et al., Sov. J. Plasma Phys. 2(4), July 1976, pp-321-323.
1215. Bushnell, D. L., Tassotto, G. R., et al., Phys. Rev., C, 14(1), July 1976, pp-75-91.
1216. Bushuev, N. N., Zh. Neorg. Khim. 22(8), Aug. 1977, pp-2215-2219.
1217. Bussiere, J., and Robson, J. M., Nucl. Instr. and Meth. 91(1), Jan. 1, 1971, pp-103-7.
1218. Bussoletti, J., Ebisawa, K., et al., Nucl. Phys. Lab., Wash. Univ., Annual Report, June 1977, pp-41-43.
1219. Bussoletti, J. E., Hasinoff, M. D., Nucl. Phys. Lab., Wash. Univ., June 1977, Annual Report, pp-48-51.
1220. Bussoletti, J. E., Ebisawa, K., et al., Nucl. Phys. Lab., Wash. Univ., June 1977, Annual Report, pp-53-59.

1221. Butler, G. W., Perry, D. G., et al., Phys. Rev. Lett. 38 (24), 13 June, 1977, pp-1380-1382.
1222. Butler, T. A., ORNL-TM-530 (Rev.), Contract W-7405-eng-26, Aug. 28, 1963, p-20.
1223. Butsev, V. S., and Chuttem, D., Phys. Lett., B, 67(1), 14 March 1977, pp-33-34.
1224. Buzhin'ski, S., Zhupran'ski, P., et al., Izv. Akad. Nauk SSSR, Ser. Fiz. ISSN 0367-6765, 43(1), Jan. 1979, pp-158-159.
1225. Bystritskii, V. M., Mekhedov, B. N., et al., Prib. Tekh. Eksp. 6, Nov. - Dec. 1968, pp-70-2.
1226. Cabot, C., Deprun, C., et al., C. R., Ser. B, 281(18), 03 Nov. 1975, pp-453-455.
1227. Cabot, C., Gauvin, H., et al., J. Phys. (Paris) Lett. 36 (12), Dec. 1975, pp-L-289-L-292.
1228. Cabot, C., Gauvin, H., et al., European Physical Soc., Geneva, Europ. Conf. on Nuclear Phys. with Heavy Ions, 6-10 September 1976.
1229. Cabot, C., Inst. de Physique Nucleaire, INKA-Conf-79-001-051, pp-120-124.
1230. Cadmus, R. R., Jr., and Haeberli, W., Nucl. Phys., A ISSN 0375-9474, 327(2), 24 Sept. 1979, pp-419-438.
1231. Cagle, T. G., Proc. of Am. Electroplaters Soc., Educational Sessions, Jenkintown, PA, 1944, pp-20-76.
1232. Cahn, L., and Schultz, H. R., Vac. Microbal. Tech. 3, 1963, NY, Plenum, p-29.
1233. Calarco, J. R., Wissink, S. W., et al., Phys. Rev. Lett. 39(15), 10 Oct. 1977, pp-925-928.
1234. Call, T., Stoner, J. O., Jr., et al., INTDS Proc., 1978, Garching, Sept. 11-14.
1235. Callaway, D. J. E., Wilets, L., et al., Nucl. Phys., A, ISSN 0375-9474, 327(1), 17 Sept. 1979, pp-250-268.
1236. Camarda, H. S., Phys. Rev., C, 16(5), Nov. 1977, pp-1803-1811.
1237. Camarda, H. S., Anderson, J. D., et al., Calif. Univ., Livermore, U. S. A., UCID-17732, 16 Feb. 1978, p-6.
1238. Cambou, F., and Rème, H., J. Phys. (Paris) 25(3), March 1964, pp-61A-3A.
1239. Cameron, C. P., Roberson, N. R., et al., Phys. Rev., C, 14(2), Aug. 1976, pp-553-562.
1240. Cammarata, J. B., and Banerjee, M. K., Phys. Rev., C, 12 (5), Nov. 1975, pp-1595-1606.
1241. Cammarata, J. B., and Donnelly, T. W., AIP Conf. Proc., 1976, 33 (Meson-Nucl. Phys.), pp-618-19.
1242. Cammarata, J. B., and Banerjee, M. K., Phys. Rev., C, 13 (1), Jan. 1976, pp-299-314.
1243. Campbell, D. S., and Blackburn, H., Trans. 7th Nat. Vac. Symp., Oxford: Pergamon, 1960, p-313.
1244. Campbell, J., and Scott, C. M., Sci. and Ind. Appl. of Small Accel., 4th Conf., 1976.

1245. Campbell, J. I., Nucl. Instr. and Meth. 142(1-2), 1-15 April, 1977, pp-263-273.
1246. Cance, M., and Grenier, G., Trans. Am. Nucl. Soc. 22, Nov. 1975, pp-664-665.
1247. Cance, M., and Grenier, G., CEA-CONF-3291, 1975, p-16.
1248. Cance, M., Gimat, D., et al., CEA-N-1969, June 1977, pp-56-63.
1249. Cannell, L. E., Ill. Univ., Urbana (U. S. A.), Thesis (Ph.D), 1976, p-143.
1250. Cannell, L. E., Zurmuehle, R. W., et al., Phys. Rev. Lett. ISSN 0031-9007, 43(12), 17 Sept. 1979, pp-837-840.
1251. Canto, L. F., European Physical Soc., Geneva, European Conf. on Nucl. Phys. with Heavy Ions, 6-10 Sept. 1976.
1952. Canty, M. J., Davison, N. E., et al., Nucl. Phys., A, 265 (1), 12 July 1976, pp-1-34.
1253. Canty, M., Busch, F., et al., Verh. Dtsch. Phys. Ges. 4, 1978, pp-894-895.
1254. Caplar, R., Fick, D., et al., Verh. Dtsch. Phys. Ges. 4, 1978, p-812.
1255. Carchon, R., Van de Vyver, R., et al., Phys. Rev., C, 14(2), Aug. 1976, pp-456-467.
1256. Caretto, A. A., Jr., Contract AT(30-1)-2897, CMV, TID-22024, p-42.
1257. Carlino, L., Musset, P., et al., Rev. Sci. Instr. 35, Oct. 1964, pp-1274-7.
1258. Carlson, A. D., Natl. Bur. of Stands. (U. S.), Spec. Publ. 1, Oct. 1975, pp-293-301.
1259. Carlson, G. W., Behrens, J. W., et al., Nucl. Sci. Eng. 63 (2), June 1977, pp-149-152.
1260. Carlson, J. D., Nucl. Instr. and Meth. 113(4), 1973, pp-541-3.
1261. Carlson, R. F., Short, T. H., et al., Nucl. Instr. and Meth. 123(3), 1 Feb. 1975, pp-509-519.
1262a. Carne, A., Ber. Kernforschungsanlage Juelich (Conf.) 1980, Jul-Conf-34.
1262b. Carroll, A. S., Chiang, I., et al., Phys. Rev., C, 14(2), Aug. 1976, pp-635-638.
1263. Carr, D. L., Rev. Sci. Instr. 40(7), 1969, pp-965-6.
1264. Carraz, L. C., Sundell, S., et al., Nucl. Instr. and Meth. 158(1), Jan. 1, 1979, pp-69-80.
1265. Carraz, L. C., Haldorsen, I. R., et al., Nucl. Instr. and Meth. 148(2), 15 Jan. 1978, pp-217-230.
1266. Carruth, R. T., Auburn Univ., ALA (U. S. A.), Thesis (Ph.D), 1975, p-76.
1267. Carswell, D. J., and Milsted, J., J. Nucl. Ener. 4, 1957, p-51.
1268. Carter, G., and Colligan, J. S., Ion Bombardment of Solids, Am. Elsevier, 1968.

1269. Carter, J., Clarkson, R. G., et al., Europ. Phys. Soc., Geneva, Europ. Conf. on Nucl. Phys. with Heavy Ions, 6-10 Sept. 1976.
1270. Carter, J., Clarkson, R. G., et al., Nucl. Phys., A, 297(3), 20 March 1978, pp-520-532.
1271. Caruana, J., Mathur, J. N., et al., Aust. Inst. of Nucl. Sci. and Eng., Lucas Heights, 6th Conf. 9-11 Feb. 1976.
1272. Carver, T. R., from Conf. on Polarized Targets and Sources, Saclay, France, Dec. 5-9, 1966. See CONF-661205.
1273. Cash, A. R., Cox, S. F. J., et al., J. Phys., E (London) Sci. Instr. ISSN 0022-3735, 13(2), Feb. 1980, pp-182-191.
1274. Casparis, R., Leemann, B., et al., Helv. Phys. Acta. 48(1), 16 June 1975, pp-45-47.
1275. Casparis, R., Leemann, B. Th., et al., Nucl. Phys., A, 263 (2), 7 June 1976, pp-285-292.
1276. Cassagnou, Y., Julien, J., et al., CEA-CONF-4094, 1977, p-1.
1277. Cassagnou, Y., Jackson, H. E., et al., CEA-CONF-4105, 1977, p-1.
1278. Cassagnou, Y., AIP Conf. Proc. ISSN 0094-243X, 47, 1978, pp-434-443.
1279. Cassapakis, C. G., New Mexico Univ., Albuquerque (U. S. A.), Thesis (Ph.D), 1975, p-175.
1280. Cassapakis, C. G., Bryant, H. C., et al., Phys. Lett., B, 63(1), 5 July 1976, pp-35-38.
1281a. Cassels, J. M., Dickson, J. M., et al., Phys. Soc. Proc. 64(379B, July 1, 1951, pp-590-4.
1281b. Cassels, J. M., Dickson, J. M., et al., Proc. Phys. Soc. (London) 64B, 1951, p-719.
1281c. Cassels, J. M., Dickson, J. M., et al., Proc. Phys. Soc. (London) 64B, 1951, pp-590-4.
1282. Cassidy, T. D., Gorka, A. J., et al., Trans. Nucl. Sci., NS-16, June 1969, pp-640-1.
1283. Casten, R. F., Greenberg, J. S., et al., Nucl. Instr. and Meth. 80, 1970, pp-296-8.
1284. Casten, R. F., Greenwood, R. C., et al., Nucl. Phys., A, 285(2), 25 July 1977, pp-235-252.
1285. Castro, J. J., and Dominguez, C. A., Phys. Rev. Lett. 39(8), 22 Aug. 1977, pp-440-442.
1286. Catillon, P., Pure Appl. Phys. 40A, 1974, pp-193-212.
1287. Catillon, P., Nucl. Spectro. and Reactions, Part A, NY; Academic Press, Inc., 1974, Cerny, J. (ed.), pp-193-212.
1288. Cattapan, G., Maglione, E., et al., Nucl. Phys., A, 296(2), 20 Feb. 1978, pp-263-277.
1289. Cauchois, Y., Ben Abdelaziz, H., et al., Univ. Paris VI, Orsay, France.
1290. Cauvin, B., Galin, J., et al., CEA-N-1861 (nd), 1 Oct. 1974 - 1 Sept. 1975, pp-98-101.

1291. Cavaignac, J. F., Vignon, B., et al., Phys. Lett., B, 67 (2), 28 March 1977, pp-148-150.
1292. Cavallaro, S., Delaunay, J., et al., Nucl. Phys., A, 293 (1-2), 12-19 Dec. 1977, pp-125-136.
1293. Cavedon, J. M., Godin, A., et al., CEA-N-1861 (nd), 1 Oct. 1974 - 1 Sept. 1975, pp-181-182.
1294. Cecchini, A., and Modena, I., Laboratori Nazionali di Frascati, Comitato Nazionale per l'Energia Nucleare (Italy), Jan. 24, 1966, p-10.
1295. Cecil, F. E., Shepard, J. R., et al., Phys. Lett., B, 64(4), 11 Oct. 1976, pp-411-413.
1296. Celenza, L. S., Nutt, W. T., et al., Phys. Lett., B, 72(1), 5 Dec. 1977, pp-23-26.
1297. Cerino, J., and Cronin, R., IEEE Trans. on Nuclear Sci., NS-26(3), June 1979.
1298. Cernigoi, C., Gabrielli, I., et al., AIP Conf. Proc. ISSN 0094-243X, 47, 1978, pp-612-613.
1299. Cester, R., Fitch, V. L., et al., Phys. Rev. Lett. 37(18), 1 Nov. 1976, pp-1178-1181.
1300. Ceulemans, H., Nuclear Energy Agency, Nuclear Data Comm., 20-22 May 1974.
1301. Chabaud, V., and Kuroda, K., Nucl. Instr. and Meth. 125(1), 1975, pp-119-24.
1302. Chackett, K. F., Ark. Fys. 36, 1967, pp-133-50.
1303. Chai, B. T., and Sinclair, D., Nucl. Phys., A, 279(3), 4 April 1977, pp-517-531.
1304. Chai, J., McClelland, J. B., et al., Am. J. Phys. 45(7), July 1977, pp-611-617.
1305. Chakrabarty, D. R., and Gupta, S. K., Dept. of Atomic Energy, Bombay (India), 20B, Dec. 26-30, 1977, Nuclear Physics.
1306. Chakraborty, P., Sargent, B. W., et al., Nucl. Instr. and Meth. 95(2), 1971, pp-209-11.
1307. Chamberlain, O., UCRL-17133, Contract W-7405-eng-48, Sept. 9, 1966, p-18.
1308. Chamberlin, E. P., Nucl. Instr. and Meth. 91, 1971, p-289.
1309. Chambers, B., Hofstadter, R., et al., from Int. Conf. on Nuclear Structure, Stanford, CA, June 1963.
1310. Chambon, B., Drain, D., et al., Nucl. Instr. and Meth. 68, Feb. 1, 1969, pp-167-8.
1311. Chambon, B., Lyon-1 Univ., 69 - Villeurbanne (France), Inst. de Physique Nucleaire, These (D. es S).
1312. Chan, K. C. D., Pitts. Univ., PA (U. S. A.), Thesis (Ph.D), 1975, p-122.
1313. Chan, T. U., Agard, M., et al., Nucl. Instr. and Meth. 146 (1), 1 Oct. 1977, p-314.
1314. Chan, Y., Cramer, J. G., et al., Wash. Univ., Seattle (U. S. A.), Nuclear Phys. Lab. Annual Report, 1977, pp-113-115.
1315. Chang, C. C., Didelez, J. P., et al., Maryland Univ., College Park (U. S. A.), June 1977, p-23.

1316. Chang, H., Phys. Lett., B, 69(3), 15 Aug. 1977, pp-272-274.
1317. Chang, H. H., Ridley, B. W., et al., Colo. Univ., Boulder.
1318. Chan, Y., Cramer, J. G., et al., Wash. Univ., Seattle, (U. S. A.), Nuclear Phys. Lab., Annual Report, 1977, June 1977, pp-112-123.
1319. Chang, J., Leeper, R. J., et al., Rev. Sci. Instr. 51(3), 1980, pp-398-9.
1320. Chandra, H., and Sauer, G., Phys. Rev., C, 13(1), Jan. 1976, pp-245-252.
1321. Chaney, D., Ferbel, T., et al., Phys. Rev. Lett. 40(2), 9 Jan. 1978, pp-71-74.
1322. Chant, N. S., Wall, N. S., et al., Phys. Rev., C, 15(1), Jan. 1977, pp-53-56.
1323. Chant, N. S., Kitching, P., et al., Phys. Rev. Lett. ISSN 0031-9007, 43(7), 13 Aug. 1979, pp-495-498.
1324. Chanut, Y., Drain, D., et al., Nucl. Instr. and Meth. 75, 1969, pp-74-6.
1325. Chanut, Y., Perrin, C., et al., Rev. Phys. Appl. 4, June 1969, pp-237-8.
1326. Chanut, Y., Perrin, C., et al., RC Accad. Naz. Lincei (Italy), 45(5), Nov. 1968, pp-237-8.
1327. Chapellier, M., Garreta, D., et al., Proc. of the 2nd Int. Symp. on Polarization Phenomena of Nucleons, Huber, P., and Schopper, H. (eds.), Basel, and Stuttgart, Birkhaeuser Verlag, 1966, pp-126-7.
1328. Chapellier, M., from Conf. on Polarized Targets and Sources, Saclay, France, Dec. 5-9, 1966.
1329. Chapman, G. T., Dickens, J. K., et al., Natl. Bur. of Stands. (U. S.), Spec. Publ., 425, Oct. 1975, pp-758-761.
1330. Chapman, K. R., Nucl. Instr. and Meth. 148(2), Jan. 15, 1978, pp-209-212.
1331. Charapatpimol, N., Fong, J. C., et al., CEA-CONF-3189, 1975, p-1.
1332. Charles, P., Auger, F., et al., Phys. Lett., B, 62(3), 7 June 1976, pp-289-292.
1333. Charles, P., Auger, F., et al., Europ. Phys. Soc., Geneva, Europ. Conf. on Nucl. Phys. with Heavy Ions, 6-10 Sept. 1976.
1334. Charlton, L. A., Phys. Rev., C, 12(2), Aug. 1975, pp-351-358.
1335. Charlton, L. A., Phys. Rev., C, 14(2), Aug. 1976, pp-506-513.
1336. Charlton, L. A., Phys., Lett., B, 72(1), 5 Dec. 1977, p-7-10.
1337. Charlton, L.A., Delic, G., LBL-6592, CONF-770968-4, 1977 Aug.
1338. Chartoire, M., Engerran, J., et al., Lyon-1 Univ., 69-Villeurbanne (Fr.), March 1975.
1339. Charvet, A., Duffait, R., et al., Phys. Rev., C, 13(6), June 1976, pp-2237-2240.
1340. Chasan, B., Milder, F. L., et al., Phys. Rev., C, ISSN 0556-2813, 20(4), Oct. 1979, pp-1603-1606.

1341. Chatterjee, A., and Gupta, S. K., Proc. Nucl. Phys. Solid State Phys. Symp. 1978, 21B, p-290.
1342. Chatoorgoon, V., and Stangeby, P. C., J. Energy 2(4), 1978, pp-254-6.
1343. Chatoorgoon, V., Univ. Toronto, Toronto, Diss. Abstr. Int. B 1979, 40(2), pp-831-2.
1344. Chaudhri, M. A., Templer, J., et al., IEEE Trans. on Nucl. Sci., NS-26(2), April 1979.
1345. Chaudhri, M. A., IEEE Trans. on Nucl. Sci., NS-26(2), April 1979.
1346. Chaudhri, M. A., Templer, J., et al., TEICCTA, IU, U. S. A., Sept. 18-21, 1978.
1347. Chaudhri, M. A., TEICCTA, IU, U. S. A., Sept. 18-21, 1978.
1348. Chaudhri, M. A., 7th Int. Conf. on Cyclotrons and their Application; Zurich, Switzerland, Aug. 19-22, 1975.
1349. Chaudhri, M. A., Nucl. Instr. and Meth. 120(2), 1 Sept. 1974, pp-357-358.
1350. Chaudhri, M. A., Chaudhri, A. J., et al., Proc. of the 6th Int. Cyc. Conf., Vancouver, Canada, July 18-21, 1972.
1351. Chaudhri, M. A., and Böttger, O., Int. Cyc. Conf., St. Catherine's Coll., Oxford, 17-20 Sept. 1969.
1352. Chaudhri, M. A., Nucl. Instr. and Meth. 62, July 1, 1968, p-316.
1353. Chaudhri, M. A., Nucl. Instr. and Meth. 45, 1966, pp-357-8.
1354. Chaudhri, R. M., and Khan, M. Y., Nature 192, Nov. 18, 1961, pp-646-7.
1355. Chaumeaux, A., CEA Centre d'Etudes Nucleaires de Saclay, 91 - Gif-sur-Yvette (France).
1356. Chaumeaux, A., Bruge, G., et al., Nucl. Phys., A, 267(3), 30 Aug. 1976, pp-413-424.
1357. Chechick, R., Fraenkel, Z., et al., Nucl. Phys., A, 287(2), 5 Sept. 1977, pp-353-361.
1358. Chechik, R., Eyal, Y., et al., Weiz. Inst. Sci., 1977, p-27.
1359. Chekh, Ya., Yanout, Z., et al., JINR-E1-3708, 1968, p-21.
1360. Chellew, N. R., Steunenberg, R. K., et al., Nucl. Instr. and Meth. 44, Sept. 1966, pp-149-50.
1361. Chellis, K. E., and Sheline, R. K., Nucl. Instr. and Meth. 54, Sept. 1967, pp-139-40.
1362. Chemarin, M., Feuvrais, L., et al., Rev. Phys. Appl. (France), 4(2), June 1969, pp-236-7.
1363. Chemarin, M., Feuvrais, L., et al., Internal Report 1968, Inst. de Physique Nucleaire, Lyon (Fr.).
1364. Chemin, J. F., and Bordeaux, Internal Report 1968, Inst. de Physique Nucleaire, Lyon (Fr.).
1365. Chemin, J. F., Andriamonje, S., et al., CENBG-7722, 1977, p-11.
1366. Chen, J. R., Dept. Phys., Univ. PA, Report, 1971, (LBL-500), pp-209-12.
1367. Chen, J. R., LBL-500, from 2nd Int. Conf. on Polarized Targets; Berkeley, CA, 30 Aug. 1971, pp-411-12.

1368. Chen, J. R., et al., Phys. Rev. Lett. 21(17), 1968, pp-1279-82.
1369. Chen, J. R., Rev. Sci. Instr. 21, 1950, pp-491-2.
1370. Chen, L. C., Phys. Rev., C, 14(6), Dec. 1976, pp-2069-81.
1371. Chen, T. W., and Hoock, D. W., AIP Conf. Proc. 33, 1976, pp-88-89.
1372. Cheon, I., AIP Conf. Proc. ISSN 0094-243X, 54(1), 15 July 1979, pp-351-352.
1373. Chermak, J., and Tinka, I., Sovet Ehkonomicheskoj Vzaimopomoshchi, Moscow (USSR).
1374a. Chermarin, M., Feuvrais, L., et al., RC Accad. Naz. Lincei (Italy), 45(5), pp-236-7.
1374b. Chernyshev, B. N., and Kavun, V. Ya., Koord. Khim. ISSN 0132-344X, 5(9), Sept. 1979, pp-1309-1313.
1375. Chestnut, C. H., III, Nelp, W. B., et al., Proc. of the 6th Int. Cyc. Conf., Vancouver, Canada, July 18-21, 1972.
1376. Chevallier, J., Haas, B., et al., Centre de Recherches Nucleaires, 1975, p-25.
1377. Chevallier, P., Disdier, D., et al., Europ. Phys. Soc., Geneva (Switzerland), 6-10 Sept. 1976.
1378. Chevarier, A., Chevarier, N., et al., Goldberg, D. A., et al., (eds.), MD Univ. (USA), 1975, pp-350-351.
1379. Chew, S. H., and Lowe, J., Nucl. Phys., A, 252(1), 3 Nov. 1975, pp-8-12.
1380. Chew, S. H., Low, J., et al., Nucl. Phys., A, 286(3), 22 Aug. 1977, pp-451-473.
1381. Chiang, T. H., Guerreau, D., et al., Phys. Rev., C, ISSN 0556-2813, 20(4), Oct. 1979, pp-1408-1418.
1382. Chida, K., and Kaneko, K., Tokyo Univ., Shitsuryo Bunseki 14, Oct. 1966, pp-149-56.
1383. Chigin', V. I., Prib. Tekh. Ehksp. 6, Nov. 1974, pp-38-40.
1384. Child, H. R., Raubenheimer, L. J., et al., Phys. Rev. 174, 1968, p-1553.
1385. Childs, J. D., Pittsburgh Univ., PA (U. S. A.), Thesis (Ph.D), 1976, p-90.
1386. Chinowsky, W., et al., UCRL-16830, 2, 1967, p-317.
1387. Chirapatpimol, N., Fong, et al., Nucl. Phys., A, 264(3), 5 July 1976, pp-379-396.
1388. Chiricj, M. D., Inst. za Nuklearne Nauke Boris Kidric, Belgrade (Yugoslavia), Internal Report 1975.
1389. Chiricj, M. D., Suboticj, M. K., et al., Zb. Rad. Prir.-Mat. Fak. 6, 1976, pp-127-129.
1390. Chirikov, B. V., Tayurski, V. A., et al., Nucl. Instr. and Meth. 144(2), 15 July 1977, pp-129-139.
1391. Chizhikov, D. M., and Shchastlivyi, V. P., Academy of Sci., 1968, USSR.
1392. Chmielewska, D., Sujkowski, Z., et al., Kernfysisch Versneller Instituut, Annual Report 1975.

1393. Chopra, K. L., Thin Film Phenomena, McGraw-Hill, NY, 1969.
1394. Chopra, K. L., and Randlett, M. R., Rev. Sci. Instr. 37, 1966, p-1421
1395. Chopra, K. L., and Randlett, M. R., Rev. Sci. Instr. 38, Aug. 1967, pp-1147-51.
1396. Choudry, A., Nucl. Instr. and Meth. 71, 1969, pp-221-5.
1397. Choudhury, A., and Mahalanabis, J., Proc. of the Nucl. Phys. and Solid State Phys. Symp., Pune, 20B, Dec. 26-30, 1977.
1398. Chretien, M., Firth, D. R., et al., Nucl. Instr. and Meth. 20, 1963, pp-120-4.
1399. Chrien, R. E., BNL-NCS-50451, March 1975, pp-147-150.
1400. Chrien, R. E., Cole, G. W., et al., Phys. Rev., C, 13(2), Feb. 1976, pp-578-594.
1401. Chrien, R. E., and Kopecky, J., BNL-NCS-22500, March 1977, pp-47-48.
1402. Chrien, R. E., AIP Conf. Proc. ISSN 0094-243X, 47, 1978, pp-628-629.
1403. Christ, A., and Gassen, A., et al., Nucl. Instr. and Meth. (Netherlands), 152(2-3), 15 June 1978, pp-367-9.
1404. Christensen, P. R., and Switkowski, Z. E., Nucl. Phys., A, 280(1), 11 April 1977, pp-205-216.
1405. Christillin, P., and Rosa-Clot, M., Phys. Lett., B, 73(1), 30 Jan. 1978, pp-23-26.
1406. Christillin, P., and Ericson, T. E. O., AIP Conf. Proc. ISSN 0094-243X, 54(1), 15 July 1979, pp-420-421.
1407. Chrusciel, E., Inst. Techniki Jacrowej AGH, Krakow, Report 1964.
1408. Chrusciel, E., Lasa, J., et al., Nukleonika 10, 1965, pp-115-21.
1409. Choquer, Y., Inst. des Sciences Nucleaires, Grenoble (France), These.
1410. Chowdhury, K. R., Chowdhury, A. R., et al., Proc. of the Nucl. Phys. and Solid State Phys. Symp., Calcutta, 18B, Dec. 22-26, 1975.
1411. Chu, L. C., and Attrep, M., Jr., J. Inorg. and Nucl. Chem. ISSN 0022-1902, 39(9), 1977, pp-1495-1496.
1412. Chu, T. L., Southern Methodist Univ., Dallas, TX (U. S. A.), Dept. of Elect. Eng., Sept. 1975, p-351.
1413. Chuang, L. S., and Lu, T. H., Nucl. Instr. and Meth. 83, 1970, pp-197-200.
1414. Chua, L. T., Becchetti, F. D., et al., Nucl. Phys., A, 273 (1), 23 Nov. 1976, pp-243-252.
1415. Chu, Y. Y., Friedlander, G., et al., Phys. Rev., C, 15(1), Jan. 1977, pp-352-360.
1416. Chuang, L. S., Nucl. Instr. and Meth. 75, 1969, pp-171-2.
1417. Chuev, V. I., Demjanova, A. S., et al., Europ. Phys. Soc., Geneva (Switzerland, 6-10 Sept. 1976.
1418. Chung, A. H., Diamond, W. T., et al., Proc. Ann. Conf. Nucl. Target Dev. Soc., 1974, pp-84-5.

1419. Chung, A. H., Ph.D. Thesis, Univ. of Toronto, 1973.
1420. Church, T. G., AECL-2750, ING Status Report, July 1967, p-197.
1421. Ciangaru, G., McGrath, R. L., et al., Phys. Lett., B, 61(1), 1 March 1976, pp-25-28.
1422. Cierjacks, S., Gupta, S. K., et al., Phys. Rev., C, 17(1), Jan. 1978, pp-12-15.
1423. Cierjacks, S., and Schouky, I., Verh. Dtsch. Phys. Ges. 4, 1978, p-952.
1424. Cierjacks, S., and Kari, K., Verh. Dtsch. Phys. Ges. 4, 1978, p-953.
1425. Cierjacks, S., Hinterberger, F., et al., Bonn Univ. (FRG.), Annual Report, Oct. 1979, pp-26-27.
1426. Cierjacks, S., and Kari, K., Bonn Univ. (FRG.), Annual Report, Oct. 1979, p-28.
1427. Cindra, N., and Holub, E., Fizika (Zagreb), 9(1), 1977, p-20.
1428. Cindro, N., Cocu, F., et al., CEA-N-1969, June 1977, pp-80-5.
1429. Cindro, N., Frehaut, J., et al., CEA-CONF-3719, 1976, p-1.
1430. Cindro, N., Basrak, Z., et al., CEA-N-1875, Apr. 1976, pp-107-108.
1431. Cindro, N., Cocu, F., et al., Phys. Rev. Lett. 39(18), 31 Oct. 1977, pp-1135-1138.
1432. Ciric, B., Fizika (Zagreb), 9(1), 1977, pp-39-40.
1433. Ciricj, D. M., Stepancicj, B. Z., et al., Zb. Rad. Prir.-Mat. Fak. 7, 1977, pp-61-65.
1434. Cisotti, M., De la Euente, F., et al., Nucl. Instr. and Meth. 159(1), 1979, pp-235-42.
1435. Citron, A., CERN-PS/AC-3, March 13, 1954, p-13.
1436. Civelekoglu, Y., Freiesleben, H., et al., Verh. Dtsch. Phys. Ges. 4, 1978, p-806.
1437. Civitarese, O., Broglia, R. A., et al., Phys. Lett., B, 72 (1), 5 Dec. 1977, pp-45-48.
1438. Cjiricj, M. D., Popicj, V. R., et al., Zb. Rad. Prir.-Mat. Fak. 6, 1976, pp-115-125.
1439. Clark, A. H., Thesis, Mat. Sci. Ctr., Cornell Univ., Ithaca, NY, 1967.
1440. Clark, A. H., Phys. Rev. 154, 1967, p-154.
1441. Clark, A. R., Holley, W. R., et al., LBL-5075, 1975, pp-193-194, Calif. Univ., Berkeley (U. S. A.).
1442. Clark, J. A., and Quinn, C. P., J. Chem. Soc. (London), Faraday Trans., I, 72(3), 1976, pp-706-714.
1443. Clark, J. C., Thakur, M. L., et al., Int. J. Appl. Radiat. Isotop. 23(7), July 1972, pp-329-35.
1444. Clack, R. W., Lambert, J. P., et al., from Tritium Symp., Las Vegas, NE, U. S. A., 30 Aug. 1971.
1445. Clarke, J. H., Report AERE-M-2950, 1978, p-51.
1446. Clarke, N. M., Nucl. Instr. and Meth. 97(2), 1971, pp-399-403.

1447. Clauser, M. J., Report SAND-78-0472C, CONF-78034303, 1978, p-23.

1448. Cleary, T. P., Ford, J. L. C., Jr., et al., Phys. Lett., B, ISSN 0370-2693, 83(1), 23 April 1979, pp-51-54.

1449. Clement, H., Graw, G., et al., Nucl. Phys., A, 285(1), 18 July 1977, pp-109-144.

1450. Clement, H., Ehrlich, D., et al., AIP Conf. Proc. ISSN 0094-243X, 47, 1978, pp-742-743.

1451. Cline, M. C., and Emigh, C. R., Report LA-UR-74-599, 1978, p-11.

1452. Close, D. A., and Bearse, R. C., Nucl. Instr. and Meth. 97 (3), 1971, pp-607-8.

1453. Close, D. A., and Malanify, J. J., LASL, U. S. A. Report 1970.

1454. Cloth, P., Darvas, J., et al., ATOMKI Kozl. 18(2), 1976, pp-439-50.

1455. Clover, M. R., DeVries, R. M., et al., AIP Conf. Proc. ISSN 0094-243X, 47, 1978, pp-598-599.

1456. Coates, M. S., Gayther, D. B., et al., Natl. Bur. of Stands (U. S.), Spec. Publ. 425, Oct. 1975, pp-568-71.

1457. Cobble, J. W., Purdue Univ., Lafayeete, Ind., Contract AT (11-1)-347, March 1958, p-98.

1458. Cobern, M. E., Lemaire, M. C., et al., CEA-CONF-3451, 1975, p-25.

1459. Cobern, M. E., Lisbona, N., et al., CEA-N-1861 (nd), pp-25-26.

1460. Cobern, M. E., Lemaire, M. C., et al., CEA-N-1861 (nd), pp-27-28.

1461. Cobern, M. E., Pisano, D. J., et al., Phys. Rev., C, 14(2), Aug. 1976, pp-491-505.

1462. Cobern, M. E., and Parker, P. D., Phys. Rev., C, 15(5), May 1977, pp-1929-1932.

1463. Coceva, C., and Giacobbe, P., Nucl. Phys., A, 293(1-2), 12-19 Dec. 1977, pp-167-188.

1464. Cochavi, S., Alster, J., et al., Bull. Isr. Phys. Soc. 21, Paper A-4, 1975, p-13.

1465. Cocking, S. J., McPhee, R. C., J. Phys. E. 1, March 1968, pp-367-8.

1466. Cocu, F., Ambrosino, G., et al., Acta Phys. Slovaca 26(1), 1976, pp-60-63.

1467. Cocu, F., Cindro, N., et al., Europ. Phys. Soc., Geneva (Switzerland), 6-10, Sept. 1976.

1468. Cocu, F., Haouat, G., et al., CEA-N-1875, April 1976, p-42-45.

1469. Cocu, F., Haouat, G., et al., CEA-R-4746, May 1976, p-28.

1470. Cocu, F., Cindro, N., et al., CEA-N-1969, June 1977, p-86-89.

1471. Cocu, F., Uzureau, J., et al., J. Phys. (Paris), Lett. 38 (21), 1 Nov. 1977, pp-L.421-L.425.

1472. Cocu, F., and Cindro, N., Fiz. (Zagreb), Suppl. 9(2), 1977, pp-26-29.
1473. Coelho, H. T., and Das, T. K., Rev. Bras. Fis. 5(3), Dec. 1975, pp-335-347.
1474. Coelho, H. T., Coutinho, F. A. B., et al., Phys. Rev., C, 14(3), Sept. 1976, pp-1280-1284.
1475. Coffin, J. P., Engelstein, P., et al., Europ. Phys. Soc., Geneva (Switzerland), Europ. Conf. on Nuclear Physics with Heavy Ions, 6-10 Sept. 1976.
1476. Coffin, J. P., Engelstein, P., et al., AIP Conf. Proc. ISSN 0094-243X, 47, 1978, pp-724-725.
1477. Coffin, J. P., Engelstein, P., et al., Phys. Rev., C, 17(5), May 1978, pp-1607-1614.
1478. Cohen, M. S., Handbook of Thin Film Tech., Ed. by L. I. Maissel, and R. Glang, McGraw-Hill, NY, 1970, Chapter 17.
1479. Coignet, G., Proc. DESY-79/48, INKA-Conf.-79-423-000, Aug. 1979.
1480. Cokinos, D., and Melkonian, E., Phys. Rev., C, 15(5), May 1977, pp-1636-1643.
1481. Cole, A. J., Longequeue, N., et al., J. Phys. (Paris), 38 (9), Sept. 1977, pp-1043-1049.
1482. Cole, B. J., Toepffer, C., et al., Phys. Rev. Lett. 39(1), 4 July 1977, pp-3-6.
1483. Cole, C. M., Gray, F. C., et al., Med. Phys. 1(6), 1974, pp-326-7.
1484. Cole, W., and Grime, G. W., INTDS Proc., 1979, Boston, Oct. 1-3.
1485. Coleman, C. F., AERE-PR/NP-13, Feb. 1968, p-69.
1486. Coleman, W. J., Proc. of 14th Ann. Conf. of the Soc. of Vacuum Coaters, Miami Beach, FL, March 4-5, 1971.
1487. Colle, R. P., and Preiss, I. L., J. Phys., E (London), 4 (12, Dec. 1971, pp-977-80.
1488. Collins, E. D., and Bigelow, J. E., ORNL CONF-761101-10, 1976, p-31.
1489. Collins, K. E., DeJesus, O. T., et al., Radiochem. Radioanal. Lett. 41(2), 1979, pp-129-32.
1490. Colombani, P., Faraggi, H., et al., Proc. of the Int. School of Phys. 'Enrico Fermi' Course 62 held at Varenna, Italy, 22 July - 3 Aug. 1974.
1491. Colombani, P., Butler, P. A., et al., Europ. Phys. Soc., Geneva (Switzerland), Europ. Conf. on Nucl. Phys. with Heavy Ions, 6-10 Sept. 1976.
1492. Comfort, J. R., Decker, J. F., et al., Phys. Rev. 150(1), 1966, p-249.
1493. Comfort, J. R., Harakeh, M. N., et al., Kernfysisch Versneller Inst. Annual Report, 1975.
1494. Comfort, J. R., Harakeh, M. N., et al., Kernfysisch Versneller Inst. Annual Report, 1975.

1495. Comiso, J. C., Boswell, J., et al., Virg. Univ., IPNO-T-75-01, ORO-4043-37, 1977, pp-23-24.
1496. Comparat, V., Inst. de Physique Nucleaire, Paris (France), 1975, p-109.
1497. Concaildi, G., George, R., et al., IEEE Trans. on Nucl. Sci., NS-18(3), June 1971.
1498. Concus, P., Nucl. Instr. and Meth. 62(2), June 15, 1968, pp-199-202.
1499. Condon, E. V., and Odishaw, H., Handbook of Physics, McGraw-Hill Book Co. Inc., 1958.
1500. Conjeaud, M., Harar, S., et al., CEA-N-1861 (nd), 1 Oct. 1974 - 1 Sept. 1975, pp-12-15.
1501. Conjeaud, M., Harar, S., et al., Nucl. Phys., A, 250(1), 29 Sept. 1975, pp-182-210.
1502. Conjeaud, M., Harar, S., et al., Europ. Phys. Soc., Geneva, (Switzerland), Europ. Conf. on Nucl. Phys. with Heavy Ions, 6-10, Sept. 1976.
1503. Conjeaud, M., Gary, S., et al., Europ. Phys. Soc., Geneva, (Switzerland), Europ. Conf. on Nucl. Phys. with Heavy Ions, 6-10, Sept. 1976.
1504. Conjeaud, M., Gary, S., et al., Europ. Phys. Soc., Geneva, (Switzerland), Europ. Conf. on Nucl. Phys. with Heavy Ions, 6-10, Sept. 1976.
1505. Conjeaud, M., Harar, S., et al., Europ. Phys. Soc., Geneva, (Switzerland), Europ. Conf. on Nucl. Phys. with Heavy Ions, 6-10, Sept. 1976.
1506. Conjeaud, M., Gary, S., et al., ANL/PHY-76-2, 2, May 1976, pp-499-507.
1507. Conjeaud, M., Gary, S., et al., CEA-N-1959 (nd), 1 Sept. 1975 - 30 Sept. 1976, pp-17-23.
1508. Conjeaud, M., Harar, S., et al., CEA-N-1959 (nd), 1 Sept. 1975 - 30 Sept. 1976, pp-51-53.
1509. Conner, W. V., and Proctor, S. G., RFP-1416, Dec. 23, 1969.
1510. Conner, W. V., and Baaso, D. L., RFP-1415, Dec. 30, 1969.
1511. Conner, W. V., from 3rd Int. Symp. on Rsch. Mat. for Nucl. Measurements, Gatlinburg, TN, 5 Oct. 1971.
1512. Conner, W. V., Nucl. Instr. and Meth. 102(3), 1972, pp-417-23.
1513. Conner, W. V., INTDS Proc., 1974, Chalk River, Canada, Oct. 1-3.
1514. Conner, W. V., and Baaso, D. L., Report RFP-2366, 1975, p-10, The Dow Chem. Co., Golden, CO, U. S. A.
1515. Conner, W. V., Proctor, S. G., et al., Report RFP-1835, 1972, p-7.
1516. Conner, W. V., and Baaso, D. L., RFP-1848, May 17, 1972.
1517. Conner, W. V., Nucl. Instr. and Meth. 102(3), 1972, pp-417-23.
1518. Constantinescu, F., Stud. Cercet. Fiz. 26(3), 1974, pp-307-16.

1519. Constantinescu, G. H., Dobrescu, S., et al., Stud. and Cercet. Fiz. (Rumania), 25(8), 1973, pp-1001-5.
1520. Conte, R. R., Liquid Helium Tech., NY, Pergamon Press, Inc., 1967, pp-395-401.
1521. Conzett, H. E., Hinterberger, F., et al., Phys. Rev. Lett. ISSN 0031-9007, 43(8), 20 Aug. 1979, pp-572-576.
1522. Cook, G. B., and Hudswell, F., J. Sci. Instr. 27, 1950, p-230-1
1523. Cook, J. H., Newbury, F. H., et al., Trans. Am. Nucl. Soc. 30, 1978, p-754.
1524. Cook, J. H., Newbury, F. H., et al., LA-UR-78-2649, Contract W-7405-eng-36, 1978, p-22.
1525. Cook, J. H., Newbury, F. H., et al., Proc. Conf. Remote Syst. Technol. 26, 1979, pp-21-5.
1526. Cook, W. B., Johns, M. W., et al., Nucl. Phys., A, 259(3), 22 March 1976, pp-461-480.
1527. Cookson, J. A., Fowler, J. L., et al., Nucl. Phys., A, 299(3), 1 May 1978, pp-365-380.
1528. Cooney, P. J., State Univ. of NY, Stony Brook (U. S. A.), Thesis (Ph.D), 1975, p-178.
1529. Coope, D. F., Ill. Univ., Urbana (U. S. A.), Thesis (Ph.D), 1975, p-90.
1530. Coope, D. F., Schell, M. C., et al., Phys. Rev. Lett. 37 (17), 25 Oct. 1976, pp-1126-1128.
1531. Coope, D. F., Cannell, L. E., et al., Phys. Rev., C, 15(6), June 1977, pp-1977-1986.
1532. Coope, D. F., Tripathi, S. N., et al., Phys. Rev., C, 16(6), Dec. 1977, pp-2223-2237.
1533. Coope, D. F., and McEllistrem, M. T., IEEE Trans. Nucl. Sci., NS-26(1 Pt. 2), 1979, pp-1721-3.
1534. Cooper, J. A., and Perkins, R. W., Nucl. Instr. and Meth. 94(1), 1971, pp-29-38.
1535. Cooper, J. W., Mich. Univ., Ann Arbor (U. S. A.), Thesis (Ph.D), 1975, p-208.
1536. Cooper, M. D., LA-UR-76-1183, CONF-760561-5, 1976, p-12.
1537. Cooper, M. D., AIP Conf. Proc. ISSN 0094-243X, 54(1), 15 July 1979, pp-222-229.
1538. Cooper, M. D., Baer, H., et al., LA-7892C, July 1979.
1539. Corcoran, M. D., Ems, S. E., et al., Stud. Nat. Sci. (NY), 1977, pp-145-61.
1540. Cordero, F., Proc. of 3rd Conf. on Accel. Targets Designed for the Prod. of Neutrons, Liege (Belgium), Sept. 18-19, 1967, pp-61-74.
1541. Cordero, F., Euratom, EUR-3895, 1968, p-14.
1542. Cordero, F., An. Fis. 64, Jan. - Feb. 1968, pp-25-32.
1543. Corelli, J. C., Livingston, M., et al., Rev. Sci. Instr. 28, June 1957, pp-471-2.
1544. Corfu, R., Egger, J. P., et al., Helv. Phys. Acta 50(2), 25 March 1977, p-197.

1545. Combe, J. C., Europ. Organization for Nuclear Rsch., Geneva, CERN-63-3, pp-111-27.
1546. Cormier, T. M., Cosman, E. R., et al., BNL-21303, CONF-760424-13, 1976, p-9.
1547. Cormier, T. M., Cosman, E. R., et al., ANL/PHY-76-2(2), May 1976, pp-509-516.
1548. Cormier, T. M., Cosman, E. R., et al., Phys. Rev., C, 14(1), July 1976, pp-334-337.
1549. Cormier, T. M., Applegate, J., et al., Phys. Rev. Lett. 38 (17), 25 April 1977, pp-940-943.
1550. Cormier, T. M., Braun-Munzinger, P., et al., Phys. Rev., C, 16(1), July 1977, pp-215-219.
1551. Correll, F. D., Mandansky, L., et al., Johns Hopkins Univ., Baltimore, MD (U. S. A.), Nuclear Moments and Nuclear Structure, Annual Progress Report, May 1, 1975 - April 30, 1976.
1552. Correy, T. B., Battelle Pacific Northwest Labs., Richland, WA (U. S. A.), BNWL-CC-496, 15 Feb. 1966.
1553. Corwin, W. C., ANL/PHY/MSD-76-1.
1554. Corwin, W. C., INTDS Proc., 1975, Argonne, U. S. A., Sept. 30 - Oct. 2.
1555. Corwin, W. C., Nucl. Instr. and Meth. 136(1), 1 July 1976, pp-41-45.
1556. Corvi, F., Rohr, G., et al., Natl. Bur. of Stands. (U. S.) Spec. Publ., 425, Oct. 1975, pp-733-737.
1557. Corvi, F., and Giacobbe, P., Natl. Bur. of Stands. (U. S.) Spec. Publ., 425(2), Oct. 1975, pp-599-602.
1558. Corwin, W. C., Ph.D. Thesis, Ill. Inst. of Tech.
1559. Cosman, E. R., Cormier, T. M., et al., Maryland Univ., College Park (U. S. A.), Dept. of Phys. and Astron., 1975, pp-528-529.
1560. Cossairt, J. D., Ind. Univ., Bloomington (U. S. A.), Thesis (Ph.D), 1975, p-182.
1561. Cossairt, J. D., Bent, R. D., et al., Maryland Univ., College Park (U. S. A.), Dept. of Phys. and Astron., 1975, pp-411-412.
1562. Cossairt, J. D., Tribble, R. E., et al., Phys. Rev., C, 15 (5), May 1977, pp-1685-1689.
1563. Cossairt, J. D., Burns, W. S., et al., Nucl. Phys., A, 287 (1), 29 Aug. 1977, pp-13-23
1564. Cossairt, J. D., Talley, S. B., et al., Phys. Rev., C, 18 (1), July 1978, pp-28-32.
1565. Cosslett, A., and Cosslett, V. E., Brit. J. Appl. Phys. 8, 1957, p-374.
1566. Cossutta, D. D., from Proc. of the 2nd Oak Ridge Conf. on the use of Small Accelerators for Teaching and Rsch., March 23-25, 1970.

1567. Cossuta, D., from Proc. of the 3rd Conf. on Accel. Tgts. Designed for the Prod. of Neutrons, Liege (Belgium), Sept. 18-19, 1967, pp-191-204.
1568. Costa, G. J., Alexander, T. K., et al., Nucl. Phys., A, 256(2), 12 Jan. 1976, pp-277-300.
1569. Cotanch, S. R., and Vincent, C. M., Phys. Rev. Lett. 36(1), 5 Jan. 1976, pp-21-25.
1570. Cotelle, S., and Haissinsky, M., C. R. Acad. Sci. Paris 206, 1938, pp-1644-6.
1571. Cottingham, J. G., and Kovarik, V. J., BNL-6940, Contract AT-30-2-GEN-16, p-9.
1572. Cottingame, W. B., Braithwaite, W. J., et al., AIP Conf. Proc. ISSN 0094-243X, 54(1), 15 July 1979, pp-272-273.
1573. Cottles, V. M., Tulane Univ., New Orleans, LA (U. S. A.), Thesis (Ph.D), 1975, p-150.
1574. Couchell, G. P., Egan, J. J., et al., Trans. Am. Nucl. Soc. 27, 1977, pp-871-872.
1575a. Couvert, P., Grenoble-1 Univ., 38 (France), These (3e Cycle), 1974, p-77.
1575b. Coupat, B., Isabelle, D. B., et al., AIP Conf. Proc. 33, 1976, pp-332-333.
1576. Coupat, B., Bertin, P. Y., et al., La Toussuire, LYCEN-7702(pt. 1), 28 Feb.-4 Mar. 1977, pp-S13.1-S13.2.
1577. Court, G. R., Crabb, D. G., et al., ANL-HEP-CP-78-57, Report No: CONF-7810119-4, 1978, p-26.
1578. Couvert, P., AIP Conf. Proc. 33, 1976, pp-310-315.
1579. Cowan, F. P., Nucl. Safety 10(6), 1969, pp-483-91.
1580. Cowan, G. R., and Holtzman, A. H., J. Appl. Phys. 34, 1963, p-928.
1581. Cowley, A. A., and Heymann, G., Maryland Univ., College Park (U. S. A.), Dept. of Phys. and Astron., ORO-4856-26, 1975, pp-323-324.
1582. Cowley, A. A., Roos, P. G., et al., Maryland Univ., College Park (U. S. A.), Dept. of Phys. and Astron., ORO-5128-5, Nov. 1976, p-34.
1583. Cowley, A. A., Roos, P. G., et al., Phys. Rev., C, 15(5), May 1977, pp-1650-1661.
1584. Cowley, A. A., Mills, S. J., et al., J. Phys., G (London) Nucl. Phys. ISSN 0305-4616, 4(6), June 1978, pp-L149-L154.
1585. Cowley, A. A., Nucl. Phys., A, ISSN 0375-9474, 325(1), 13 Aug. 1979, pp-83-92.
1586. Cox, J., Martin, F., et al., Nucl. Instr. and Meth. 69, March 1969, pp-77-88.
1587. Cox, S. F. J., Report 75, RL-75-019, p-15.
1588. Cracknell, P. J., Gettings, M., et al., Nucl. Instr. and Meth. 92(4), 1971, pp-465-9.
1589. Craig, J. N., Maryland Univ., College Park (U. S. A.), Thesis (Ph.D), 1975, p-237.

1590. Cramer, D. S., and Cranberg, L., Nucl. Instr. and Meth. 93 (2), 1971, pp-405-7.
1591. Cramer, J. G., DeVries, R. M., et al., Phys. Rev., C, 14(6), Dec. 1976, pp-2158-2161.
1592. Cramer, J. G., Goldberg, D. A., et al., ORNL-5306, Sept. 1977, pp-33-34.
1593. Cramer, J. G., Lynch, W. J., et al., Wash. Univ., Seattle (U. S. A.), Nuclear Phys. Lab. Annual Report, RLO/1388-362, June 1977, pp-128-129.
1594. Cranberg, L., Phys. Med. and Biol. (GB) 17(6), Nov. 1972, p-881.
1595. Cranfill, B., Gursky, J. C., et al., LA-6850-C, June 1977, pp-130-137, INTDS Proc., 1976, Los Alamos, Oct. 19-21.
1596. Crannell, H., Finn, J. M., et al., Nucl. Phys., A, 278(2), 7 March 1977, pp-253-260.
1597. Crase, K., Farley, W. E., et al., Report UCRL-52459, 1977, p-286.
1598. Crawley, G. M., Steele, W. F., et al., Phys. Lett., B, 64(2), 13 Sept. 1976, pp-143-146.
1599. Crawley, G. M., Benenson, W., et al., Phys. Rev. Lett. 39 (23), 5 Dec. 1977, pp-1451-1454.
1600. Crawley, R. L., and Steinman, D. A., Rev. Sci. Instr. 51(5), 1980, pp-673-4.
1601. Cress, L. W., and Rydin, R. A., Phys. Med. Biol. 18(5), Sept. 1973, pp-742-743.
1602. Criegee, L., Garrell, M., et al., Phys. Lett. 28B, Nov. 11, 1968, pp-140-2.
1603. Crinean, G. B., Melbourne Univ., Parkville (Aust.), Thesis (Ph.D), 1974.
1604. Croall, I. F., and Cunninghame, J. G., Uses of Cyc. in Chem., Metall., and Biol., Proc. of a Conf. held at St. Catherine's Coll., Oxford, 22-23 Sept. 1969.
1605. Croall, I. F., and Cunninghame, J. G., Uses of Cyc. in Chem., Metall., and Biol./Amphlett, C. B. (ed.), London, Butterworth and Co. Ltd., 1970.
1606. Cross, W. G., and Ing, H., Nucl. Sci. and Eng. 58(4), Dec. 1975, pp-377-386.
1607. Crouzel, C., Comar, D, et al., IEEE Trans. on Nucl. Sci., NS-26(2), April 1979.
1608. Crouzel, C., Comar, D., et al., 7th Int. Conf. on Cyc. and their Application, Zurich, (Switz.), Aug. 19-22, 1975.
1609. Crouzel, C., Comar, D., et al., TEICCTA, IU, U. S. A., Sept. 18-21, 1978.
1610. Crumpton, D., Nucl. Instr. and Meth. 55(1), 1967, pp-198-200.
1611. Csajka, M., Csoeke, A., et al., Isotopenpraxis 2, June 1966, pp-224-6.
1612. Csepregi, L., Kennedy, E. F., et al., J. Appl. Phys. 49(7), July 1978, pp-3906-3911.
1613. Csihas, L., Nucl. Instr. and Meth. 167(1), 1979, pp-171-2.

1614. Csihas, L., INTDS Proc., 1978, Garching, Sept. 11-14.
1615. Csihas, L., INTDS Proc., 1978, Garching, Sept. 11-14.
1616. Csöke, A., Istvan, P., et al., Magy. Tud. Akad. Kozp. Fiz. Kut. Int. Kozlemen 12, 1964, pp-385-93.
1617. Csonka, P. L., Nucl. Instr. and Meth. 91, 1971, pp-469-76.
1618. Cugnon, J., and DaSilveira, R., Europ. Phys. Soc., Geneva, (Switz.), Europ. Conf. on Nucl. Phys. with Heavy Ions, 6-10 Sept. 1976.
1619. Cujec B., and Barnes, C. A., Nucl. Phys., A, 266(2), 9 Aug. 1976, pp-461-493.
1620. Cummings, C. E., Ogard, A. E., et al., Trans. Am. Nucl. Soc. 30, 1978, p-781.
1621. Cummings, C. E., Ogard, A. E., et al., Proc. Conf. Remote Syst. Technol. 26, 1979, pp-201-6.
1622. Cuninghame, J. G., Morris, B., et al., Int. J. Appl. Rad. Isot. 27(11, Nov. 1976, pp-597-603.
1623. Cunnane, J. C., Piiparinen, M., et al., Phys. Rev., C, 13 (6), June 1976, pp-2197-2207.
1624. Cunsolo, A., Foti, A., et al., Europ. Phys. Soc., Geneva (Switz.), Europ. Conf. on Nucl. Phys. with Heavy Ions, 6-10 Sept. 1976.
1625. Cunsolo, A., Foti, A., et al., CEA-N-1959 (nd), 1 Sept. 1975-30 Sept. 1976, pp-10-13.
1626. Cunsolo, A., Foti, A., et al., Nuovo Cim., A, 40(3), 1 Aug. 1977, pp-293-312.
1627. Currie, W. M., and Johnson, C. H., Nucl. Instr. and Meth. 63, Aug. 1968, pp-221-9.
1628. Cusson, R. Y., and Maruhn, J., Phys. Lett., B, 62(2), 24 May 1976, pp-134-138.
1629. Cusson, R. Y., CONF-770602, Oct. 1977, pp-99-114.
1630. Cusson, R. Y., and Meldner, H. W., Phys. Rev. Lett. ISSN 0031-9007, 42(11), 12 March 1979, pp-694-696.
1631. Cutrona, G., Lanzano, G., et al., Lett. Nuovo Cim. 21(10), 11 March 1978, pp-368-372.
1632. Cuypers, M., Weber, G., et al., Liege Univ. (Belgium), Laboratoire des Radioelements, Feb. 20, 1969, p-52.
1633. Cuypers, M., Peters, J. M., et al., Nucl. Instr. and Meth. 68, Feb. 15, 1969, pp-245-50.
1634. Cverna, F., Baer, H., et al., LA-7892C, July 1979.
1635. Czirr, J. B., and Sidhu, G. S., Natl. Bur. of Stands. (U. S.), Spec. Publ. 425, Oct. 1975, p-546-548.
1636. Czirr, J. B., and Sidhu, G. S., Nucl. Sci. and Eng. 60(4), Aug. 1976, pp-383-389.
1637. Dabbs, J. W. T., Walker, W., et al., from 3rd Int. Symp. on Polarization Phenomena in Nucl. Reactions, Madison, WI, Aug.-Sept. 1970.
1638. Dabbs, J. W. T., Hill, N. W., et al., Natl. Bur. of Stands. (U. S.), Spec. Publ., 1, Oct. 1975, pp-81-82.
1639. Dadayan, N. A., Yad. Fiz. 24(2), 1976, pp-417-420.

1640. Daehnick, W. W., Spisak, M. J., et al., Phys. Rev., C, 14 (5), Nov. 1976, pp-2011-2012.
1641. Dagenhart, W. K., and Whitehead, T. W., Jr., Nucl. Instr. and Meth. 85, 1970, p-215.
1642. Dagenhart, W. K., Stelson, P. H., et al., Nucl. Phys., A, 284(3), 11 July 1977, pp-484-500.
1643. Dagostino, B. M., Massa, I., et al., Nuovo Cim., A, 27(1), 1 May 1975, p-1-26.
1644. Dahl, J. R., and Tilbury, R. S., Int. J. Appl. Radiat. Isotop. 23(9), 1972, pp-431-7.
1645. Dakowski, M., Piekarz, H., et al., Inst. of Nuclear Rsch., Warsaw, Jan. 1965, p-15.
1646. Dalitz, R. H., Polarized Targets and Ion Sources, Gif-sur-Yvette, France, Centre d'Etudes Nucleaires de Saclay, 1967, pp-261-7.
1647. Dally, E. B., and Croissiaux, M., Rev. Sci. Instr. 38, May 1967, pp-646-8.
1648. Danagulyan, A. S., Demekhina, N. A., et al., Nucl. Phys., A, 285(3), 1 Aug. 1977, pp-482-492.
1649. Daneshvar, K., and Kovar, D. G., ANL-78-66, 1 April 1977 - 31 March 1978, pp-41-42.
1650. Daniel, H., and Leon, M., Nucl. Instr. and Meth. 158(1), 1979, pp-233-6.
1651. Daniels, J. M., from 2nd Int. Conf. on Polarized Targets, Berkeley, CA, 30 Aug. 1971.
1652. Daniels, J. M., Univ. of Toronto, Report LBL-500, 1971, pp-88-83.
1653. Daniels, J. M., Kiang, A. K. C., et al., Can. J. Phys. 49, 1 March 1971, pp-576-81.
1654. Daniels, J. M., Kirkby, P., et al., from Int. Conf. on Instrumentation for High Energy Phys., Dubna, USSR, 8 Sept. 1970.
1655. Danilkin, N. P., and Denisenko, P. F., Geomagn. Aeron. (USSR) (Engl. Transl.), 13(5), 1973, pp-786-788.
1656. Danilyan, G. V., Dronyaev, V. P., et al., TsNIIAtominform., 1977, pp-295-299.
1657. Danilyan, G. V., Novitskij, V. V., et al., TsNIIAtominform., 1977, pp-300-304.
1658. Dantzig, R. van (ed.), Inst. voor Kernphysisch Onderzoek, Amsterdam, Annual Report, 1975.
1659. Dar, Y., Gerber, J., et al., Phys. Rev., C, 12(6), Dec. 1975, pp-1771-1779.
1660. Darden, S. E., Murillo, G., et al., Nucl. Phys., A, 266(1), 2 Aug. 1976, pp-29-52.
1661. Daruga, V. K., Matusevich, E. S., Sov. At. Energy 33(5), Nov. 1972, pp-1091-1093.
1662. Dassie, D., CENBG-7601, 1976, p-137, Bordeau-1 Univ., These.
1663. Datta, S. K., Falk, W. R., et al., AIP Conf. Proc. ISSN 0094-243X, 47, 1978, pp-690-691.

1664. Datta, S. K., Berg, G. P. A., et al., Nucl. Phys., A, ISSN 0375-9474, 332(1-2), Dec. 1979, pp-125-143.
1665. Datz, S., Moak, C. D., et al., Phys. Rev. 179, 1969, p-315.
1666. Daum, C., Proc. Triangle Semin., 3rd, 1973, pp-127-54.
1667. David, T. P., AEET-214, 1965, p-90.
1668. David, P., Debrus, J., et al., Phys. Lett., B, 61(2), 15 March 1976, pp-158-160.
1669. David, P., Debrus, J., et al., Verh. Dtsch. Phys. Ges. 6, 1977, pp-973-974.
1670. David, P., Bisplinghoff, J., Blann, M., et al., Nucl. Phys., A, 287(1), 29 Aug. 1977, pp-179-188.
1671. Davidson, J. M., Taylor, T., et al., Nucl. Phys., A, 250(2), 6 Oct. 1975, pp-221-226.
1672. Davidson, J. M., Sheppard, D. M., et al., Phys. Rev., C, 15(1), Jan. 1977, pp-104-113.
1673. Davies, K. T. R., ORNL/TM-7000, Aug. 1979, p-46.
1674. Davies, K. T. R., Devi, K. R. S., et al., Phys. Rev., C, ISSN 0556-2813, 20(4), Oct. 1979, pp-1372-1382.
1675. Davies, J. A., J. Electrochem. Soc. 112, 1965, pp-675-80.
1676. Davies, W. G., Devries, R. M., et al., Nucl. Phys., A, 269 (2), 5 Oct. 1976, pp-477-492.
1677. Davis, J., Kaellne, J., et al., Phys. Rev., C, ISSN 0556-2813, 20(5), Nov. 1979, pp-1946-1948.
1678. Davis, J. C., and Anderson, J. D., J. Vac. Sci. Technol. 12(1), Jan. 1975, pp-358-360.
1679. Davis, J. C., and Anderson, J. D., Report UCRL-75757, 1974, p-13.
1680. Davis, J. R., and Din, G. U., Aust. J. Phys. 30(2), Apr. 1977, pp-167-175.
1681. Davis, M. C., Knoll, G. F., et al., ANL-76-90, 1976, pp-225-236.
1682. Davis, M. C., and Knoll, G. F., BNL-NCS-22500, March 1977, pp-165-167.
1683. Davis, S., Glashausser, C., et al., Nucl. Phys., A, 270(2), 19 Oct. 1976, pp-285-299.
1684. Davison, A., and Allen, K. J. F., Aust. Inst. of Nucl. Sci. and Eng., Lucas Heights, 6th AINSE Nucl. Phys. Conf. 9-11 Feb. 1976.
1685. Davison, A., Allen, K. J. F., et al., Melbourne Univ., Parkville (Aust.), School of Phys., 1977, p-32.
1686. Davison, N. E., Hasell, D. K., et al., Nucl. Phys., A, 290 (1), 24 Oct. 1977, pp-45-54.
1687. Davletshin, A. N., Tipunkov, A. O., et al., Tsentral'nyj. Nauchno-Issledovatel'skij Inst. Informatsii i Tekhniko-Ehkonomicheskikh Issledovanij po Atomnoj Nauke i Tekhnike, Moscow (USSR), Neutron Physics, Part 4, 1976, No. 4, pp-99-103.
1688. Dawson, J., Cash, A. R., et al., J. Phys. E, 13(2), 1980, pp-182-91.

1689. Dayton, R. W., from Am. Power Conf. 26th Annual Meeting, Chicago, April 1964.

1690. Dayras, R., Cujec, B., et al., Nucl. Phys., A, 257(1), 26 Jan. 1976, pp-118-134.

1691. Dayras, R. A., Switkowski, Z. E., et al., Nucl. Phys., A, 261(3), 3 May 1976, pp-365-372.

1692. Dayras, R. A., Stokstad, R. G., et al., Nucl. Phys., A, 261(3), 3 May 1976, pp-478-490.

1693. Dayras, R., Switkowski, Z. E., et al., Nucl. Phys., A, 279 (1), 21 March 1977, pp-70-84.

1694. Dayras, R. A., Fulmer, C. B., et al., ORNL-5306, Sept. 1977, pp-41-43.

1695. De, J. N., Gross, D. H. E., et al., Europ. Phys. Soc., Geneva (Switz.), Europ. Conf. on Nucl. Phys. with Heavy Ions, 6-10 Sept. 1976.

1696. Dealler, B. F. J., and Ettinger, V. K., Proc. of a Conf. held at St. Catherine's Coll., Oxford, 22-23 Sept. 1969.

1697. Dearnaley, G., Rev. Sci. Instr. 31, 1960, pp-197-8.

1698. Dearnaley, G., Phil. Mag. 8(1), 1956 Sept., pp-821-34.

1699. Debrun, L. J., Barrandon, N. J., et al., Proc. of a Conf. held at St. Catherine's Coll., Oxford, 22-23 Sept. 1969.

1700. Debus, G. H., AERE-R-5097, Paper 1, p-4, Geel, Belgium.

1701. Debus, G. H., Proc. of the Seminar on the Prep. and Stand. of Isotopic Targets and Foils, Atomic Energy Rsch. Estab., Oct. 20 & 21, 1965.

1702. Deckman, H. W., and Halpern, G. M., Inertial Confinement Fusion, Dig. Tech. Pap. Top. Meet. 1, 1977, TuE7, p-5.

1703. Decowski, P., Benenson, W., et al., Nucl. Phys., A, ISSN 0029-5582, 302(1), 12 June 1978, pp-186-204.

1704. Dedov, V. B., and Kosyakov, V. N., Proc. of the Int. Conf. on Peaceful Uses of Atomic Energy 7, 1956, p-369.

1705. Degre, A., Centre de Recherches Nucleaires, Strasbourg-1 Univ., 67 (France), These (D. es S.), CRN-PN-76-24, 1976, p-214.

1706. Dehesa, J. S., Faessler, A., et al., Verh. Dtsch. Phys. Ges. 6, 1977, pp-984-985.

1707. Dehnhard, D., Shkolnik, V., et al., Phys. Rev. Lett. 40(24), 12 June 1978, pp-1549-1552.

1708. Delagrange, H., Fleury, A., et al., Europ. Phys. Soc., Geneva (Switz.), Europ. Conf. on Nucl. Phys. with Heavy Ions, 6-10 Sept. 1976.

1709. Delagrange, H., Lin, S. Y., et al., Phys. Rev. Lett. 39(14), 3 Oct. 1977, pp-867-870.

1710. Delagrange, H., Vaz, L. C., et al., Phys. Rev., C, ISSN 0556-2813, 20(5), Nov. 1979, pp-1731-1742.

1711. Delbar, T., Gregoire, G., et al., J. Phys. (Paris), Colloq. 5, 1975, pp-C5.113-C5.115.

1712. DelBianco, W., Kundu, S., et al., Nucl. Phys., A, 270(1), 12 Oct. 1976, pp-45-60.
1713. Deleeuw, J. H., Haasz, A. A., et al., Nucl. Instr. and Meth. 145(1), 1977, pp-119-25.
1714. Deleplanque, M. A., Gerschel, C., et al., Nucl. Phys., A, 249(2), 15 Sept. 1975, pp-366-378.
1715. Deleplanque, M. A., Husson, J. P., et al., Phys. Rev. Lett. ISSN 0031-9007, 43(14), 1 Oct. 1979, pp-1001-1005.
1716. DelFiore, G., and Peters, J. M., Radiochem. Radioanal. Lett. 43(1), 1980, pp-9-18.
1717. Delheij, P. P. J., and Ligthart, H., Nederlandse Natuurkundige Vereniging, Amsterdam, Spring Meeting, 26 and 27 April 1976, Amsterdam.
1718. Delic, G., and Kurath, D., Phys. Rev., C, 14(2), Aug. 1976, pp-619-624.
1719. Deloff, A., Nucl. Phys. A, A251(1), 1975, pp-76-92.
1720. Delvecchio, R. M., Bacco, W. L., et al., Nucl. Instr. and Meth. 144(3), 1977, pp-429-35.
1721. Demond, F. J., Albrecht, R., et al., Verh. Dtsch. Phys. Ges. 6, 1977, pp-899-900.
1722. Dem'yanov, A. V., Dzhelepov, V. P., et al., JINR-P-9-8222, 1974, p-21.
1723. Dem'yanova, A. S., Man'ko, V. I., et al., Nauka, 1978, p-210.
1724. Denisov, F. P., Milovanov, V. P., et al., Yadern. Fiz. 9, May 1969, pp-930-2.
1725. Denisov, S. P., Nekrasov, A. N., et al., Instr. Exp. Tech. 6, Nov. - Dec. 1970, pp-1569-71.
1726. Denisov, S. P., Nekrasov, A. N., et al., IFVE-SEF-70-22, 1970, p-28.
1727. Denkin, N. M., Kavanagh, R. W., Phys. Rev., C, 15(1), Jan. 1977, pp-303-306.
1728. Dennert, K., Graw, G., et al., Verh. Dtsch. Phys. Ges. 10 (7), 1975, pp-709-710.
1729. Denschlag, H. O., Braun, H., et al., Verh. Dtsch. Phys. Ges. 6, 1977, p-971.
1730. Deprun, C, Gauvin, H., et al., Nucl. Instr. and Meth., 138 (1), 1 Oct. 1976, pp-111-24.
1731. Deregel, J., and Lottin, J. C., Bull. Inf. Sci. and Tech. 214, May 1976, pp-31-9.
1732. Dergobuzov, K. A., Evdokimov, O. B., et al., Vsesoyuznyj Nauchno-Issledovatel'skij Inst. Fiziko-Tekhnicheskikh i Radiotekhnicheskikh Izmerenij, Moscow, 1974, pp-27-8.
1733a. Derkach, A. Ya., Karnaukhov, I. M., et al., Vopr. Atom. Nauki i Tekhn. Ser. Obshch. i Yader. Fiz. (1/1), 1978, pp-68-70.
1733b. Dermendzhiev, E. G., and Delchev, M. K., Pribory i Tekhn. Eksperim. 4, July - Aug. 1963, pp-170-3.

1734. Dermendzhiev, E., Kalinkova, N., et al., Int. Symp. on the Interactions of Fast Neutrons with Nuclei, Nov. 17-21, 1975 in Gaussig (GDR).

1735. Derrien, H., and Lucas, B., Natl. Bur. of Stands. (U. S.), Spec. Publ. 425, Oct. 1975, pp-637-641.

1736. Derrien, H., and Edvardson, L., Natl. Bur. of Stands. (U. S.), Spec. Publ. 493, Oct. 1977, pp-14-29.

1737. Deruytter, A. J., Spaepen, J., et al., Neutron Cross Section Tech. I, NBS special Pub. 299, 1968, p-491.

1738. Deruytter, A. J., and Pelfer, P., J. Nucl. Energy 21, 1967, p-833.

1739. Deruytter, A. J., Nuclear Data for Reactor I, IAEA, Vienna, 1970, p-127.

1740. Desgrolard, P., Chambon, B., et al., Phys. Rev., C, 12(6), Dec. 1975, pp-1711-1716.

1741. Desportes, H., ANL-LBL-500, 1971, pp-57-68.

1742. Desreumaux, S., Perrin, P., et al., CEA-CONF-3869, 1975, p-3.

1743. Detaint, J., CEA-R-4212, Thesis, submitted to Grenoble Univ. (France), Faculte des Sciences, Nov. 1971, p-140.

1744. Detaint, J., SC-T-722545, CEA-R-4212, p-186.

1745. Detaint, J., Raffin, M., et al., Vide 26(154), July-Oct. 1971, pp-143-9.

1746. Detaint, M., Proc. of the 3rd Conf. on Accel. Targets Designed for the Prod. of Neutrons, Liege(Belgium), Sept. 18-19, 1967, pp-205-12.

1747. Detraz, C., Barreto, J., et al., Europ. Phys. Soc., Geneva (Switz.), Europ. Conf. on Nucl. Phys. with Heavy Ions, 6-10 Sept. 1976.

1748. Detraz, C., Naulin, F., et al., Phys. Rev., C, 15(5), May 1977, pp-1738-1741.

1749. Deubler, H. H., and Dietrich, K., Europ. Phys. Soc., Geneva (Switz.), Europ. Conf. on Nucl. Phys. with Heavy Ions, 6-10 Sept. 1976.

1750. Deutsch, J., Favart, D., et al., AIP Conf. Proc. 33, 1976, pp-630-631.

1751. Deutchman, P. A., AIP Conf. Proc. ISSN 0094-243X, 54(1), 15 July 1979, p-363.

1752. Deutchmann, S., Sci. Foundations of Vacuum Tech., 2nd Ed., edited by J. M. Lafferty (John Wiley and Sons, Inc., NY, 1962), p-14.

1753. Devan, K., and Brient, C., BNL-NCS-22500, March 1977, pp-229-238.

1754. Devaney, J. J., Report LA-7790-MS, 1979, p-10.

1755. Devaux, A., Landaud, G., et al., AIP Conf. Proc. ISSN 0094-243X, 47, 1978, pp-634-635.

1756. Devlin, T. J., McFadden, J. T., et al., PURC-4159-17, 1970, p-14.

1757. Devous, M. D., Sr., Texas Agricultural and Mech. Univ., College Station (U. S. A.), Thesis (Ph.D), 1976, p-191.

1758. Diaz, J. D., Gonzalez, C. E. G., et al., JEN-60, 1959, p-10.
1759. Dickens, J. K., Morgan, G. L., et al., Natl. Bur. of Stands. (U. S.), Spec. Publ. 425, Oct. 1975, pp-762-765.
1760. Dickens, J. K., Nucl. Sci. and Eng. 58(3), Nov. 1975, pp-331-338.
1761. Dickens, J. K., Nucl. Sci. and Eng. 63(1), 1 May 1977, pp-101-109.
1762a. Dickson, J. M., Proc. Phys. Soc. (London) 64B, 1951, pp-615-16.
1762b. Didelez, J. P., Frascaria, R., et al., Phys. Rev., C, 12(6), Dec. 1975, pp-1974-1977.
1763. Didelez, J. P., Djaloeis, A., et al., Verh. Dtsch. Phys. Ges. 6, 1977, pp-962-963.
1764. Didelez, J. P., Chant, N. S., et al., Verh. Dtsch. Phys. Ges. 6, 1977, p-989.
1765. Didelez, J. P., Djaloeis, A., et al., Nucl. Phys., A, ISSN 0375-9474, 325(1), 13 Aug. 1979, pp-93-99.
1766. Didriksson, R., Einarsson, L., et al., INTDS Proc., 1978, Garching, Sept. 11-14.
1767. Dieterle, B., Denes, P., et al., Nucl. Instr. and Meth. ISSN 0029-554X, 165(2), 1 Oct. 1979, pp-351-353.
1768. Dieterle, B. D., Berkeley, CA, Univ. of Calif., Thesis, 1967, p-53.
1769. Dietrich, W., Baecklin, A., et al., Nucl. Phys., A, 253(2), 24 Nov. 1975, pp-429-447.
1770. Dietrich, F. S., Hansen, L. F., et al., Nucl. Sci. and Eng. 61(2), Oct. 1976, pp-267-268.
1771. Dietrich, F. S., Heikkinen, D. W., et al., Phys. Rev. Lett. 38(4), 24 Jan. 1977, pp-156-158.
1772. Difilippo, F. C., Perez, R. B., et al., CONF-760622-28, ORNL, 1976, p-5.
1773. Difilippo, F. C., Perez, R. B., et al., Nucl. Sci. and Eng. 63(2), June 1977, pp-153-166.
1774. Digiacomo, N. J., Rosenthal, A. S., et al., Phys. Lett., B, 66(5), 28 Feb. 1977, pp-421-424.
1775. Dikshit, J. J., and Singh, B. P., J. Phys., G (London), 2(4), April 1976, pp-219-232.
1776. Diksic, M., Galinier, J. L., et al., Int. J. Appl. Radiat. and Isot. (GB), 28(10-11), Oct.-Nov. 1977, pp-885-8.
1777. Dillig, M., and Huber, M. G., Proc. of the June Workshop in Intermediate Energy Electromagnetic Interactions with Nuclei, held at MIT, June 13-24, 1977.
1778. Dimbylow, P. J., Nat. Radio. Prot. Board, Harwell (UK), Annual Rsch. and Develop. Report 1977.
1779. Dimov, G. I., NP-tr-1669, Akademiya Nauk SSSR, Novosibirsk, Institut Yadernoi Fiziki, p-14.
1780. Din, G. U., and Heusch, B., CRN-PN-75-46, 1975, p-16.
1781. Dittrich, T. R., Gould, C. R., et al., Nucl. Phys., A, 279 (3), 4 April 1977, pp-430-444.

1782. Diven, B. C., Annual Review of Nucl. Sci. 20, 1970, p-79.
1783. Dixon, W. R., Storey, R. S., et al., Phys. Rev., C, 15(5), May 1977, pp-1896-1910.
1784. Dixon, W. R., and Storey, R. S., Nucl. Phys., A, 284(1), 27 June 1977, pp-97-113.
1785. Djaloeis, A., Jahn, P., et al., Nucl. Phys., A, 250(1), 29 Sept. 1975, pp-149-162.
1786. Djaloeis, A., Bojowald, J., et al., Nucl. Phys., A, 273(1), 23 Nov. 1976, pp-29-44.
1787. Dmitrevskii, Yu. P., Cryogenics 16(4), 1976, pp-225-30.
1788. Dmitrevskii, Yu. P., Mel'nik, Yu. M., et al., Cryogenics 14(10, Oct. 1974, pp-564-567.
1789. Dmitrevskii, Yu. P., Elistratov, V. V., et al., Instr. Exp. Tech. 14(4), 1972, pp-1256-1257.
1790. Dmitriev, P. P., Konstantinov, I. O., et al., JINR, July 24, 1967.
1791. Dmitriev, P. P., Konstantinov, I. O., et al., Soviet Atomic Energy 23(1), 1967, pp-733-4
1792. Dmitriev, P. P., Konstantinov, I. O., et al., J. of Atomic Energy 26(5), May 1969.
1793. Dmitriev, P. P., Konstantinov, I. O., et al., Atomnaya Energiya 29(3), Sept. 1970, p-205.
1794. Dmitriev, P. P., Molin, G. A., et al., At. Ehnerg. 40(1), Jan. 1976, pp-66-68.
1795. Dmitriev, P. P., Molin, G. A., et al., At. Ehnerg. 41(6), Dec. 1976, pp-431-434.
1796. Do Dang, G., Publ. Lett. B, Paris-11 Univ., 91-Orsay.
1797. Do, H. P., Chery, R., et al., Verh. Dtsch. Phys. Ges. 6, 1977, p-846.
1798. Dobberstein, P., and Henke, L., Nucl. Instr. and Meth. 119, 1974, p-611.
1799. Dobiasch, H., Fischer, R., et al., Verh. Dtsch. Phys. Ges. 6, 1977, pp-870-871.
1800. Dodder, D. C., LA-UR-75-1447, 1975, p-7.
1801. Dodge, W. R., and Molen, H. V., Maryland Univ., College Park (U. S. A.), ORO-4856-26, 1975, pp-325-26.
1802. Doering, R. R., Galonsky, A., et al., Phys. Rev. Lett. 35 (25), 22 Dec. 1975, pp-1691-1693.
1803. Doering, R., Schweizer, T. C., et al., Phys. Rev. Lett. 40 (22), 29 May 1978, pp-1433-1435.
1804. Doering, R. R., Taddeucci, T. N., et al., Phys. Rev., C, ISSN 0556-2813, 20(5), Nov. 1979, pp-1627-1630.
1805. Dogotar, G. E., Eramzhyan, R. A., et al., AIP Conf. Proc. ISSN 0094-243X, 54(1), 15 July 1979, pp-355-356.
1806. Doi, Y., Fujii, T., et al., Jap. J. Appl. Phys. 10, April 1971, pp-468-471.
1807. Dolbilov, G. V., Ivanov, I. N., et al., JINR-P-9-4737, from 7th Int. Conf. on High Energy Accelerators, Moscow, USSR, UCRL-TRANS-1408.

1808. Doldeman, J. W., Natl. Bur. of Stand. (U. S.), Spec. Publ. 493, Oct. 1977, pp-182-193.
1809. Dolinskij, Eh. I., Mukhamedzhanov, A. M., et al., Izv. Akad. Nauk SSSR, Ser. Fiz. ISSN 0367-6765, 43(1), Jan. 1979, pp-167-169.
1810. Dolinskij, Eh. I., Mukhamedzhanov, A. M., et al., Nauka, 1978, p-205.
1811. Dolinskij, Eh. I., Krekoten', S. P., et al., Yad. Fiz. ISSN 0044-0027, 29(1), Jan. 1979, pp-71-80.
1812. Dodson, R. W., Graves, A. C., et al., Misc. Physical and Chemical Tech. of the Los Alamos Project, Natl. Nuclear Energy Ser. 3, pp-1-46.
1813. Dollard, H., Erdman, K. L., et al., Phys. Lett., B, 63(4), 16 Aug. 1976, pp-416-418.
1814. Donahue, D. J., McCullen, J. D., et al., Arizona Univ., Tucson, Contract AT(11-1)-1468, Jan. 1969, p-14.
1815. Donahue, D. J., Hausser, O., et al., Phys. Rev., C, 12(5), Nov. 1975, pp-1547-1556.
1816. Donohue, J. T., Laboratoire de Physique Theorique Et des Hautes Energies, Nice, Report No.: N-TH-71/1, Feb. 1971, p-8.
1817. Donichkin, A. G., Smirnov, A. N., et al., Int. Atomic Energy Agency, Vienna (Austria), Int. Nucl. Data Comm., Feb. 1977, p-16.
1818. Donichkin, A. G., Smirnov, A. N., et al., Int. Atomic Energy Agency, Vienna (Austria), Int. Nucl. Data Comm., Nuclear Physics Research in the USSR, 1977 Feb., p-18.
1819. Donnelly, T. W., Proc. of the June Workshop in Intermediate Energy Electromagnetic Interactions with Nuclei, held at MIT, June 13-24, 1977.
1820. Donnelly, T. W., Orden, J. W. van, et al., Phys. Lett., B, ISSN 0031-9163, 76(4), 19 June 1978, pp-393-396.
1821. Donoghue, T. R., Doyle, S. M. A., et al., Phys. Rev. Lett. 37(15), 11 Oct. 1976, pp-981-984.
1822. Doornbos, J., Krijgsman, W., et al., Nederlandse Natuurkundige Vereniging, Amsterdam, Spring Meeting, 26 and 27 April 1976, Amsterdam.
1823. Doornbos, J., Krijgsman, W., et al., Nucl. Phys., A, 297(3), 20 March 1978, pp-412-428.
1824. Doron, T. A., and Blann, M., Nucl. Phys. A, A161(1), 25 Jan. 1971, pp-12-48.
1825. Doron, A., Moinester, M. A., et al., Bull. Isr. Phys. Soc. 21, 1975, pp-12-13.
1826. Doron, A., Alster, J., et al., LAMPF Workshops on Pion Single Charge Exchange, LA-7892C, July 1979.
1827. Doss, K. G. R., Dytman, S. A., et al., AIP Conf. Proc. 33, 1976, pp-344-345.
1828. Doss, K. G. R., Eisenstein, R. A., et al., AIP Conf. Proc. ISSN 0094-243X, 54(1), 15 July 1979, pp-533-534.

1829. Doubre, H., Roynette, J. C., et al., IPNO-PhN-77-16, 1977, p-31.
1830. Doubre, H., and Marty, C., Paris-11 Univ., 91 - Orsay (France), Inst. de Physique Nucleaire, IPNO-TH-77-28, May 1977, p-18.
1831. Doubre, H., Roynette, J. C., et al., Phys. Rev., C, 17(1), Jan. 1978, pp-131-142.
1832. Doubre, H., Gamp, A., et al., Phys. Lett., B, 73(2), 13 Feb. 1978, pp-135-138.
1833. Doubt, H. A., Fechner, J. B., et al., Nucl. Instr. and Meth. 109(3), 1973, pp-571-572.
1834. Douglas, D. A., ORNL-3470, pp-240-7.
1835. Douglas, D. A., Jr., ORNL-3970, pp-179-84.
1836. Douglas, R. A., et al., Can. J. Phys. 34, 1956, p-1097.
1837. Doukellis, G. C., Ohio Univ., Athens (U. S. A.), Thesis (Ph. D), 1975, p-163.
1838. Dover, C. B., AIP Conf. Proc. 33, 1976, pp-249-259.
1839. Dow, P. A., Briers, G. W., et al., Nucl. Instr. and Meth. 60, April 1968, pp-293-6.
1840. Dracoulis, G. D., Ferguson, S. M., et al., Aust. Nat. Univ., Canberra, Rsch. School of Physical Sciences, 1976, p-30.
1841. Dracoulis, G. D., Ferguson, S. M., et al., Nucl. Phys., A, 279(2), 28 March 1977, pp-251-268.
1842. Dracoulis, G. D., Walker, P. M., et al., Aust. Nat. Univ., Canberra, Rsch. School of Physical Sciences, ANU-P-674, Nov. 1977, p-61.
1843. Dragichesku, P., Lushchikov, V. I., et al., JINR-P-1797, 1964, p-9.
1844. Dragoun, O., Brabec, V., et al., Z. Phys., A, 279(1), Oct. 1976, pp-107-111.
1845. Drake, D., Grenier, G., et al., CEA-CONF-3960, 1977, p-14.
1846. Drake, D. M., Auchampaugh, G. F., et al., LA-6257, Aug. 1976, p-13.
1847. Drake, D. M., Arthur, E. D., et al., Nucl. Sci. and Eng. 65(1), Jan. 1978, pp-49-64.
1848. Drake, M. K., Sargis, D. A., et al., Sci. Appl., Inc. LaJolla, CA, PB-261828, Nov. 1976, p-88.
1849. Drake, M. K., Sargis, D. A., et al., Sci. Appl., Inc., LaJolla, CA, EPRI-NP-250, Nov. 1976, p-84.
1850. Drake, T. E., and Pai, H. L., Nucl. Phys., A, 259(2), 15 March 1976, pp-317-323.
1851a. Draper, J. E., and McDonald, R. J., Nucl. Instr. and Meth. 171(2), 1980, pp-215-17.
1851b. Draper, J. E., McDonald, R. J., et al., Phys. Rev., C, 16 (4), Oct. 1977, pp-1594-1604.
1852. Drenckhahn, W., Feigel, A., et al., Verh. Dtsch. Phys. Ges. 6, 1977, pp-960-961.
1853. Drenckhahn, W., Feigel, A., et al., Verh. Dtsch. Phys. Ges. 6, 1977, p-964.

1854. Dressler, E., Proc. of the June Workshop in Intermediate Energy Electromagnetic Interactions with Nuclei, held at MIT, June 13-24, 1977.
1855. Dressel, R. W., Phys. Rev. 144, Apr. 8, 1966, pp-332-43.
1856. Dressel, R., AD Rep. 66, (AD 640634), p-4, U. S. Govt. Res. Develop. Rep. 41(23), 1966, p-161.
1857. Dreyer, I., Fritsch, T., et al., Gesellschaft fuer Schwerionenforschung m.b.H., Darmstadt (F.R. Germany), Nov. 1975, pp-61-63.
1858. Dreze, C., and Duquesne, H., Nucl. Instr. and Meth. 28, Aug. 1964, pp-321-4.
1859. Driel, M. A. van, Eggenhuisen, H. H., et al., Nederlandse Natuurkundige Vereniging, Amsterdam, Spring meeting, 26 and 27 April 1976, Amsterdam.
1860. Driel, J. van, Kamermans, R., et al., Rijksuniversiteit Groningen (Netherlands), Kernfysisch Versneller Inst., pp-41-43.
1861. Drooks, L. J., Yale Univ., New Haven, CT (U. S. A.), Thesis (Ph.D), 1976, p-139.
1862. Dropesky, B. J., Butler, G. W., et al., Phys. Rev., C, ISSN 0556-2813, 20(5), Nov. 1979, pp-1844-1856.
1863. Drosg, M., Smith, R. K., et al.,LA-6262-MS, Feb. 1976, p-3.
1864. Drosg, M., and Auchampaugh, G. F., Nucl. Instr. and Meth. 140(3), 1 Feb. 1977, pp-515-518.
1865. Drosg, M., Nucl. Sci. and Eng. 65(3), March 1978, pp-553-4.
1866. Druzhinin, A. A., Grigor'ev, V. K., et al., At. Ehnerg. 42 (4), April 1977, pp-314-315.
1867. Duarte, L. R., and Spinelli, D., Rev. Brazil Technol. 2(1), March 1971, pp-21-6.
1868. Dubal, L., Eaton, G. H., et al., Verh. Dtsch. Phys. Ges. 6, 1977, p-920.
1869. Dubler, T., Kaeser, K., et al., Nucl. Phys., A, 294(3), 16 Jan. 1978, pp-397-416.
1870. Duboc, J., Minten, A. G., et al., CERN-65-2, Jan. 4, 1965, p-21.
1871. Dubovikov, M. S., Zh. Eksp. Teor. Fiz., Pis'ma Redkat. 3, Feb. 15, 1966, pp-156-9.
1872. Dudley, N. D., and Kennerley, R., BNL-NCS-50446, April 1975, pp-80-87.
1873. Dudley, N. D., and Kennerley, R., BNL-NCS-50446, April 1975, pp-34-41.
1874. Duclos, J., CEA-N-1861 (nd), 1 Oct. 1974 - 1 Sept. 1975, pp-153-156.
1875. Dueck, P., Froelich, H., et al., Europ. Phys. Soc., Geneva (Switz.), Europ. Conf. on Nucl. Phys. with Heavy Ions, 6-10 Sept. 1976.
1876. Dufour, C., Vide 3, 1948, p-480.
1877. Dufour, C., and Zega, B., Vide 18, 1963. 180.
1878. Duggan, J. L., (ed.), Wells, G. W., ORNL, TN, Proc. of 4th Conf. on Sci. and Ind. Appl. of Small Accelerators.

1879. LeDuigou, Y., and Lauer, K. F., Nucl. Instr. and Meth. 97, 1971, p-199.
1880. Dumail, M., Nucl. Instr. and Meth. 163(1), 1 July 1979, pp-61-5.
1881. Dumazet, G., Institut d'Etudes Nucleaires, Algiers, Thesis, June 1976, p-122.
1882. Dumont, P. D., Livingston, A. E., et al., Phys. Sci. 14, 1976, p-122.
1883. Dupetit, G. A., J. Inorg. and Nucl. Chem. 24, 1962, p-1297.
1884. Dupont, F., and Gremont, B., Int. Cyc. Conf. St. Catherine's Coll., Oxford, 17-20 Sept. 1967.
1885. Dupzyk, R. J., Henderson, C. M., et al., Nucl. Instr. and Meth. 153(1), 1978, pp-53-8.
1886. Durand, G., Grenoble-1 Univ., 38 (France), Inst. des Sciences Nucleaires, 1974, p-113.
1887. Durell, J. L., Christy, A., et al., Univ. of Manchester, England, 1975.
1888. Durell, J. L., Buttle, P. J. A., et al., Nucl. Phys., A, 269(2), 5 Oct. 1976, pp-443-459.
1889. Durisch, J. E., Neumann, W., et al., Nucl. Instr. and Meth. 80, 1970, pp-1-12.
1890. Dushman, S., Sci. Fnd. of Vacuum Technique (John Wiley & Sons, Inc., NY), 1962.
1891. Dushman, S., Sci. Fnd. of Vacuum Technique (John Wiley & Sons, Inc., NY), 1949.
1892. Dworschak, F., Neuhaeuser, J., et al., Rev. Sci. Instr. 41, Jan. 1970, pp-64-7.
1893. Dyer, F. F., Bate. L. C., et al., Anal. Chem. 39, Dec. 1967, pp-1907-9.
1894. Dyer, P., Puigh, R. J., et al., LBL-1851, 1978.
1895. Dyer, P., Puigh, R. J., et al., Wash. Univ., Seattle (U. S. A.), Nucl. Phys. Lab. Annual Report, 1977.
1896. Dymarz, R., and Malecki, A., Phys. Lett., B, ISSN 0370-2693, 83(1), 23 April 1979, pp-15-18.
1897. Dytman, S. A.Amann, J. F., et al., Phys. Rev. Lett. 38(19), 9 May 1977, pp-1059-1062.
1898. Dzhavadov, A. V., Mirabutalybov, M. M., et al., Izv. Akad. Nauk SSSR, Ser. Fiz. ISSN 0367-6765, 42(9), Sept. 1978, pp-1875-1882.
1899. Dzhavadov, A. V., and Mirabutalybov, M. M., Izv. Akad. Nauk SSSR, Ser. Fiz. ISSN 0367-6765, 42(9), Sept. 1978, pp-1869-1874.
1900. Dzhibuti, R. I., Kezerashvili, R. Ya., et al., Yad. Fiz. ISSN 0044-0027, 29(1), Jan. 1979, pp-65-70.
1901. Eagg, L. W., Jones, E. C., Jr., et al., Nucl. Instr. and Meth. 77(1), 1970, pp-136-40.
1902. Eagle, R., Clarke, N. M., et al., Phys. Rev., C, 16(4), Oct. 1977, pp-1314-1321.
1903. Earle, E. D., Knowles, J. W., et al., Nucl. Phys., A, 257 (3), 9 Feb. 1976, pp-365-377.

1904. East, M. T., Kean, D. C., et al., Aust. Inst. of Nuclear Sci. and Eng., Lucas Heights, 6th AINSE Nuclear Physics Conf., 9-11 February, 1976.
1905. Eaton, E. E., Ehart, E. P., et al., Contract W-7405-eng-36, June 1972, p-6.
1906. Eaton, E. E., Ehart, E. P., et al., LASL, Report LA-4862, 1971, p-6.
1907. Eberhard, K. A., and Bernhardt, K. G., Phys. Rev., C, 13(1), Jan. 1976, pp-440-443.
1908. Eberhard, K. A., Wit, M., et al., Phys. Rev., C, 14(6), Dec. 1976, pp-2332-2334.
1909. Eberhard, K. A., and Cuengco, B. D., Verh. Dtsch. Phys. Ges. 6, 1977, p-963.
1910. Ebersold, P., Aas, B., et al., Nucl. Phys., A, 296(3), 27 Feb. 1978, pp-493-518.
1911a. Ebert, H. G., EUR-3895, Proc. of the 3rd Conf. on Accel. Targets Designed for the Prod. of Neutrons, Liege (Belgium), Sept. 18-19, 1967.
1911b. Ebert, K., Wild, W., et al., Facultad de Ciencias Exactas, Sao Paulo Univ. (Brazil), Inst. de Fisica, IFUSP-P-175 (nd), p-52.
1912. Eberth, U., Eberth, J., et al., Nucl. Phys., A, 257(2), 2 Feb. 1976, pp-285-302.
1913. Ebinghaus, H., Steffens, E., et al., Nucl. Instr. and Meth. 125(1), 1975, pp-73-8.
1914. Ebisawa, K., and Snover, K. A., Wash. Univ., Seattle (U. S. A.), Nuclear Phys. Lab. Annual Report, 1977.
1915. Eccleston, G. W., and Woodruff, G. L., Wash. Univ., Seattle (U. S. A.), Dept. of Nuclear Eng., 1976, p-37.
1916. Eccleston, G. W., and Woodruff, G. L., Nucl. Sci. and Eng. 62(4), April 1977, pp-636-651.
1917. Eck, J. S., Gray, T. J., et al., Phys. Rev., C, 16(5), Nov. 1977, pp-1873-1877.
1918. Eckroad, S. W., Leber, R. E., et al., Nucl. Instr. and Meth. 140(3), 1977, pp-447-51.
1919. Eckstein, P., Helfer, H., et al., Sektion Physik, Technische Univ., Dresden, Sektion Physik, Zentralinstitut fuer Kernforschung, Rossendorf bei Dresden, Annual Report 1975, ZfK-295, Aug. 1975, pp-15-16.
1920. Edelstein, R. M., Makuchowski, E. J., et al., Phys. Rev. Lett. 38(5), 31 Jan. 1977, pp-185-188.
1921. Eckstein, W., and Verbeek, H., Vacuum 23(5), May 1973, pp-159-162.
1922. L'Ecuyer, J., Volders, R., et al., Phys. Rev., C, 12(6), Dec. 1975, pp-1878-1887.
1923. Edeskuty, F. J., Hwang, C. F., et al., LA-4465, pp-33.1-2.
1924. Edeskuty, F. J., Hwang, C. F., et al., Polarization Phenomena in Nuclear Reactions, Barschall, H. H. (ed.), Madison, WI, Univ. of Wisconsin Press, 1971, pp-890-2.

1925. Edguer, E., Fac. Chem., Hacettepe Univ., Ankara, Turk., 1979.
1926. Edwards, F. M., Univ. of Colo., 1976.
1927. Eeghem, W. P. Th. M. van, and Verhaar, B. J., Nucl. Phys., A, 258(1), 16 Feb. 1976, pp-70-82.
1928. Egan, J. J., Kegel, G. H. R., et al., Natl. Bur. of Stands. (U. S.), Spec. Publ. 425, Oct. 1975, pp-950-952.
1929. Egerton, R. F., and Juharez, C., Thin Solid Films 4, 1969, p-239.
1930. Eggenhuisen, H. H., Engelbertink, G. A. P., et al., Europ. Phys. Soc., Geneva (Switz.), Europ. Conf. on Nuclear Phys. with Heavy Ions, 6-10 Sept. 1976.
1931. Egger, C., Gunten, H. R. von., et al., Radiochim. Acta 21 (3/4), 1974, pp-200-202.
1932. Egger, J. P., Burman, R. L., et al., LA-6926-C, Aug. 1977, pp-45-70.
1933. Egger, J. P., Rebel, H., et al., Proc. KFK-2830, INKA-Conf.-79-431-000.
1934. Eggermann, J., Ernst, J., et al., Z. Phys., A, 273(4), 1975, pp-381-384.
1935. Eggers, R., Namboodiri, M. N., et al., Europ. Physical Soc., Geneva (Switz.), Europ. Conf. on Nucl. Phys. with Heavy Ions, 6-10, Sept. 1976.
1936. Eggers, R., Namboodiri, M. N., et al., ANL/PHY-76-2(2), May 1976, pp-531-540.
1937. Eggers, R. C., Ribbe, W. S., et al., Nucl. Phys., A, 263(3), 14 June 1976, pp-500-510.
1938. Ehl'-Ashri, F. M., Vinogradov, L. I., et al., Nauka 1978, p-129.
1939. Ehl'-Ashri, F. M., Chubinskij, O. V., et al., Nauka 1978, p-132.
1940. Ehlers, D. H., Knowles, H. B., et al., Rev. Sci. Instr. 37, Dec. 1966, pp-1708-10.
1941. Eickhoff, H., Gaul, G., et al., Verh. Dtsch. Phys. Ges. 6, 1977, p-999.
1942. Eickmeyer, J., Michalowski, S., et al., Phys. Rev. Lett. 36(6), 9 Feb. 1976, pp-289-291.
1943. Eijk, W. van der, Oldenhof, W., et al., Nucl. Instr. and Meth. 112, 1973, pp-343-351.
1944. Eijk, W. van der, and Vaninbrounx, R., TISRMNM, ORNL, Oct. 5-8, 1971.
1945. Eijk, W. van der, Source Improvements in the Prep. of Thin Radioactive Sources, Ph. D Thesis to be submitted at the Univ. of Amsterdam.
1946. Eijkern, F. E. H. van, Middelkoop, G. van, et al., Nucl. Phys., A, 260(1), 29 March 1976, pp-124-140.
1947. Eijkern, F. E. H. van, Middelkoop, G. van, et al., Internal Report, Rijksuniversiteit Utrecht, Neth., 1975.
1948. Eisen, Y., Fortune, H. T., et al., ANL-75-75, 1975, pp-18-9.
1949. Eisen, Y., Erskine, J. R., et al., ANL-75-75, 1975, pp-20, 22, 23.

1950. Eisen, Y., Fortune, H. T., et al., ANL-75-75, 1975, pp-22, 24-26.
1951. Eisen, Y., Fortune, H. T., et al., Phys. Rev., C, 13(2), Feb. 1976, pp-699-711.
1952. Eisen, Y., and Day, B., ANL/PHY-76-2(2), May 1976, pp-541-8.
1953. Eisenberg, Y., Haber, B., et al., Phys. Rev. Lett. 38(3), 17 Jan. 1977, pp-108-111.
1954. Eisend, M., Frank, K., et al., Nucl. Instr. and Meth. 140 (2), 1977, pp-227-30.
1955. Eisenstein, R. A., AIP Conf. Proc. 33, 1976, pp-55-64.
1956. Eisenstein, R. A., AIP Conf. Proc. ISSN 0094-243X, 54(1), 15 July 1979, pp-440-454.
1957. Ekstrom, C., and Lindgren, I., At. Phys. 5, 1977, pp-201-13.
1958. Ekstroem, L. P., Eggenhuisen, H. H., et al., Europ. Physical Soc., Geneva (Switz.), Europ. Conf. on Nucl. Phys. with Heavy Ions, 6-10 Sept. 1976.
1959. Eggenhuisen, H. H., Ekstroem, L. P., et al., Nederlandse Natuurkundige Vereniging, Amsterdam, Spring Meeting of the Dutch Physical Soc., NNV voorjaarsvergadering, March 1978, p-14.
1960. Ekstroem, L. P., Scherpenzeel, D. E. C., et al., Nucl. Phys., A, 295(3), 6 Feb. 1978, pp-525-531.
1961. Elbek, B., Nucl. Instr. and Meth. 38, 1965, pp-314-18.
1962. El-Kazzaz, S., Lien, J. R., et al., Nucl. Phys., A, 280(1), April 11, 1977, pp-1-12.
1963. Ellegaard, C., Garrett, J. D., et al., Verh. Dtsch. Phys. Ges. 6, 1977, p-868.
1964. Ellegaard, C., Julin, R., et al., Nucl. Phys., A, ISSN 0029-5582, 302(1), 12 June 1978, pp-125-133.
1965. Elliot, A. G., Ph. D Thesis, Mat. Sci. Dept., Stanford Univ.
1966. Elliott, D. O., Jr., Lee, Y. K., et al., Nuclear Moments and Nuclear Structure, Annual Progress Report, May 1, 1975-April 30, 1976.
1967. Ellsworth, C. E., INTDS Proc., 1977, Berkeley, Oct. 19-20.
1968. Elmore, D., and Alford, W. P., Phys. Rev., C, 14(2), Aug. 1976, pp-583-589.
1969. Elo, D., Morris, D., et al., LBL-7735, Contract W-7405-eng-48, Sept. 1978, p-6.
1970. Elsenaar, R. J., Leun, C. van der, et al., Rijksuniversiteit Utrecht (Netherlands), Robert van de Graaff Lab., Annual Report 1976, Feb., 1977, pp-13-14.
1971. Elsner, G., Reuter, W., et al., Verh. Dtsch. Phys. Ges. 6, 1977, p-1002.
1972. Elsenaar, R. J., Graber, H. D., et al., Rijksuniversiteit Utrecht (Netherlands), Robert van de Graaff Lab., Annual Report 1976, Feb. 1977, pp-15-16.
1973. Elsenaar, R. J., Lancman, H., et al., Rijksuniversiteit Utrecht (Netherlands), Robert van de Graaff Lab., Annual Report 1977, March 1978, pp-15-16.

1974. Elwyn, A. J., Holland, R. E., et al., Natl. Bur. of Stands. (U. S.), Spec. Publ. 425, Oct. 1975, pp-692-696.
1975. Elwyn, A. J., Holland, R. E., et al., ANL-78-66, 1 April 1977-31 March 1978.
1976. Emel'yanenko, G. A., Kopaleishvili, T. I., et al., JINR-E-4-12320, 1979, p-22.
1977. Emery, G. T., Phys. Lett., B, 60(4), 2 Feb. 1976, pp-351-54.
1978. Emigh, R. A., Lind, D. A., et al., Colo. Univ., Boulder (U. S. A.), Technical Progress Report TID-27909, 1 Nov. 1977, pp-19-21.
1979. Emigh, C. R., LA-4097, Contract W-7405-eng-36, Dec. 1969, p-72.
1980. Emigh, R. A., and Anderson, R. E., Colo. Univ., Boulder (U. S. A.), Tech. Prog. Report TID-27909, 1 Nov. 1977, pp-16-19.
1981. Emigh, R. A., Anderson, R. E., Nucl. Phys., A, 293(3), 26 Dec. 1977, pp-379-396.
1982. Emsallem, A., Lyon-1 Univ., 69-Villeurbanne (France), Inst. de Physique Nucleaire.
1983. Emsallem, A., and Asghar, M., Z. Phys., A, 275(2), Nov. 1975, pp-157-16
1984. Emsallem, A., Asghar, M., et al., Z. Phys., A, 286(4), 1978, pp-411-414.
1985. Enayet, U. M., Atomic Energy Centre, Dacca (Pakistan), Nov. 1971, p-99.
1986. Engelbertink, G. A. P., and Olness, J. W., Phys. Rev. C, 3(1), Jan. 1971.
1987. Engelbrecht, C. A., Nucl. Instr. and Meth. 80, 1970, pp-187-91.
1988. Engelmann, Ch., Nucl. Instr. and Meth. 91, 1971, pp-189-94.
1989. Engelmohr, G. O., Mueller, R. M., et al., Nucl. Instr. and Meth. 83, 1970, pp-160-4.
1990. Engfer, R., Hartmann, R., et al., AIP Conf. Proc. ISSN 0094-243X, 54(1), 15 July 1979, pp-176-177.
1991. England, J. B. A., Nucl. Instr. and Meth. 98(2), 1972, pp-237-54.
1992. England, J. B. A., Casal, E., et al., Nucl. Phys., A, 284(1), 27 June 1977, pp-29-40.
1993a. England, J. B. A., Nucl. Instr. and Meth. 147(2), 1 Dec. 1977, pp-441-446.
1993b. England, J. B. A., Vlastou, R., et al., AIP Conf. Proc. ISSN 0094-243X, 47, 1978, pp-610-611.
1994. Ensslin, N., Bendel, W. L., et al., Colo. Univ., Boulder (U. S. A.), Dept. of Phys. and Astrophys., Technical Progress Report and Proposal for Continuation of Contract, 1 Nov. 1975.
1995. Epaneshnikov, V. D., Kuznetsov, V. M., et al., Conf. on Nucl. Spectro. and Nucl. Structure, Baku, 3-6 Feb. 1976.

1996. Epaneshnikov, V. D., Kuznetsov, V. M., et al., Conf. on Nucl. Spectro. and Nucl. Structure, Baku, 3-6 Feb. 1976.
1997. Eponeshnikov, V. N., Krechetov, Yu. F., et al., Leningrad Nauka 1978, p-255.
1998. Epstein, G. N., Singham, M. K., et al., Phys. Rev., C, 17(2), Feb. 1978, pp-702-709.
1999. Epstein, G. N., Tabakin, F., et al., Phys. Rev., C, 17(4), April 1978, pp-1501-1504.
2000. Epstein, G. N., Singham, M. K., et al., AIP Conf. Proc. ISSN 0094-243X, 54(1), 15 July 1979, pp-432-433.
2001. Erb, K. A., Betts, R. R., et al., ANL/PHY-76-2(2), May 1976, pp-549-556.
2002. Erb, K. A., Hanson, D. L., et al., Europ. Physical Soc., Geneva (Switz.), Europ. Conf. on Nucl. Phys. with Heavy Ions, 6-10 Sept. 1976.
2003. Erb, K. A., Betts, R. R., et al., Europ. Conf. on Nucl. Phys. with Heavy Ions, 6-10 Sept. 1976.
2004. Erb, K. A., Betts, R. R., et al., Phys. Rev. Lett. 37(11), 13 Sept. 1976, pp-670-673.
2005. Erber, T., Z. Agnew. Phys. 24, 1968, pp-188-91.
2006. Erlandsson, B., and Marcinkowski, A., Phys. Scr. 12(1-2), Aug. 1975, pp-95-102.
2007. Ermolov, P. F., Lepilov, V. I., et al., JINR-P-2711, p-5.
2008. Ermolov, P. F., Lepilov, V. I., et al., Prib. Tekh. Eksp. 3, May-June 1967, pp-41-3.
2009. Erne, F. C., AERE-R-5097, Paper 6, p-3.
2010. Erne, F. C., PSPSITF, AERE, Oct. 20-21, 1965.
2011. Erskine, J. R., and Browne, C. P., Nucl. Instr. and Meth. 13, Oct. 1961, pp-359-60.
2012. Erskine, J. C., Jr., Cleveland, Western Reserve Univ., Thesis, 1966, p-71.
2013. Erskine, J. R., INTDS Proc., 1975, Argonne, Sept. 30-Oct. 2, U. S. A.
2014. Erskine, J. R., ANL Report CONF-750968-5, 1975, p-30.
2015. Erskine, J. R., and Gemmell, D. S., Nucl. Instr. and Meth. 24, 1963, p-397.
2016. Ernst, J., Loehr, R., et al., Verh. Dtsch. Phys. Ges. 6, 1977, p-1017.
2017. Ernst, D. J., and McLeod, R. J., AIP Conf. Proc. ISSN 0094-243X, 54(1), 15 July 1979, pp-71-72.
2018. Erskine, J. R., INTDS Proc., 1975, Argonne, U. S. A., Sept. 30-Oct. 2.
2019. Erskine, J. R., Internal Report, ANL, September 30 - Oct. 2, 1974.
2020. Erskine, J. R., Phys. Rev., C, 17(3), March 1978, pp-934-38.
2021. Ertek, C., Nucl. Sci. and Eng. 64(4), Dec. 1977, pp-889-894.
2022. Esat, M. T., Kean, D. C., et al., ANU-P-623, Feb. 1976, p-13.
2023. Esat, M. T., Kean, D. C., et al., Phys. Lett., B, 61(3), March 29, 1976, pp-242-244.

2024. Esat, M. T., Kean, D. C., Aust. Nat. Univ., Canberra, Rsch. School of Physical Sciences, Nov. 1976, p-10.
2025. Esat, M. T., Kean, D. C., Nucl. Instr. and Meth. 141, 1977, pp-405-7.
2026. Eschbach, H. L., and Verheyen, F., Thin Solid Films 21, 1974, pp-237-243.
2027. Eschbach, H. L., Lycke, W., et al., Vakuumtechnik 22(8), 1973, pp-233-238.
2028. Eschbach, H. L., Nucl. Instr. and Meth. 102(3), 1972, pp-469-75.
2029. Eschbach, H. L., 3rd Symp. on Rsch. Mat. for Nucl. Measurements, Gatlinburg, TN, 1971, pp-122-133.
2030. Eschbach, H. L., and Grillot, A., LeVide 140, Construction et Caractéristiques de sources miniatures d'évaporation, April, 1969.
2031. Eschbach, H. L., and Kruidhof, E., A Direct Calibration Method for a Crystal Oscillator Vacuum Microbalance Techniques 5, 1966, p-207.
2032. Eschbach, H. L., PSPSITF, AERE, Oct. 20-21, 1965.
2033. Eschbach, H. L., and Kruedhof, E. W., 5th Conf. on Vacuum Microbalance Techniques, Princeton, NJ, 1965.
2034. Escovitz, W. H., IEEE Trans. on Nucl. Sci., NS-26(1), Feb. 1979.
2035. Esmore, L., Prod. Finish. 30(11), 1966, p-70.
2036. Esposito, A., Lucci, F., et al., Nucl. Instr. and Meth. 138 (2), 15 Oct. 1976, pp-209-212.
2037. Eswaran, M. A., Gove, H. E., et al., Nucl. Phys., A, ISSN 0375-9474, 325(1), 13 Aug. 1979, pp-269-282.
2038. Etemad, M. A., Natl. Bur. of Stands. (U. S.), Spec. Publ. 425, Oct. 1975, pp-871-874.
2039. Etkin, A., Lindenbaum, S. J., et al., BNL-19347, 1974, p-5.
2040. Etkin, A., BNL-20415, Dec. 1975, pp-331-339.
2041. Etkin, A., BNL-20415, Dec. 1975, pp-379-384.
2042. Ettinger, R. V., Fremlin, J. H., et al., 7th Int. Conf. on Cyc. and Their Application, Zurich, Switz., 19-22 Aug. 1975.
2043. Eubanks, I. D., and Thompson, M. C., Inorg. and Nucl. Chem. Letters 5, p-187
2044. Euteneuer, H., Friedrich, J., et al., Phys. Rev. Lett. 36(3), 19 Jan. 1976, pp-129-132.
2045. Evans, A. E., and Krick, M. S., Trans. Am. Nucl. Soc. 23, June 1976, pp-491-492.
2046. Evans, A. E., Krick, M. S., Nucl. Sci. and Eng. 62(4), April 1977, pp-652-659.
2047. Evans, C., AERE-I/M-38, Sept. 26, 1955, p-6.
2048. Evans, J. E., Lougheed, R. W., et al., U. S. At. Energy Comm., UCRL-73121, 1971, p-35.
2049. Evans, J. E., Lougheed, R. W., et al., Nucl. Instr. and Meth. 102(3), 1972, pp-389-401.

2050. Evans, J. E., Lougheed, R. W., et al., UCRL-73121, July 1, 1971
2051. Evans, J. E., Lougheed, R. W., et al., 3rd Int. Symp. on Rsch. Mat. for Nucl. Measurements, 1971, pp-21-38.
2052. Evans, J. E., Lougheed, R. W., et al., UCRL-73121, July 1, 1971, p-35.
2053. Evans, J. E., Lougheed, R. W., et al., U. S. At. Energy Comm., UCRL-73121, 1971, p-35.
2054. Evans, P. A. R., Huxtable, G. B., et al., ANL-76-90, 1976, pp-149-153.
2055. Evans, R. B., III., and Rutherford, J. L., J. Appl. Phys. 38, July 1967, pp-3127-34.
2056. Evans, W. H., and Hall, S. J., Nucl. Instr. and Meth. 24, Oct. 1963, pp-345-8.
2057. Evans, W. S., New England Nuclear, U. S. A., Internal Publication.
2058. Evers, D., Harasim, A., et al., Phys. Rev., C, 15(5), May 1977, pp-1690-1697.
2059. Ewen, K., and Gonsior, B., Nucl. Instr. and Meth. 105(3), 1972, p-599.
2060. Eyal, Y., Backerman, M., et al., Weizmann Inst. of Sci., Rehovoth (Israel), 1975, p-30.
2061. Eyal, Y., Europ. Conf. on Nucl. Phys. with Heavy Ions, 6-10 Sept. 1976.
2062. Eyrich, W., Erlangen-Nuernberg Univ., Erlangen (Germany, F. R.), Naturwissenschaftliche Fakultaet Diss. (D. Sc.), May 1976, p-129.
2063. Eyrich, W., Hofmann, A., et al., Phys. Lett., B, 63(4), 16 Aug. 1976, pp-406-408.
2064. Eyrich, W., Hofmann, A., et al., Erlangen-Nuernberg Univ., Erlangen (Germany, F. R.), Physikalisches Inst., Rebel, H, Oct. 1979, pp-48-49.
2065. Eyrich, W., Hofmann, A., et al., Erlangen-Nuernberg Univ., Erlangen (Germany, F. R.), Physikalisches Inst., Rebel, H, Oct. 1979, pp-50-51.
2066. Eyrich, W., Hofmann, A., et al., Erlangen-Nuernberg Univ., Erlangen (Germany, F. R.), Physikalisches Inst., Rebel, H, Oct. 1979, pp-52-53.
2067. Fabian, F., Accel. Targets Designed for the Production of Neutrons, EUR-1815.e(Rev.), 1964, p-209.
2068. Fabian, H., and Hanau, W. B., Report No. BMVG-FBWT-77-9, 1977, p-37, Nuc.-Chemie und Metallurgie, G. m. b. H.
2069. Fabian, H., and Muenzer, H., EUR-1815.e(Rev.), pp-190-200.
2070. Fabian, H., EUR-1815.e(Rev.), pp-201-28. See also 2068.
2071. Fabian, H., from Seminar on the Monitoring of Radioactive Effluents, Karlsruhe, F. R. Germany, 14 May 1974.
2072. Fabian, H., and Hanau, W. B., Report No. BMVG-FBWT-71-1, 1971, p-24. See also 2068.
2073. Fabian, H., and Hanau, W. B., Report No. BMVG-FBWT-73-22, 1973, p-47. See also 2068.

2074. Fabian, H. Kerntech, Atomprax 13(4), April 1971, pp-176-8.

2075. Fabian, H., and Hanau, W. B., Report No. BMVG-FBWT-71-8, Nov. 1970, p-34. See also 2068.

2076. Fabian, H., Report No. NASA-TT-F-14337; BMVG-FBWT-F1-8, July 1972, p-24.

2077. Fabian, H., Proc. of the 3rd Conf. on Accel. Targets Designed for the Prod. of Neutrons, Liege (Belgium), Sept. 18-19, 1967, pp-239-46.

2078. Fabian, H., Oesterr. Akad. Wiss., Math.-Naturwiss. Kl., Anz. 101, 1964, pp-248-76.

2079. Fagg, L. W., Jones, E. C., Jr., et al., Nucl. Instr. and Meth. 77, 1970, pp-136-40.

2080. Farnum, E. H., Fries, R. J., et al., J. Nucl. Mater. 85-86(A), 1979, pp-99-102.

2081. Farouk, M. A., Nassef, M. H., et al., Nucl. Instr. and Meth. 35, 1965, pp-210-12.

2082. Farrar, R. L., Jr., and Smith, D. F., Report K-L-3054, ORNL, 1972.

2083. Farrell, K., King, R. T., et al., Report ORNL-TM-4139, 1973, p-65.

2084. Fasoli, U., Galeazzi, G., et al., Lett. Nuovo Cimento 5 Ser. 2(3), 16 Sept. 1972, pp-209-12.

2085. Fasso, A., and Hoefert, M., Nucl. Instr. and Meth. 133(2), 1976, pp-213-18.

2086. Faubel, M., Dissertation, Inst. of Nuclear Physics, Mainz Univ., 1969.

2087. Faubel, M., Inst. of Nuclear Physics, Mainz, FRG

2088. Fagot, J., Lucas, R., et al., CEA-N-1791 (nd), 1 Oct. 1973-30 Sept. 1974, pp-211-212.

2089. Fagan, M. J., and Shite, G. G., INIS-mf-3205, 1976, p-60.

2090. Feher, I., and Biro, J., Fizikai Kutato Intezete, Budapest, 1970, p-12.

2091. Fehr, E. B., and Peschel, R. E., Nucl. Instr. and Meth. 112(3), 1973, pp-617-18.

2092. Fehr, E. B., INTDS Newsletter, July 1979.

2093. Fain, J., Gardes, J., et al., Nucl. Phys., A, 262(3), 24 May 1976, pp-413-432.

2094. Falk, W. R., Djaloeis, A., et al., Z. Phys., A, 273(3), July 1975, pp-265-267.

2095. Falk, W. R., Djaloeis, A., et al., Nucl. Phys., A, 252(2), 10 Nov. 1975, pp-452-476.

2096. Falomkin, I. V., Lyashenko, V. I., et al., AIP Conf. Proc. 33, 1976, pp-326-327.

2097. Falomkin, I. V., Nichitiu, F., et al., Nuovo Cim., A, 43(4), 21 Feb. 1978, pp-604-614.

2098. Fedorets, I. D., Popov, A. I., et al., Conf. on Nucl. Spectro. and Nucl. Structure, Baku, 3-6 Feb. 1976.

2099. Fedorets, I. D., Mishchenko, V. M., et al., Nauka. 1978, p-170.

2100. Fedotov, P. I., Tsentral'nyj Nauchno-Issledovatel'skij Inst. Informatsii i Tekhniko-Ehkonomicheskikh Issledovanij po Atomnoj Nauke i Tekhnike, Moscow (USSR), Neutron Phys., Part 6, 1976, pp-366-373.
2101a. Feenstra, S. J., Ockels, W. J., et al., Nederlandse Natuurkundige Vereniging, Utrecht, Sectie Kernfysica Autumn Meeting, 1977.
2101b. Fehl, D. L., and Chang, J., Tropical Meeting on Inertial Confinement Fusion, San Diego, CA (U. S. A.), 26-28 Feb. 1980.
2102. Feigel, A., Drenckhahn, W., et al., Verh. Dtsch. Phys. Ges. 4, 1978, p-835.
2103. Felber, F. S., and Bodner, S. E., Naval Rsch. Lab., Washington, D. C. (U. S. A.), NRL-MR-3574, Aug. 1977, p-16.
2104. Felder, R. D., Witten, T. R., et al., Nucl. Phys., A, 264 (3), 5 July 1976, pp-397-408.
2105. Feldl, E. J., Nucl. Instr. and Meth. 117(1), 1974, pp-5-7.
2106. Feldl, E. J., and Leachman, R. B., Nucl. Instr. and Meth. 101(3), 1972, pp-563-5.
2107. Feldl, E. J., and Umbarger, C. J., Nucl. Instr. and Meth. 103(2), 1972, pp-341-3.
2108. Feldl, E. J., and Eck, J. S., Nucl. Instr. and Meth. 95(2), 1971, pp-233-5.
2109. Feltman, A. V., and Cottingham, J. G., Rev. Sci. Instr. 35, July 1964, pp-814-15.
2110. Feng, D. H., Udagawa, T., et al., ANL/PHY-76-2(2), May 1976, p-557.
2111. Feng, D. H., Tamura, T., et al., Phys. Rev., C, 14(4), Oct. 1976, pp-1484-1487.
2112. Feng, D. H., Geller, F. N., et al., Phys. Rev., C, 18(1), July 1978, pp-33-41.
2113. Fenyes, T., Magy. Fiz. Foly. 18, 1970, pp-110-49.
2114. Ferdinande, H., Knuyt, G., et al., Nucl. Instr. and Meth. 91, 1971, pp-135-40.
2115. Ferdinande, H., Jacobs, R., et al., from Prof. Comm. on Nucl. Phys., Bad Neuenahr, Germany, 20 April 1968, p-19.
2116. Ferguson, J. M., Phys. Rev., C, 17(5), May 1978, pp-1888-90.
2117. Fernandez, F., and Nalda, J., AIP Conf. Proc. ISSN 0094-243X, 47, 1978, pp-558-559.
2118. Fernow, R. C., Nucl. Instr. and Meth. 148(2), 1978, pp-311-16.
2119. Ferreira, E. M., Rosa, L. P., et al., AIP Conf. Proc. 33, 1976, pp-452-453.
2120. Ferrero, A., Gadioli, E., et al., Z. Phys., A, 293(2), Nov. 1979, pp-123-134.
2121. Eschbach, H., TISRMNM; ORNL, Oct. 5-8, 1971.
2122. Eschbach, H., J. Phys. (Paris), Colloq. 5, 1976, pp-C5.177-C5.194.

2123. Festag, J. G., Max-Planck-Inst. fuer Kernphysik, Heidelberg, CONF-690540, pp-596-604.
2124. Fettweis, P., and Marmol, P. del., Z. Phys., A, 275(4), Dec. 1975, pp-359-367.
2125. Fewel, M. P., Baxter, A. M., et al., Aust. Nat. Univ., Canberra, Rsch. School of Physical Sci., ANU-P-660, Nov. 1976, p-15.
2126. Fewell, T. R., Nucl. Instr. and Meth. 61, April 15, 1968, pp-61-71.
2127. Fewell, T. R., Sandia Corp., Albuq., NM, Cont. AT(29-1)-789.
2128. Fick, D., Phys. Lett. 24B, Jan. 9, 1967, pp-13-14.
2129. Fick, D., and Hofmann, H. M., Phys. Lett. 20, March 1, 1966, pp-416-18.
2130. Fick, D., Huther, P., et al., Max-Planck-Inst. fuer Kernphysik, Heidelberg, 1975, pp-201-202.
2131. Fick, D., Amakawa, A., et al., Europ. Physical Soc., Geneva (Switz.), Europ. Conf. on Nucl. Phys. with Heavy Ions, 6-10 Sept. 1976.
2132. Fielding, H. W., Wyoming Univ., Laramie (U. S. A.), Thesis (Ph. D), 1974, p-244.
2133. Fielding, H. W., Anderson, R. E., et al., Colo. Univ., Boulder (U. S. A.), Dept. of Phys. and Astrophys., Nov. 1975, pp-76-81.
2134. Fifield, L. K., Zurmuhle, R. W., et al., Phys. Rev., C, 14(3), Sept. 1976, pp-1010-1015.
2135. Figureau, A., AIP Conf. Proc. 33, 1976, pp-616-619.
2136. Filevich, A., Rensfelt, K. G., et al., Nucl. Instr. and Meth. 98(3), 1972, pp-601-3.
2137. Filss, P., Guldbakke, S., et al., Nucl. Instr. and Meth. 91, 1971, pp-1-4.
2138. Fink, R. W., Hamburg Univ., I. Inst. für Experimentalphysik, EUR-1815.e(Rev.), pp-229-34.
2139. Findlay, D. J. S., and Owens, R. O., Phys. Rev. Lett. 37 (11), 13 Sept. 1976, pp-674-675.
2140. Findlay, D. J. S., and Owens, R. O., Nucl. Phys., A, 292 (2), 21-28 Nov. 1977, pp-53-60.
2141. Finlay, R. W., McKenna, C., et al., Nucl. Phys., A, 261(3), 3 May 1976, pp-413-426.
2142. Fintz, P., Guillaume, G., et al., CRN-PN-75-44, 1975, p-6.
2143. Firk, F. W. K., Bond, J. E., et al., Natl. Bur. of Stands. (U. S.), Spec. Publ. 425, Oct. 1975, pp-875-878.
2144. Fischbach, G., and Denschlag, H. O., Verh. Dtsch. Phys. Ges. 4, 1978, p-928.
2145. Fischer, H., Sizmann, R., et al., Z. Phys. 224, 1969, pp-135-43.
2146. Fischer, P., Harz, U., et al., Verh. Dtsch. Phys. Ges. 4, 1978, p-953.
2147. Fischer, V. K., Columbia Univ., NY, Contract AT-30-1-GEN-72, March 2, 1959, p-29.

2148. Fishbane, P. M., and Trefil, J. S., Phys. Rev. D, 9(1), 1974, pp-168-77.
2149. Fisher, T. R., Healey, D. C., et al., Rev. Sci. Instr. 41, May 1970, pp-684-7.
2150. Fitz, W., Kienle, F., et al., Phys. Rev., C, 14(2), Aug. 1976, pp-755-757.
2151. Fitzgerald, D. H., Lilley, J. S., et al., John H. Williams Lab. of Nucl. Phys., Univ. of Minn., Annual Report, 1975, pp-103-110.
2152. Fitzgerald, D. H., Lilley, J. S., et al., John H. Williams Lab. of Nucl. Phys., Univ. of Minn., Internal Report, 1974.
2153. Fitzgerald, D. H., Lilley, J. S., et al., John H. Williams Lab. of Nucl. Phys., Univ. of Minn., Annual Report, 1975, pp-111-114.
2154. Fitzgerald, D. H., Minn. Univ., Minneapolis (U. S. A.), Thesis (Ph. D), 1976, p-141.
2155. Fitzsimmons, W. A., Rice Univ., Houston, TX, Thesis, 1968, p-101.
2156. Fivel, H. J., Lang, G. P., et al., McDonnell Douglas Astronautics Co., St. Louis, MO (U. S. A.), Dec. 1976, p-104.
2157. Flach, S., Inst. fuer Radiochemie; Karlsruhe Univ. (TH), KFK-2279, March 1976, p-64.
2158. Flatau, C. R., BNL-6512, pp-188-91.
2159. Flatte, S. M., Heusch, C. A., et al., Nucl. Instr. and Meth. 119(2), 15 July 1974, pp-333-45.
2160. Flaum, C., Rochester Univ., NY (U. S. A.), Thesis (Ph. D), 1975, p-120.
2161. Fleissner, J. G., Rakel, D. A., et al., Phys. Rev., C, 17(3), March 1978, pp-1001-1007.
2162. Flerov, G. N., Oganessian, Yu. Ts., et al., Nucl. Phys., A, 267(2), 23 Aug. 1976, pp-359-364.
2163. Flerov, N. N., Lipatov, V. P., et al., IAE-614, 1964, p-24.
2164. Flocard, H., Koonin, S. E., et al., Phys. Rev., C, 17(5), May 1978, pp-1682-1699.
2165. Flocard, H., and Weiss, M. S., Phys. Rev., C, 18(1), July 1978, pp-573-575.
2166. Florent, R., from Int. Colloq. on Intense Magnetic Fields, Their Production and Their Applications, Grenoble, France, CEN, 1967, pp-247-253.
2167. Florent, R., Geles, C., et al., Nucl. Instr. and Meth. 56 (1), 1967, pp-160-4.
2168. Flowers, A. G., Shotter, A. C., et al., Phys. Rev. Lett. 40(11), 13 March 1978, pp-709-712.
2169. Flowers, A. G., Univ. of Edinburgh, U. K., Report 1977.
2170. Fluri, L. D., Freudenriech, K., et al., Helv. Phys. Acta 45(6), 31 Dec. 1972, p-950.
2171. Flynn, D. S., Duke Univ., Durham, NC (U. S. A.), Thesis (Ph. D), 1976, p-214.

2172. Flynn, D. S., Hershberger, R. L., et al., Phys. Rev., C, ISSN 0556-2813, 20(5), Nov. 1979, pp-1700-1705.
2173. Flynn, E. R., Hardekopf, R. A., et al., Phys. Rev. Lett. 36(2), 12 Jan. 1976, pp-79-81.
2174. Flynn, E. R., Sherman, J. D., et al., Phys. Rev., C, 13(2), Feb. 1976, pp-568-577.
2175. Flynn, E. R., Hardekopf, et al., Phys. Lett., B, 61(5), 26 April 1976, pp-433-436.
2176. Flynn, E. R., Anderson, R. E., et al., Phys. Rev., C, 16(1), July 1977, pp-139-141.
2177. Flynn, E. R., and Burke, D. G., Phys. Rev., C, 17(2), Feb. 1978, pp-501-507.
2178. Flynn, K. F., Gindler, J. E., et al., Phys. Rev., C, 12(5), Nov. 1975, pp-1478-1482.
2179. Fodor, G., and Cohen, B. L., Rev. Sci. Instr. 31, 1960, pp-73-4.
2180. Folda, D., Steuer, H., et al., Verh. Dtsch. Phys. Ges. 4, 1978, p-856.
2181. Foley, K. J., BNL-20415, Dec. 1975, pp-361-378.
2182. Folger, H., and Klemm, J., INTDS Proc., Berkeley, U. S. A., Oct. 19-20, 1977.
2183. Folger, H., INTDS Proc., Boston, U. S. A., Oct. 1-3, 1979.
2184. Folger, H., and Richter, U., Nucl. Instr. and Meth. 167(1), 1979, 85-9.
2185. Folger, H., and Richter, U., INTDS Proc., Garching, FRG, Sept. 11-14, 1978.
2186. Folger, H., Ges. Schwerionenforsch, Darmstadt, Report 1978.
2187. Fomushkin, Eh. F., Maslennikov, B. K., et al., Int. Atomic Energy Agency, Vienna (Austria), Int. Nucl. Data Comm., Nucl. Phys. Rsch. in the USSR.
2188. Fontell, A., and Arminen, E., Can. J. Phys. 47, 1969, p-2405.
2189. Fontes, P., Paris-11 Univ., 91-Orsay (France), Centre de Spectrometrie Nucleaire et de Spectrometrie de Masse, These (D. es S.).
2190. Fontes, P., Phys. Rev., C, 15(6), June 1977, pp-2159-2168.
2191. Ford, J. L. C., Jr., Gomez, del Campo, J., et al., ORNL-5025, May 1975, pp-82-85.
2192. Ford, G. P., and Norris, A. E., LA-6129, Oct. 1975, p-43.
2193. Formann, E., and Viehboeck, F. P., Nucl. Instr. and Meth. 42(2), 1966, pp-331-2.
2194. Formann, E., Hechtl, E., et al., Nucl. Instr. and Meth. 38, 1965, pp-144-7.
2195. Foroughi, F., Nussbaum, C., et al., Helv. Phys. Acta. 49(5), 29 Oct. 1976, p-788.
2196. Forsling, W., Svensk Kem. Tidskr. 73, 1961, pp-210-32.
2197. Fort, E., and Huet, J. L., Proc. of the 3rd Conf. on Accel. Targets Designed for the Prod. of Neutrons, Liege (Belgium), Sept. 18-19, 1967, pp-21-30.

2198. Forte, M., Jt. Nucl. Res. Cent., Euratom, Ispra, Italy, Euratom (Rep.) 74, (EUR 5054), p-17.
2199. Forte, M., Nuovo Cimento Soc. Ital. Fis. A, 18(4), 1973, pp-726-36.
2200. Forterre, M., Gerber, J., et al., Phys. Lett. 55B, 1975, p-56.
2201. Fortier, S., Paris-11 Univ., 91-Orsay (France), Inst. de Physique Nucleaire, These (D. es S.), 1976.
2202. Fortune, H. T., Medsker, L. R., et al., Phys. Rev., C, 12 (6), Dec. 1975, pp-1723-1725.
2203. Fortune, H. T., Betts, R. R., et al., Phys. Lett., B, 62(3), 7 June 1976, pp-287-288.
2204. Fortune, H. T., Braid, T. H., et al., Phys. Lett., B, 63(4), 16 Aug. 1976, pp-403-405.
2205. Fortune, H. T., Headley, S. C., et al., Phys. Rev., C, 14 (3), Sept. 1976, pp-1271-1272.
2206. Fortune, H. T., and Garrett, J. D., Phys. Rev., C, 14(5), Nov. 1976, pp-1695-1701.
2207. Fortune, H. T., Greenwood, L. R., et al., Phys. Rev., C, 15(1), Jan. 1977, pp-439-443.
2208. Fortune, H. T., Headley, S. C., et al., Phys. Lett., B, 72(2), 19 Dec. 1977, pp-173-175.
2209. Fortune, H. T., and Bishop, J. N., Nucl. Phys., A, 293(1-2), 12-19 Dec. 1977, pp-221-229.
2210. Fortune, H. T., and Bingham, H. G., Nucl. Phys., A, 293 (1-2), 12-19 Dec. 1977, pp-197-206.
2211. Fortune, H. T., Betts, R. R., et al., Phys. Rev., C, 17(1), Jan. 1978, pp-401-402.
2212. Fortune, H. T., Cobern, M. E., et al., Phys. Rev., C, 17(3), March 1978, pp-888-891.
2213. Fortune, H. T., Phys. Rev., C, 17(3), March 1978, pp-861-64.
2214. Fortune, H. T., Cobern, M. E., et al., Phys. Rev. Lett. 40(19), 8 May 1978, pp-1236-1239.
2215. Fortune, H. T., Courtney, W. J., et al., Phys. Rev., C, 17(6), June 1978, pp-1955-1960.
2216. Fortune, H. T., and Kurath, D., Phys. Rev., C, 18(1), July 1978, pp-236-238.
2217. Foster, C. C., Greenebaum, B., et al., Trans. Nucl. Sci. NS-16, June 1969, pp-620-1.
2218. Foster, J. L., Jr., Krmpotic, F., et al., Phys. Rev., C, 17(5), May 1978, pp-1602-1606.
2219. Foti, A., Pappalardo, G., et al., Goldberg, D. A., et al., Maryland Univ., College Park (U. S. A.), Dept. of Phys. and Astronomy, 1975, pp-435-436.
2220. Fou, C., Balamuth, D. P., et al., Phys. Rev., C, ISSN 0556-2813, 20(5), Nov. 1979, pp-1754-1758.
2221. Fourmond, M., Paris-11 Univ., 91-Orsay (France), These (3e Cycle), 1978.

2222. Fox, G., ANL/HEP-7208 (Vol. 3), pp-1072-83.
2223. Fox, J. D., Courtney, W. J., et al., Phys. Rev. Lett. 37 (10), 6 Sept. 1976, pp-629-631.
2224. Fox, M., Morrison, I., et al., Phys. Lett., B, ISSN 0370-2693, 86(2), 24 Sept. 1979, pp-121-124.
2225. Fraenkel, Z., Mayk, I., et al., Phys. Rev., C, 12(6), Dec. 1975, pp-1809-1825.
2226. Frahn, W. E., and Rehn, K. E., Lett. Nuovo Cim. 17(10), 6 Nov. 1976, pp-339-346.
2227. Frair, J. L., Heisenberg, J., et al., Proc. of the June Workshop in Intermediate Energy Electromagnetic Interactions with Nuclei, held at MIT, June 13-24, 1977.
2228. LaFrance, S., Fortune, H. T., et al., Phys. Rev., C, ISSN 0556-2813, 20(5), Nov. 1979, pp-1673-1679.
2229. Franco, V., and Nutt, W. T., Nucl. Phys., A, 292(3), 5 Dec. 1977, pp-506-522.
2230. Francombe, M. H., Basic Problems in Thin Film Physics (Göttingen: Vandenhoeck and Ruprecht), 1966, p-52.
2231. Franey, M. A., and Lilley, J. S., John H. Williams Lab. of Nucl. Phys., Univ. of Minn., Annual Report, 1975.
2232. Franey, M. A., Lilley, J. S., et al., Minn. Univ., Minneapolis (U. S. A.), John H. Williams Lab. of Nucl. Phys., Annual Report, 1977.
2233. Frank, G. G., Robertson, B. C., et al., Nucl. Phys., A, 255(2), 22 Dec. 1975, pp-351-359.
2234. Franke, P. R., Jr., Trans. Am. Nucl. Soc. 22, Nov. 1975, pp-732-733.
2235. Frankel, J. P., and Old, C. C., Calif. Rsch. and Dev. Co., Livermore, CA, CRD-T2B-45, Aug. 3, 1951, Decl. Feb. 14, 1957, p-9.
2236. Frankel, S., Frati, W., et al., Phys. Rev., C, 13(2), Feb. 1976, pp-737-741.
2237. Frankel, K., and Stevenson, J., Phys. Rev., C, 14(4), Oct. 1976, pp-1455-1457.
2238. Frankel, S., Phys. Rev. Lett. 38(23), 6 June 1977, pp-1338-1341.
2239. Frankl, D. R., and Venables, J. A., Adv. Phys. 19, 1970, p-409.
2240. Franz, J., Hamann, N., et al., Verh. Dtsch. Phys. Ges. 4, 1978, p-834.
2241. Frascaria, N., Didelez, J. P., et al., Phys. Rev., C, 16(2), Aug. 1977, pp-603-612.
2242. Frascaria, N., Stephan, C., et al., Phys. Rev. Lett. 39 (15), 10 Oct. 1977, pp-918-921.
2243. Fraser, J. S., AECL-2177, Paper 3, p-8.
2244. Fraser, J. S., Proc. Inf. Meet. Accel.-Breed, AECL, 162-76.
2245. Frawley, A. D., Bray, K. H., et al., Aust. Nat. Univ., Canberra, Rsch. School of Physical Sciences, ANU-P-669, June 1977, p-28.

2246. Frawley, A. D., Crawley, G. M., et al., Z. Phys., A, 286(3), May 1978, pp-307-311.
2247. Frawley, A. D., Bray, K. H., Nucl. Phys., A, 294(1-2), 2-9 Jan. 1978, pp-161-176.
2248. Freeman, J. R., Baker, L., et al., Sandia Lab., Albuquerque, NM (U. S. A.), Report 1979.
2249. Freeman, J. H., Nucl. Instr. and Meth. 38, 1965, pp-97-102.
2250. Freeman, J. H., and Gard, G. A., AERE-R-6330, March 1970, p-15.
2251. Freeman, J. R., Goldstein, S. A., et al., Plasma Phys. Controlled Nucl. Fusion Res. 1, 1977, pp-167-175.
2252. Freeman, R. M., Haas, F., et al., Centre de Recherches Nucleaires, CRN-PN-75-45, 1975, p-12.
2253. Freeman, R. M., Haas, F., et al., Centre de Recherches Nucleaires, CRN-PN-76-35, 1976, p-6.
2254. Freeman, R. M., Gallmann, A., et al., European Physical Soc., Geneva (Switz.), Europ. Conf. on Nucl. Phys. with Heavy Ions, 6-10 Sept. 1976, p-119.
2255. Freeman, R. M., and Haas, F., Phys. Rev. Lett. 40(14), 3 April 1978, pp-927-930.
2256. Freer, C. M., Nucl. Instr. and Meth. 86, 1970, p-311.
2257. Frehaut, J., and Mosinski, G., Acta Phys. Slovaca 25(2-3), 1975, pp-195-198.
2258. Frehaut, J., and Mosinski, G., Natl. Bur. of Stands. (U. S.), Spec. Publ. 425, Oct. 1975, pp-855-858.
2259. Frehaut, J., and Mosinski, G., CEA-CONF-3311, 1975, p-16.
2260. Freiesleben, H., Hildenbrand, K. D., et al., European Physical Soc., Geneva (Switz.), Europ. Conf. on Nucl. Phys. with Heavy Ions, 6-10 Sept. 1976.
2261. Freiesleben, H., Hildenbrand, K. D., et al., Z. Phys. A, 292(2), Sept. 1979, pp-171-189.
2262. Freitag, K., Heising, C., et al., Nucl. Instr. and Meth. 139(1), 1976, pp-83-5.
2263. Frekers, D., and Langanke, K., Fiz. (Zagreb), Suppl. 9(2), 1977, pp-17-21.
2264. DeFrenne, D., Thierens, H., et al., Phys. Rev., C, 18(1), July 1978, pp-486-492.
2265. Frick, G., Chaki, V., et al., Centre de Recherches Nucleaires, Strasbourg, France, CRN/PN 77-9, 1977.
2266. Friebel, A., Manakos, P., et al., Nucl. Phys., A, 294(1-2), 2-9 Jan. 1978, pp-129-140.
2267. Friebel, H. U., Frischke, D., et al., INTDS Proc., Garching, FRG, Sept. 11-14, 1978.
2268. Friebel, H. U., Frischke, D., et al., INTDS Proc., Berkeley, U. S. A., Oct. 19-20, 1977.
2269. Friebel, H. U., Frischke, D., et al., Nucl. Instr. and Meth. 167(1), 1979, pp-9-11.
2270. Friebel, H. U., Frischke, D., et al., INTDS Proc., Berkeley, U. S. A., Oct. 19-20, 1977.

2271. Friebel, U. H., Frischke, D., et al., INTDS Proc., Los Alamos, Oct. 19-21, 1976, U. S. A.
2272. Friebel, U. H., Frischke, D., et al., INTDS Proc., Garching, FRG, Sept. 11-14, 1978.
2273. Friedenberg, R. A., and Weiss, D. L., Phys. Rev., C, 14(1), July 1976, pp-204-210.
2274. Friederichs, H., Gelberg, A., et al., Phys. Rev., C, 13(6), June 1976, pp-2247-2256.
2275. Friedland, E., Goldschmidt, M., et al., Nucl. Phys., A, 256(1), 5 Jan. 1976, pp-93-105.
2276. Freedman, S. J., Nero, A. V., et al., Phys. Rev. Lett. 36 (5), 2 Feb. 1976, pp-279-281.
2277. Freedman, S. J., Gagliardi, C. A., et al., Phys. Rev. Lett. 37(24), 13 Dec. 1976, pp-1606-1609.
2278. Freedman, S. J., Gagliardi, C. A., et al., Phys. Rev., C, 17(6), June 1978, pp-2071-2075.
2279. Friedman, A. M., and Mohr, W. C., Nucl. Instr. and Meth. 17, Sept. 1962, pp-78-80.
2280. Friedman, E., Moalem, A., et al., Phys. Rev., C, 15(1), Jan. 1977, pp-456-458.
2281. Friedman, E., Moalem, A., et al., Phys. Rev., C, 14(6), Dec. 1976, pp-2082-2088.
2282. Friedman, E., and Batty, C. J., Phys. Rev., C, 16(4), Oct. 1977, pp-1425-1430.
2283. Friedman, E., Gils, H. J., et al., Inst. fuer Angewandte Kernphysik, Annual Report, KFK-2868, Oct. 1979, Phys. Rev. Lett. 41, 1978, p-1220.
2284. Friedman, W. A., McVoy, K. W., et al., AIP Conf. Proc. 33, 1976, pp-278-279.
2285. Friedrich, J., Voegler, N., et al., Phys. Lett., B, 64(3), 27 Sept. 1976, pp-269-272.
2286. Fries, J. R., INTDS Proc., Los Alamos, Oct. 19-21, 1976, U. S. A.
2287. Friskney, C. A., and Simpson, K. A., J. Nucl. Mater. 57(3), Sept. 1975, pp-341-347.
2288. Froelich, H., Dueck, P., et al., Phys. Lett., B, 64(4), 11 Oct. 1976, pp-408-410.
2289. Frois, B., and Heisenberg, J., Proc. of the June workshop in Intermediate Energy Electromagnetic Interactions with Nuclei, held at MIT, June 13-24, 1977.
2290. Frois, B., Bellicard, J. B., et al., Phys. Rev. Lett. 38(4), 24 Jan. 1977, pp-152-155.
2291. Fry, E. M., Jr., Lasalle, R. A., et al., Nucl. Instr. and Meth. 166(2), 1979, pp-191-6.
2292. Fu, C. Y., and Perey, F. G., At. Data Nucl. Data Tables 16 (5), Nov. 1975, pp-409-450.
2293. Fu, C. Y., At. Data Nucl. Data Tables 17(2), Feb. 1976, pp-127-156.
2294. Fu, C. Y., ORNL, 1975 Report.

2295. Fujiwara, N., Hourany, E., et al., Phys. Rev., C, 15(1), Jan. 1977, pp-4-9.
2296. Fukushima, M., Horikawa, N., et al., Nucl. Instr. and Meth. 140(2), 1977, pp-275-8.
2297. Fulbright, H. W., and Freiesleben, H., Nucl. Instr. and Meth. 115, 1974, pp-83-4.
2298. Fulbright, H. W., Nucl. Instr. and Meth. 72, 1969, pp-229-30.
2299. Fulbright, H. W., Robbins, J. A., et al., Phys. Rev. 184, 1969, p-1068.
2300. Fuller, R. C., and Moffa, P. J., Phys. Rev., C, 14(5), Nov. 1976, pp-1721-1726.
2301. Fuller, R. C., and Moffa, P. J., Phys. Rev., C, 15(1), Jan. 1977, pp-266-280.
2302. Fullwood, R. R., Steadman, B. L., et al., Contract AT(30-3)-328, Nucl. Instr. and Meth. 46, 1967, pp-194-6.
2303. Fullwood, R. R., Gaerttner, E. R., et al., Trans. Nucl. Sci., NS-12(3), June 1965, pp-705-7.
2304. Fullwood, R. R., Gaerttner, E. R., et al., Contract AT (30-3)-328, from Particle Accelerator Conf., Washington, DC, p-5, 1964.
2305. Fulmer, C. B., and Kindred, G., ORNL-TM-2834, Contract W-7405-eng-26, Aug. 1970, p-28.
2306. Fulmer, C. B., Hensley, D. C., et al., Phys. Rev., C, 13(3), March 1976, pp-937-943.
2307. Fulmer, C. B., Butler, H. M., et al., from Conf. on Cyc. and Their Applications; Bloomington, IN, U. S. A., 18 Sept. 1978.
2308. Funke, L., Doering, J., et al., Zentralinstitut fuer Kernforschung, Rossendorf bei Dresden (German Democratic Republic), Annual Report 1976.
2309. Funke, L., Dubbers, F., et al., Nauka, 1978, p-222.
2310. Funsten, H. O., Kossler, W. J., et al., Phys. Rev., C, 16(4), Oct. 1977, pp-1521-1525.
2311. Furber, R. D., Minn. Univ., Minneapolis (U. S. A.), Thesis (Ph. D), 1976.
2312. Furyi, S., AIP Conf. Proc. 33, 1976, pp-320-321.
2313. Furui, S., AIP Conf. Proc. 33, 1976, pp-620-621.
2314. Furui, S., Prog. Theor. Phys. (Kyoto), 58(3), Sept. 1977, pp-864-878.
2315. Furutani, H., Horiuchi, H., et al., AIP Conf. Proc. ISSN 0094-243X, 47, 1978, pp-556-557.
2316. Fusco, A. M., Peek, F. N., et al., J. Nucl. Med. 13, 1972, p-729.
2317. Futami, Y., Prog. Theor. Phys. (Kyoto), 53(3), March 1975, pp-725-731.
2318. Gabric, A., and Amos, K., Aust. Inst. of Nucl. Sci. and Eng., Lucas Heights, 6th AINSE Nucl. Phys. Conf. 9-11 Feb. 1976.

2319. Gabriel, T. A., Alsmiller, R. G., Jr., et al., ORNL-4599, 1970, p-27.
2320. Gabriel, T. A., and Santoro, R. T., Nucl. Instr. and Meth. 95(2), 1971, pp-275-83.
2321. Gabriel, T. A., Nucl. Instr. and Meth. 91(1), 1 Jan. 1971, pp-67-72.
2322. Gabriel, T. A., Phys. Rev., C, 13(1), Jan. 1976, pp-240-44.
2323. Gabrysh, F. A., Leyring, H., et al., J. Appl. Phys. 13(10), Oct. 1960, pp-1785-91.
2324. Gadioli, E., Gadioli, E. E., et al., Phys. Rev., C, 16(4), Oct. 1977, pp-1404-1424.
2325. Gaeggeler, H., Von Gunten, H. R., et al., J. Inorg. and Nucl. Chem. 38(2), 1976, pp-205-210.
2326. Gaerttner, E. R., Rensselaer Polytechnic Inst., Troy, NY, Contract AT(11-1)-3058, 1973, p-52.
2327. Gaerttner, E. R., RPI, Troy, NY, Contract AT(30-1)-328, Oct. 1 1970-Sept. 30, 1971, p-159.
2328. Gaerttner, E. R., Greenspan, E., et al., Trans. Am. Nucl. Soc. 11, June 1968, p-211.
2329. Gäggeler, H., Brüchle, W., et al., Z. Physik A 286, 1978, p-419.
2330. Gaillard, Y. R., Grenoble-1 Univ., 38 (France), Inst. des Sciences Nucleaires, These (3e Cycle), 1975.
2331. Gaillard, Y. R., Martin, P., et al., Nucl. Phys., A, ISSN 0375-9474, 327(2), 24 Sept. 1979, pp-349-372.
2332. Gaille, F., Mureramanzi, S., et al., Nuovo Cim., A, 40(1), 1 July 1977, pp-31-40.
2333. Gakh, G. I., and Rekalo, A. P., Fizika Vysokikh Ehnergij i Atomnogo Yadra, KFTI-77-9, Voprosy Atomnoj Nauki i Tekhniki No. 2(19), 1977.
2334. Gakh, G. I., AN Ukrainskoj SSR, Kharkov, Fiziko-Tekhnicheskij Inst. KFTI-79-19, 1979, p-23.
2335. Gal, A., AIP Conf. Proc. 33, 1976, pp-694-704.
2336. Gal, A., and Eisenberg, J. M., Phys. Rev., C, 14(3), Sept. 1976, pp-1273-1276.
2337. Gales, S., Bartol, F., et al., Rev. Phys. Appl. 5, Dec. 1970, pp-835-9.
2338. Gales, S., Fortier, S., et al., Nucl. Phys., A, 259(2), 15 March 1976, pp-189-212.
2339. Gales, S., Fortier, S., et al., Nucl. Phys., A, 265(2), 19 July 1976, pp-213-219.
2340. Gales, S., Fortier, S., et al., Nucl. Phys., A, 268(2), 14 Sept. 1976, pp-257-292.
2341. Gales, S., ElHage, Y., et al., et al., Paris-11 Univ., 91-Orsay (France), Inst. de Physique Nucleaire, IPNO-PhN-79-23, 1979, p-49.
2342. Galin, J., Moretto, L. G., et al., Nucl. Phys., A, 255(2), 22 Dec. 1975, pp-472-490.

2343. Galinier, J. L., and Yaffe, L., J. Inorg. and Nucl. Chem. ISSN 0022-1902, 39(9), 1977, pp-1497-1508.
2344. Galinier, J. L., Diksic, M., et al., Can. J. Chem. 55(20), 15 Oct. 1977, pp-3609-3615.
2345. Gallant, J. L., ANL/PHY/MSD-76-1, 1975, pp-52-59.
2346. Gallant, J. L., AECL-5503, April 1976, pp-169-178.
2347. Gallant, J. L., Yaraskavitch, D. J., et al., Nucl. Instr. and Meth. 167(1), 1979, pp-55-9.
2348. Gallant, J. L., Chalk River Nuclear Labs., Canada
2349. Gallant, J. L., Yaraskavitch, D. J., et al., INTDS Proc., Boston, U. S. A., Oct. 1-3, 1979.
2350. Gallant, J. L., INTDS Newsletter, July 1979.
2351. Gallant, J. L., INTDS Proc., Boston, U. S. A., Oct. 1-3, 1979.
2352. Gallant, J. L., Yaraskavitch, D. J., et al., INTDS Proc., Garching, FRG, Sept. 11-14, 1978.
2353. Gallant, J. L., INTDS Newsletter, Feb. 1980.
2354. Gallant, J. L., and Yaraskavitch, D. J., INTDS Proc., Berkeley, U. S. A., Oct. 19-20, 1977.
2355. Gallant, J. L., INTDS Proc., Los Alamos, U. S. A., Oct. 19-21, 1976.
2356. Gallant, J. L., INTDS Proc., ANL, U. S. A., Sept. 30-Oct. 2, 1975.
2357. Gallant, J. L., INTDS Proc., Chalk River, Canada, Oct. 1-3, 1974.
2358. Gallant, J. L., Nucl. Instr. and Meth. 102(3), 1972, pp-477-83.
2359. Gallant, J. L., Proc. 3rd Int. Symp. on Rsch. Mat. for Nucl. Measurements, Gatlinburg, TN, 1971, p-138.
2360. Gallant, J. L., Chalk River Nuclear Labs., AECL, Oct. 1971.
2361. Gallant, J. L., Nucl. Instr. and Meth. 81, 1970, pp-27-8.
2362. Galonsky, A., Branson, J. G., et al., Phys. Rev. Lett. 35 (18), 3 Nov. 1975, pp-1208-1211.
2363. Galonsky, A., Doering, R. R., et al., Phys. Rev., C, 14(2), Aug. 1976, pp-748-752.
2364. Galonsky, A., Didelez, J. P., et al., Phys. Lett., B, 74(3), 10 April 1978, pp-176-178.
2365. Galster, W., Treu, W., et al., Phys. Lett., B, 67(3), 11 April 1977, pp-262-264.
2366. Gamalij, E. G., Isakov, A. I., et al., Chast' 6, Materialy 3, Vsesoyuznoj Konferetsii po nejtronnoj fizike, INIS-mf-4270, Moscow, 1976, pp-166-169.
2367. Gamp, A., Bohlen, H. G., et al., Verh. Dtsch. Phys. Ges. 10(7), 1975, pp-764-765.
2368. Gamp, A., Braun-Munzinger, P., et al., Nucl. Phys., A, 250(2), 6 Oct. 1975, pp-341-350.
2369. Gamp, A., Jacmart, J. C., et al., Phys. Lett., B, 74(3), 10 April 1978, pp-215-218.
2370. Gamp, A., Fuchs, H., et al., Inst. de Physique Nucleaire, 1979, p-18.

2371. Gangulee, A., and Taylor, R. C., J. Appl. Phys. 49(3), March 1978, pp-1762-1764.
2372. Garcilazo, H., AIP Conf. Proc. 33, 1976, pp-454-455.
2373. Gardner, D. G., and Dietrich, F. S., Calif. Univ., Livermore, U. S. A., Lawrence Livermore Lab., UCRL-82998.
2374. Gardner, M. A., and Gardner, D. G., Nucl. Phys., A, 265(1), 12 July 1976, pp-77-92.
2375. Garg, J. B., State Univ. of New York, Albany, U. S. A., Dept. of Physics, 1 April 1975-31 March 1976.
2376. Garg, U., Sjoreen, T. P., et al., Phys. Rev. Lett. 40(13), 27 March 1978, pp-831-834.
2377. Gari, M., Hyuga, H., et al., Phys. Rev., C, 14(6), Dec. 1976, pp-2196-2210.
2378. Garlea, I., Miron, C., et al., Rev. Roum. Phys. ISSN 0035-4090, 23(4), 1978, pp-409-417.
2379. Garrett, C., Leigh, J. R., et al., Aust. Nat. Univ., Canberra, Rsch. School of Physical Sciences, March 1976, p-40.
2380. Garreta, D., from Conf. on Polarized Targets and Sources, Saclay, France, Dec. 5-9, 1966.
2381. Garrett, J. D., Wegner, H. E., et al., Phys. Rev., C, 12 (2), Aug. 1975, pp-481-488.
2382. Garrett, J. D., Wegner, H. E., et al., Phys. Rev., C, 12 (2), Aug. 1975, pp-489-498.
2383. Garrett, J. D., Hansen, O., Nucl. Phys., A, 276(1), 17 Jan. 1977, pp-93-100.
2384. Garvey, P. M., Fraser, J. S., et al., Trans. Am. Nucl. Soc. 28, 1978, pp-754-6.
2385. Gary, A. S., Paris-11 Univ., 91-Orsay (France), These (3e Cycle).
2386. Gasior, M., Lizurej, H. I., et al., Nukleonika 13, 1968, pp-635-43.
2387. Gass, J., and Mueller, H. H., Nucl. Instr. and Meth. 136 (3), 1 Aug. 1976, pp-559-561.
2388. Gassen, H. J., and Reichelt, T., Vak. Tek. 28(7), 1979, pp-204-8.
2389. Gastebois, J., CEA-CONF-3990, 1977, p-17.
2390. Gatty, B., Guerreau, D., et al., Nucl. Phys., A, 253(2), 24 Nov. 1975, pp-511-532.
2391. Gavrilov, Yu. K., Kim Zi Khvan, et al., Yad. Fiz. 24(2), 1976, pp-241-245.
2392. Gavron, A., Phys. Rev., C, 13(1), Jan. 1976, pp-98-104.
2393. Gavron, A., Britt, H. C., et al., Phys. Rev. Lett. 38(25), 20 June 1977, pp-1457-1460.
2394. Gebauer, B. E., Heidelberg Univ. (F. R. Germany) Naturwissenschaftliche Gesamtfakultaet. Diss. (D. Sc.), July 1974, p-98.
2395. Geesaman, D. F., State Univ. of New York, Stony Brook, U. S. A., Thesis (Ph. D), 1976, p-161.

2396. Geesaman, D. F., Henning, W., et al., ANL-78-66, April 1, 1, 1977-March 31, 1978, pp-38-39.
2397. Geesaman, D. F., McGrath, R. L., et al., Phys. Rev., C, 15(5), May 1977, pp-1835-1838.
2398. Gehrke, R. J., Int. J. Appl. Radiat. Isot. ISSN 0020-708X, 31(1), Jan. 1980, pp-37-40.
2399. Geibel, J. A., Ranft, J., et al., Tr. Mezhdunar. Konf. po Uskoritelyam, Dubna, 1963-1964, pp-759-66.
2400. Geist, J. J., Société Grenobloise d'Etude et d'Applications Hydrauliques, Grenoble, France, 1964, p-81.
2401. Gelbke, C. K., Nucl. Instr. and Meth. 128(1), 15 Sept. 1975, pp-175-177.
2402. Gelbke, C. K., and Cramer, J. G., Phys. Rev., C, 14(3), Sept. 1976, pp-1048-1057.
2403. Gelbke, C. K., Buenerd, M., et al., Phys. Lett., B, 65(3), 22 Nov. 1976, pp-227-230.
2404. Gelbke, C. K., Scott, D. K., et al., Phys. Lett., B, 70(4), 24 Oct. 1977, pp-415-417.
2405. Gelbke, C. K., Bini, M., et al., Phys. Lett., B, 71(1), 7 Nov. 1977, pp-83-86.
2406. Geles, C., Harigel, G., et al., Nucl. Instr. and Meth. 56, Nov. 1967, pp-160-4.
2407. Gel'fand, E. K., Man'ko, B. V., et al., INIS-mf-5287, 1977, pp-187-195.
2408. Geller, K. N., and Kollarits, R. V., ORO-4856-26, 1975, pp-538-539.
2409. Geller, K. N., and Kollarits, R. V., Phys. Rev. Lett. 37(5), 2 Aug. 1976, pp-279-282.
2410. Geller, R., and Yerouchalmi, F., CEA-R-3139, Jan. 1967, p-16.
2411. Gelletly, W., Chapman, R., et al., J. Phys., G (London), 2(1), Jan. 1976, pp-L1-L9.
2412. Gelletly, W., Kane, W. R., et al., Proc. of the 2nd Int. Symp. on Neutron Capture Gamma Ray Spectroscopy and Related Topics held at Petten, the Netherlands, 2-6 Sept. 1975, pp-526-531.
2413. Gemmell, D. S., and Worthington, J. N., Nucl. Instr. and Meth. 91, 1971, pp-15-28.
2414. Genin, J. P., Julien, J., et al., CEN, Saclay, France, ORO-4856-26, 1975, pp-303-304.
2415. Genin, J. P., Julien, J., et al., CEA-N-1861 (nd), 1 Oct. 1974-1 Sept. 1975, pp-152-153.
2416. Genty, C., INTDS Newsletter, Jan. 1979.
2417. Geoffroy, K. A., and Natowitz, J. B., Phys. Rev. Lett. 37(18), 1 Nov. 1976, pp-1198-1201.
2418. Geramb, H. V., Amos, K., et al., Phys. Rev., C, 12(6), Dec. 1975, pp-1697-1710.
2419. Gerasimov, V. F., and Knizhnikov, Yu. N., Akademiya Nauk SSSR, Moscow, Institut Atomnoi Energii, IAE-1737, 1968, p-27.

2420. Gerber, J., Goldberg, M. B., et al., Phys. Lett. 60B, 1976, p-338.
2421. Gerbino, J. J. G., Cohen, I. M., et al., J. Inorg. and Nucl. Chem. 38(7), 1976, pp-1386-1387.
2422. Gerlic, F., Langevin-Joliot, H., et al., Phys. Rev., C, 12(6), Dec. 1975, pp-2106-2110.
2423. Gerlic, E., Berrier-Ronsin, G., et al., Inst. de Physique Nucleaire, IPNO-PhN-79-20, 1979, p-59.
2424. Germai, G., Guillaume, M., Int. J. Appl. Radiat. Isot. 21, Oct. 1970, pp-595-8.
2425. Germond, J. F., and Wilkin, C., Nucl. Phys., A, 249(3), 22 Sept. 1975, pp-457-465.
2426. Germond, J. F., and Wilkin, C., Phys. Lett., B, 68(3), 6 June 1977, pp-229-233.
2427. Germond, J. F., Wilkin, C., et al., ISBN 0 444 85256 5, Volume 2, Amsterdam, Netherlands, North-Holland, 1979.
2428. Gersch, H. U., Grambole, D., et al., Zentralinstitut fuer Kernforschung, Rossendorf bei Dresden, Annual Report, 1975.
2429. Gersch, H. U., Grambole, D., et al., Zentralinstitut fuer Kernforschung, Rossendorf bei Dresden, Annual Report, 1975.
2430. Gerstenkorn, S., and Stroke, H. H., J. Opt. Soc. Am. 66(2), Feb. 1976, p-155.
2431. Gervé, A., and Schatz, G., 7th Int. Conf. on Cyclotrons and Their Applications, Zurich, Switz., 19-22 Aug. 1975.
2432. Van Gestle, J., and Pauwels, J., INTDS Proc., Boston, U. S. A., Oct. 1-3, 1979.
2433. Getoff, N., and Bildstein, H., Nucl. Instr. and Meth. 36, 1965, pp-173-5.
2434. Getoff, N., Bildstein, H., et al., Nucl. Instr. and Meth. 46, 1967, pp-305-8.
2435. Ghiorso, A., et al., UCRL, private communication.
2436. Ghiorso, A., UCRL Report.
2437. Ghiorso, A., Nitschke, J. M., et al., AED-Conf.-75-416-042, 1975, p-1.
2438. Ghorai, S. K., Hudson, C. G., et al., Nucl. Phys., A, 266 (1), 2 Aug. 1976, pp-53-60.
2439. Giaettli, H., Abragam, A., et al., Phys. Rev. Lett. 40(12), 20 March 1978, pp-748-750.
2440. Gibbs, W. R., Gibson, B. F., et al., AIP Conf. Proc. 33, 1976, pp-464-465.
2441. Gibbs, W. R., Gibson, B. F., et al., AIP Conf. Proc. 33, 1976, pp-622-623.
2442. Gibbs, W. R., Hess, A. T., et al., AIP Conf. Proc. 33, 1976, pp-324-325.
2443. Gibbs, W. R., Gibson, B. F., et al., Phys. Rev. Lett. 39 (21), 21 Nov. 1977, pp-1316-1319.

2444. Gibbs, W. R., Baer, H., et al., LA-7892C, July 1979, pp-100-107.
2445. Gibson, B. F., AIP Conf. Proc. 33, 1976, pp-418-442.
2446. Gibson, B. F., and Hess, A. T., LA-6320-MS, April 1976, p-14.
2447. Gibson, B. F., AIP Conf. Proc. ISSN 0094-243X, 54(1), 15 July 1979, pp-691-702.
2448. Gibson, W. M., Rasmussen, J. B., et al., Can. J. Phys. 46, 1968, pp-551.
2449. Gibson, J. R., J. Vac. Sci. and Tech. 8(1), Jan.-Feb. 1971, p-332.
2450. Gibson, J. R., ORNL-TM-3209, Contract W-7405-eng-26, p-11.
2451. Giedd, G. R., and Perkins, M. H., Rev. Sci. Instr. 31, 1960, p-773.
2452. Giffon, M., Guichon, P., et al., LYCEN-7702(pt. 1), 1977, pp-S11.1-S11.6. See also 1338.
2453. Gilat, J., Fleury, A., et al., Phys. Rev., C, 16(2), Aug. 1977, pp-694-705.
2454. Gilbert, K. E., Mich. Univ., Ann Arbor, U. S. A., Thesis (Ph. D), 1976, p-254.
2455. Gilbreath, J. R., and Simpson, O. C., ANL-4613(Del.), Contract W-31-109-eng-38, Oct. 3, 1951, p-75.
2456. Gill, R. L., Iowa State Univ., Ames, IO, Diss. Abstr. Int. B, 38(6), 1977, p-2668.
2457. Gillespie, F. C., Vacuum, 1972, pp-255-6.
2458. Gillespie, W. A., Singhal, R. P., et al., J. Phys., G (London), 2(3), March 1976, pp-185-197.
2459. Gillette, J. H., ORNL-TM-3621, Contract W-7405-eng-26, Nov. 1971, p-17.
2460. Gillette, J. H., ORNL-TM-3529, Contract W-7405-eng-26, Aug. 1971, p-12.
2461. Gillette, J. H., ORNL-TM-3089, Contract W-7405-eng-26, Aug. 1970, p-10.
2462. Gillette, J. H., ORNL-TM-2889, Contract W-7405-eng-26, March 1970, p-16.
2463. Gillette, J. H., and Lane, I. B., ORNL-3802, Contract W-7405-eng-26, May 1965, p-71.
2464. Gillette, J. H., ORNL-TM-428, Contract W-7405-eng-26, Jan. 9, 1963, p-20.
2465. Gillette, J. H., ORNL-TM-198, Contract W-7405-eng-26, June 5, 1962, p-39.
2466. Gilliam, D. M., and Knoll, G. F., Natl. Bur. of Stands. (U. S.), Spec. Publ. 425, Oct. 1975, pp-635-636.
2467. Gilliam, D. M., and Grundl, J. A., BNL-NCS-22500, March 1977, pp-186-188.
2468. Gilliam, D. M., Natl. Bur. of Stands. (U. S.), Spec. Publ. 493, Oct. 1977, pp-299-303.
2469. Gilot, J. F., Bosman, M., et al., Nucl. Instr. and Meth. 171(3), 1980, pp-607-8.

2470. Gils, H. J., and Rebel, H., Phys. Rev., C, 13(6), June 1976, pp-2159-2165.
2471. Gils, H. J., Rebel, H., et al., Phys. Lett., B, 68(5), 4 July 1977, pp-427-428.
2472. Gils, H. J., and Rebel, H., Inst. fuer Angewandte Kernphysik, Annual Report, KFK-2868, Oct. 1979, pp-34-36.
2473. Gils, H. J., Inst. fuer Angewandte Kernphysik, 1979, pp-123-169
2474. Gindler, J. E., Oselka, M. C., et al., Int. J. Appl. Radiat. Isot. 27(5-6), May 1976, pp-330-332.
2475. Giorni, A., Glasser, F., et al., Inst. des Sciences Nucleaires, June 1977, p-19.
2476. Giorgi, A. L., and Matthias, B. T., Phys. Rev., B, 17(5), 1 March 1978, pp-2160-2162.
2477. Giorni, A., Glasser, F., et al., Nucl. Phys., A, 292(1-2), 21-28 Nov. 1977, pp-213-224.
2478. Giovagnoli, A., Valladon, M., et al., Anal. Chim. Acta. ISSN 0003-2670, 109(2), 1 Sept. 1979, pp-411-418.
2479. Girija, V., and Devanathan, V., Proc. of the Nucl. Phys. and Solid State Phys. Symp., held at Pune, Dec 26-30, 1977, Vol. 20B, pp-114-116.
2480. Girnius, R. J., and Anderson, L. W., Nucl. Instr. and Meth. 137(2), 1 Sept. 1976, 373-8.
2481. Gismatullin, Yu. R., and Lantsev, I. A., Nauka, 1978, p-265.
2482. Gismatullin, Yu. R., Kosmach, V. F., et al., Yad. Fiz. 27 (1), 1978, pp-37-41.
2483. Gizon, J., Gizon, A., et al., Phys. Rev., C, 17(2), Feb. 1978, pp-596-600.
2484. Gjovig, A. J., Little, J. D., et al., LA-UR-782721, Contract W-7405-eng-36, 1978, p-4.
2485. Gladyshev, V. A., Katsaurov, L. N., et al., Pribory i Tekh. Ekspt. 7, 1962, pp-20-2.
2486. Glaessel, P., Roesler, H., et al., Verh. Dtsch. Phys. Ges. 10(7), 1975, pp-835-836.
2487. Glaettli, H., LBL-500, from 2nd Int. Conf. on Polarized Targets, Berkeley, CA, 30 Aug. 1971, pp-281-7.
2488. Glassel, P., Cauvin, B., et al., LBL-5075, 1975, pp-140-41.
2489. Glasgow, McDaniel, et al., BNL-NCS-22500, March 1977, pp-323-326.
2490. Glavanakov, I. V., and Stibunov, V. N., Yad. Fiz. ISSN 0044-0027, 29(6), 1979, pp-1455-1461.
2491. Glavanakov, I. V., and Stibunov, V. N., Izv. Akad. Nauk SSSR, Ser. Fiz. ISSN 0367-6765, 43(1), Jan. 1979, pp-141-144.
2492. Glawe, U., Strohbusch, U., et al., DESY-79/46, July 1979, p-7.
2493. Glendenning, N. K., and Wolschin, G., LBL-5007, CONF-760424-9, March 1976, p-8.

2494. Glendenning, N. K., and Wolschin, G., Nucl. Phys., A, 269 (1), 28 Sept. 1976, pp-223-236.
2495. Glendenning, N. K., and Karant, Y., LBL-8151, 1978, pp-209-11.
2496. Glor, M., Naegele, H. P., et al., Nucl. Phys., A, 286(1), 8 Aug. 1977, pp-31-41.
2497. Glover, K. M., Rogers, F. J. G., et al., Nucl. Instr. and Meth. 102(3), 1972, pp-443-50.
2498. Glover, K. M., and Robinson, P. S., AERE-R-5097, Paper 11, p-8.
2499. Glover, K. M., INTDS Proc., Boston, U. S. A., Oct. 1-3, 1979.
2500. Glover, K. M., Rogers, F. J. G., et al., Nucl. Instr. and Meth. 102(3), 1972, pp-443-50.
2501. Glover, K. M., Rogers, F. J. G., et al., AERE, CONF-711002, Oct. 1971, pp-91-100.
2502. Glover, K. M., Rogers, F. J. G., et al., TISRMNM-ORNL, Oct. 5-8, 1971
2503. Glover, K. M., and Robinson, P. S., PSPSITF, AERE, Oct. 20 & 21, 1965.
2504. Glover, K. M., and Borrell, P., J. Nucl. Eng. 1, 1955, p-214.
2505. Glowacka, L., Jaskola, M., et al., Nucl. Phys., A, 262(2), 17 May 1976, pp-205-213.
2506. Glowacka, L., Jaskola, M., et al., Nucl. Phys., A, 284(2), 4 July 1977, pp-257-263.
2507. Glukhov, Yu. A, Dem'yanova, A. S., et al., Leningrad Nauka, 1978, p-208.
2508. Glukhov, Yu. A., Dem'yanova, A. S., et al., Leningrad Nauka, 1978, p-209.
2509. Gmitro, M., Tosunyan, L. A., et al., JINR-R-2-11549, 1978, p-9.
2510. Gmitro, M., Korenman, G. Ya., et al., JINR-R-2-11991, 1978, p-14.
2511. Gnidak, N. L., Kirilyuk, A. L., et al., Inst. Informatsii i Tekhniko-Ehkonomicheskikh Issledovanij po Atomnoj Nauke i Tekhnike, Moscow (USSR), Neutron Physics 2 Pt. Proceeding of the 4, 22 April, 1977, pp-223-226.
2512. Gobbi, B., Rosen, J. L., et al., Phys. Lett. B, 58B(2), 1975, pp-219-22.
2513. Gobert, G., Mani, G. S., et al., Nucl. Instr. and Meth. 42, July 1966, pp-250-7.
2514. Godin, A., and Montenon, M., Nucl. Instr. and Meth. 79, 1970, pp-349-52.
2515. Goebel, J., McCaslin, J. B., et al., UCRL-20664, Contract W-7405-eng-48, April 1971, p-11.
2516. Goel, B., Kuesters, H., et al., Natl. Bur. of Stands. (U. S.), Spec. Publ. 425(1), Oct. 1975, pp-313-316.
2517. Goel, B., and Weller, F., Kernforschungszentrum Karlsruhe (Germany, F. R.), KFK-2386/III, March 1977.

2518. Goerlach, U., Habs, D., et al., Max-Planck-Institut fuer Kernphysik, Heidelberg (Germany, F. R.), Annual Report, 1976, pp-56-59.

2519. Goff, R. F., and Hendee, C. F., 27th Annual Conf. on Physical Electronics, Cambridge, MA, March 20-22, 1967, pp-231-8.

2520. Goggi, G., Conta, C., et al., Lett. Nuovo Cim. ISSN 0024-1318, 24(11), 17 March 1979, pp-381-386.

2521. Gogny, D., and Maire, M., Int. Centre for Theoretical Physics, Trieste (Italy), Proc. of the Int. Conf. held at Trieste, Italy, 24-28 Feb. 1975, pp-149-150.

2522. Goin, G., TEICOCTA, IU, U. S. A., Sept. 18-21, 1978.

2523. Gokhberg, B. M., Dubrovina, S. M., et al., Int. Atomic Energy Agency, Vienna (Austria), Int. Nucl. Data Comm., Nucl. Phys. Rsch. in the USSR, Feb. 1977, p-8.

2524. Golam Mostafa, A. B. M., and Fremlin, J. H., Nucl. Instr. and Meth. 82, 1970, pp-313-15.

2525. Goland, N. A., Snead, L. C., Jr., IEEE Trans. on Nucl. Sci., NS-22(3), June 1975.

2526. Gold, R., Westinghouse Hanford Co., Richland, WA (U. S. A.), HEDL-SA-901, Aug. 1975, p-33.

2527. Goldberg, D. A., Roos, P. G., et al., ORO-4856-26, 1975, pp-311-312.

2528. Goldring, B. G., and Scharenberg, R. P., Rev. Sci. Instr. 29, June 1958, p-532.

2529. Gol'dshtejn, V. A., Lubyanyi, V. V., et al., Prib. Tekh. Eksp. (USSR), 15(4), Pt. 1, July-Aug. 1972, pp-37-40.

2530. Gol'dshtejn, V. A., Vlasenko, V. G., et al., Kharkov. Fiziko-Tekhnicheskij Inst., Nuclear Sci. and Eng. Problems, KFTI-76-27, 1976, pp-41-42.

2531. Gol'dshtejn, V. A., Kuplennikov, Eh. L., et al., Fizika Vysokikh Ehnergij i Atomnogo Yadra, KFTI-77-9, 1977, pp-47-48.

2532. Gol'dshtejn, V. A., Afanas'ev, N. G., et al., Leningrad Nauka 1978, p-247.

2533. Gol'dshtejn, V. A., Afanas'ev, N. G., et al., Leningrad Nauka 1978, p-248.

2534. Gol'dshtejn, V. A., Kuplennikov, Eh. I., et al., Leningrad Nauka 1978, p-249.

2535. Goldsmith, M., and Ullo, J. J., Natl. Bur. of Stands. (U. S.), Spec. Publ. 425, Oct. 1975, pp-557-559.

2536. Goldsmith, M., and Ullo, J. J., Nucl. Sci. and Eng. 60(3), July 1976, pp-251-261.

2537. Goldstone, P. D., Hopkins, F., et al., Phys. Rev. Lett. 35(17), 27 Oct. 1975, pp-1141-1143.

2538. Goldstone, P. D., Hopkins, F., et al., Phys. Lett., B, 62 (3), 7 June 1976, pp-280-282.

2539. Goldstone, P. D., Britt, H. C., et al., Phys. Rev. Lett. 38(22), 30 May 1977, pp-1262-1265.

2540. Golikov, I. G., Zhukov, M. N., et al., Yad. Fiz. 27(1), 1978, pp-7-9.
2541. Golovanov, L. B., Mazarskii, V. L., et al., JINR-8-11957, 1978, p-7.
2542. Golovanov, L. B., and Mazarskij, V. L., Prib. Tekh. Ehksp. 5, Sept. 1975, p-252.
2543. Golovanov, L. B., Phys. and Techniques of Low Temperatures, Prague, Pub. House of the Czech. Acad. of Sci., 1964, pp-229-31.
2544. Golovanov, L. B., Mazarskii, V. L., et al., Report 1979, JINR-8-11956, p-7.
2545. Golovanov, L. B., and Mazarskii, V. L., Prob. Tekh. Eksp. (USSR), 18(5), Pt. 2, Sept.-Oct. 1975, p-252.
2546. Golovanov, L. B., Probl. Fiz. Elem. Chastits At. Yadra 2(3), 1972, pp-717-62.
2547. Golovanov, L. B., Cryogenics 12(4), 1972, pp-307-8.
2548. Golovanov, L. B., Mazarskii, V. L., et al., Instr. Exp. Tech. (USSR), 14(5), Sept.-Oct. 1971, pp-1310-13.
2549. Golovanov, L. B., Instr. Exp. Tech. (USSR), 14(3), May-June 1971, pp-715-16.
2550. Golovkov, M. S., Kondrat'ev, S. N., et al., Leningrad Nauka 1978, p-201.
2551. Golovnya, V. Ya., and Moldovanov, V. P., Pribory i Tekh. Ekspt. 6(3), May-June 1961, pp-185-6.
2552. Gomez del Campo, J., Ortiz, M. E., et al., Nucl. Phys., A, 262(1), 10 May 1976, pp-125-136.
2553. Gomez del Campo, J., Ford, J. L. C., Jr., et al., ORNL-5306, Sept. 1977, pp-55-57.
2554. Gomez del Campo, J., Dayras, R. A., et al., ORNL-5306, Sept. 1977, pp-50-51.
2555. Gomez del Campo, J., Ford, J. L. C., et al., Phys. Lett., B, 69(4), 29 Aug. 1977, pp-415-418.
2556. Gomez del Campo, J., Ford, , J. L. C., Jr., et al., Nucl. Phys., A, 297(1), 6 March 1978, pp-125-135.
2557. Gomez del Campo, J., ORNL, TN (U. S. A.), CONF-790743-7, 1979, p-24.
2558. Goncharova, N. G., Korenman, G. Ya., et al., Leningrad Nauka 1978, p-369.
2559. Goncharova, N. G., Zhivopistsev, F. A., et al., Leningrad Nauka 1978, p-399.
2560. Gono, Y., Lieder, R. M., et al., Nucl. Phys., A, ISSN 0375-9474, 327(2), 24 Sept. 1979, pp-269-287.
2561. Gonsior, B., Hort, M., et al., Verh. Dtsch. Phys. Ges. 10(7), 1975, p-710.
2562. Goode, P. D., Nucl. Instr. and Meth. 92(4), 1971, pp-447-53.
2563. Goodman, C. D., Hanson, D. L., et al., Phys. Rev., C, 17 (2), Feb. 1978, pp-493-500.
2564. Goodman, C. D., Bainum, D. E., et al., ORNL, TN (U. S. A.), CONF-790847-4, 1979, p-10.

2565. Goodman, L. J., Marino, S. A., et al., Contract AT-(30-1)-2740, 1970, Columbia Univ., Upton, NY (U. S. A.).
2566. Goosman, D. R., Nucl. Instr. and Meth. 116(3), 1974, pp-445-449.
2567. Gorbenko, V. G., Zhebrovskii, Yu. V., et al., Instr. Exp. Tech. (USSR) (Eng. Transl.), 16(2), 1973, pp-396-98.
2568. Gordon, S., INTDS Proc., Berkeley, U. S. A., Oct. 19-20, 1977.
2569. Gorlov, G. V., Kirillov, A. I., et al., Pribory i Tekhn. Eksperim. 4, July-Aug. 1965, pp-221-2.
2570. Gorlovoi, G. D., and Stepanenko, V. A., Moscow, Atomizdat, 1965, p-116.
2571. Gorodetzky, S., and Drouin, R. S., Strasbourg Univ. (France), AERE-R-5097, Paper 14, p-8.
2572. Gorodetzky, S., and Drouin, R. S., PSPSITF, AERE, Oct. 20 & 21, 1965.
2573. Gorodetzky, S., and Drouin, R. S., AEC Access. Nos. 65, AERE-R-5097, Paper 14, 1966, p-8.
2574. Gorodetzky, S., Denimal, J., et al., Nucl. Instr. and Meth. 38, 1965, pp-79-81.
2575. Gorodkov, Yu. V, Eliseev, G. P., et al., Cryogenics 16(8), 1976, pp-503-4.
2576. Gorpinich, O. K., Stryuk, Yu. S., et al., AN Ukrainskoj SSR, Kiev Inst. Yadernykh Issledovanij, 1976, p-39.
2577. Goryachev, Yu. M., Zenkevitch, P. R., et al., CEAL-2000, Gosudarstvennyi Komitet po Ispol'zovaniyu Atomnoi Energii SSSR, Moscow, Inst. Teoreticheskoi i Eksperimental'noi Fiziki, pp-A106-8.
2578. Goryachev, Yu. M., Kanavets, V. P., et al., Instr. Exp. Tech. (USSR), 4, July-Aug. 1970, pp-979-81.
2579. Goryunov, O. Yu., Dobrikov, V. N., et al., Ukr. Fiz. Zh. 20(11), Nov. 1975, pp-1775-1780.
2580. Goryunov, O. Yu., Dobrikov, V. N., et al., Leningrad Nauka 1978, p-193.
2581. Goss, J. D., Jolivette, P. L., et al., Phys. Rev., C, 12 (6), Dec. 1975, pp-1730-1738.
2582. Gosset, J., Paris-11 Univ., 91-Orsay (France), These (D es S), FRNC-TH-586, 1975, p-87.
2583. Gosset, J., Mayer, B., et al., CEA-CONF-3370, 1975, p-2.
2584. Gosset, J., Mayer, B., et al., Phys. Rev., C, 14(3), Sept. 1976, pp-878-905.
2585. Gould, C. R., Joyce, J. M., et al., Natl. Bur. of Stands. (U. S.), Spec. Publ. 425, Oct. 1975, pp-697-700.
2586. Goulding, C. A., Murdoch, B. T., et al., Nucl. Instr. and Meth. (Netherlands), 148(1), 1 Jan. 1978, pp-11-12.
2587. Govaerts, J., Liege Univ., EUR-1815.e(Rev.), pp-182-5.
2588. Govorov, V. V., Dobretsov, Yu. P., et al., JINR-13-12341, Report 1979, p-8.
2589. Grabez, B., Todorovic, Z., et al., Z. Phys., A, 292(1), Aug. 1979, pp-67-72.

2590. Graham, M. J., and Bray, C. S., J. of Physics, E Series 2(2), 1969, p-706.
2591. Gram, P. A. M., Contract W-7405-eng-36, LA-4612, Contract W-7405-eng-36, Dec. 1970, p-11.
2592. Grand, P., and Goland, A. N., Nucl. Instr. and Meth. 145, 1977, pp-49-76.
2593. Grand, P., Report 1975, BNL-20159, p-160.
2594. Grantsev, V. I., Dryapachenko, I. P., et al., AN Ukrainskoj SSR, Kiev Inst. Yadernykh Issledovanij (In Russian), 1976, p-18.
2595. Grau, J. A., Purdue Univ., Lafayette, Ind. (U. S. A.), Thesis (Ph. D), 1976, p-233.
2596. Graves, E. R., Rodriques, A. A., et al., Rev. Sci. Instr. 20, 1949, pp-579-82.
2597. Grawe, H., Heidinger, F., et al., Nucl. Instr. and Meth. 127(2), 1 Aug. 1975, pp-311-316.
2598. Gray, W. H., Hassenzah, W. V., et al., Advances in Cryogenic Eng. 22, NY, Plenum Press, 1977, pp-526-534.
2599. Gray, W. H., and Hassenzah, W. V., Int. Cryogenic Mat. Conf., Kingston, Ontario, Canada, 22 July 1975
2600. Grechukhin, D. P., and Soldatov, A. A., Yad. Fiz. ISSN 0044-0027, 28(5), 1978, pp-1206-1222.
2601. Green, A., Bauer, E., et al., Tech. 5, 1970, p-345.
2602. Green, A. K., Bauer, E., et al., J. Appl. Phys. 41, 1970, p-4736, J. Vac. Sci. Tech. 7, 1970, p-159.
2603. Green, A. K., Bauer, E., et al., Molecular Processes on Solid Surfaces, McGraw-Hill, NY, 1968.
2604. Green, A. K., Dancy, J., et al., J. Vac. Sci. Tech. 8, 1971, p-165.
2605. Green, L., Bettis Atomic Power Lab., West Mifflin, PA (U. S.A.), WAPD-TM-1279, Nov. 1976, p-19.
2606. Green, L., Nucl. Sci. and Eng. 66(1), April 1978, pp-127-134.
2607. Green, P. W., and Sheppard, D. M., Nucl. Phys., A, 274 (1-2), 7 Dec. 1976, pp-125-140.
2608. Green, S. J., Holtkamp, D. B., et al., LA-UR-79-2430, CONF-790847-10, 1979, p-12.
2609. Greene, M. W., and Lebowitz, E., Int. J. of Appl. Radiat. Isot. 23(7), July 1972, pp-342-4.
2610. Greenspan, E., and Miley, G. H., Trans. Am. Nucl. Soc. 34, 1980, pp-61-2.
2611. Greenwood, R. C., Helmer, R. G., et al., Idaho Nat. Eng. Lab., BNL, Upton, NY (U. S. A.), CONF-790968-1, 1979, p-13.
2612. Gregor, L. V., Phys. Thin Films 3, 1966, p-131.
2613. Greil, A., Truetzschler, K., et al., Kernenergie 12, Sept. 1969, pp-294-7.
2614. Grek, B., Pepin, H., et al., Nucl. Fusion 17(6), 1977, pp-1165-1185.

2615. Gridasov, V. I., Kardash, A. A., et al., At. Energ. (USSR), 30(6), June 1971, pp-520-5.
2616. Gridasov, V. I., Myznikov, K. P., et al., Sov. Phys. Tech. Phys. 19(7), Jan. 1975, pp-922-925.
2617. Gridnev, K. A., Darvish, N. Z., et al., Leningrad. Nauka 1978, p-211.
2618. Gridnev, K. A., Semenov, V. M., et al., Leningrad. Nauka 1978, p-231.
2619. Griess, J. C., J. Electrochem. Soc. 100, 1953, pp-429-33.
2620. Griffioen, R. D., Thompson, R. C., et al., Progress Report, Roch. Univ., NY, 1 June 1976-31 Aug. 1977, pp-50-62.
2621. Griffith, J. A. R., and Burge, E. J., Rev. Sci. Instr. 37, Feb. 1966, pp-147-54.
2622. Griffiths, J., BNL-NCS-50451, March 1975, pp-81-86.
2623. Grigoryan, L. A., and Shakhbazyan, V. A., Erevanskij Fizicheskij Inst. (USSR), 1976, p-18.
2624. Grigoryan, L. A., and Shakhbazyan, V. A., Yad. Fiz. ISSN 0044-0027, 28(5), 1978, pp-1372-1378.
2625. Grilly, E. R., LA-UR-77-1523, 1977, p-13.
2626. Grilly, E. R., LASL Report 1978.
2627. Grilly, E. R., Adv. Cryog. Eng. 23, 1978, pp-676-81.
2628. Grilly, E. R., Rev. Sci. Instr. 48(2), 1977, pp-148-51.
2629. Grimes, S. M., Anderson, J. D., et al., Rev., C, 13(6), June 1976, pp-2224-2236.
2630. Grimes, S. M., Haight, R. C., et al., Nucl. Sci. and Eng. 62(2), Feb. 1977, pp-187-194.
2631. Grimes, S. M., Haight, R. C., et al., Phys. Rev., C, 17(2), Feb. 1978, pp-508-515.
2632. Grimson, J. H., Lindberg, J. F., et al., Proc. of the 18th Conf. on Remote Systems Technology, Washington, D. C., November 1970, pp-135-42.
2633. Grimson, J. H., Lindberg, J. F., et al., Trans. Am. Nucl. Soc. 13(2), 1970, p-847.
2634. Große-Kreul, B., Sander, V., et al., INTDS Newsletter, Feb. 1980.
2635. Groce, D. C., Gulf General Atomic, Inc., San Diego, CA, Contract AT(04-3)-167, June 30, 1967, p-60.
2636. Gross, D. A., and Melissinos, A., UR-875-365, Contract AT(30-1)-875, 2 Dec. 1971, p-52.
2637. Gross, D. A., and Melissinos, A., Nucl. Instr. and Meth. 130(1), 1 Dec. 1975, pp-1-13.
2638. Gross, E. E., Cleary, T. P., et al., Phys. Rev., C, 17(5), May 1978, pp-1665-1671.
2639. Grossiord, J. Y., Guichard, A., et al., ORO-4856-26, 1975, pp-305-306, Maryland Univ.
2640. Grossiord, J. Y., Bedjidian, M., et al., J. Phys. (Paris). Colloq. No. 5, 1975, pp-C5.123-C5.125.
2641. Groves, J. L., Holloway, L. E., et al., Phys. Rev., D, 15 (1), 1 Jan. 1977, pp-47-58.

2642. Gruppelaar, H., Dragt, J. B., et al., Natl. Bur. of Stands. (U. S.), Spec. Publ. 1, Oct. 1975, pp-165-168.
2643. Gruverman, I. J., Int. J. of Appl. Radiat. and Isot. 13, 1962, pp-223-228.
2644. Gruzensky, P. M., and Block, F. E., U. S. Bureau of Mines Report of Investigation No. 5305, 1957.
2645. Guang-chang, Xia, Wu Li Hsueh Pao 20., Oct. 1964, pp-1056-7.
2646. Gubernator, K., and Moret, H., EUR-3950e, CNMB, Geel, Belgium, 1968
2647. Gubler, H. P., Plattner, G. R., et al., Nucl. Phys., A, 284(1), 27 June 1977, pp-114-122.
2648. Guenther, P., Havel, D., et al., ANL/NDM-22, Sept. 1976, p-37.
2649. Guenther, P. T., Havel, D. G., et al., Nucl. Sci. and Eng. 65(1), Jan. 1978, pp-174-180.
2650. Guerreau, D., Galin, J., et al., Inst. de Physique Nucleaire, IPNO-RC-79-07, 1979, p-33.
2651. Gurevich, G. M., Lazareva, L. E., et al., Nucl. Phys., A, 273(2), 30 Nov. 1976, pp-326-340.
2652. Guichard, A., Chevallier, M., et al., Nucl. Phys. A164, 1971, pp-56-68.
2653. Guichard, A., Benenson, W., et al., Phys. Rev., C, 12(6), Dec. 1975, pp-1762-1770.
2654. Guichard, A., Benenson, W., et al., Phys. Rev., C, 12(6), Dec. 1975, pp-1780-1788.
2655. Guichard, A., Benenson, W., et al., Phys. Rev., C, 13(2), Feb. 1976, pp-540-547.
2656. Guichard, A., Inst. de Physique Nucleaire, 4 Biennial Session of Nuclear Physics, La Toussuire, 28 Feb. - 4 March 1977, LYCEN-7702(pt. 1), pp-C11.1-C11.20.
2657. Guidry, M. W., Eichler, E., et al., Phys. Rev., C, 12(6), Dec. 1975, pp-1937-1944.
2658. Guidry, M. W., Sturm, R. J., et al., Phys. Rev., C, 13(3), March 1976, pp-1164-1172.
2659. Guidry, M. W., Butler, P. A., et al., Phys. Rev. Lett. 40 (15), 10 April 1978, pp-1016-1019.
2660. Guidry, M. W., Lee, I. Y., et al., Phys. Rev., C, ISSN 0556-2813, 20(5), Nov. 1979, pp-1814-1818.
2661. Guillaume, G., Rastegar, B., et al., Nucl. Phys., A, 272 (2), 16 Nov. 1976, pp-338-352.
2662. Guillaume, M. A., and Peeters, J. M., Liege Univ. (Belgium), Report No: EUR-3474.F, May 1967, p-76.
2663. Guillaume, M., Delfiore, G., et al., Nucl. Instr. and Meth. 92(4), 1971, pp-571-6.
2664. Guillaume, M., Delfiore, G., et al., Liege Univ. (Belgium), EUR-4286(pt. 2), Nov. 1970, p-60.
2665. Guillaume, M., Lambrecht, R. M., et al., Int. J. Appl. Radiat. Isot. 26(12), 1975, pp-703-7.

2666. Guillot, J., Van de Wiele, J., et al., Inst. des Sciences Nucleaires, ISN-79-03, Jan. 1979, p-50, Grenoble-1 Univ.
2667. Guinn, V. P., General Atomic Div., General Dynamics Corp., San Diego, CA, A/CONF.28/P/197, 1964, p-13.
2668. Guisan, O., LBL-500, from 2nd Int. Conf. on Polarized Targets, Berkeley, CA, 30 Aug. 1971, pp-187-90
2669. Gul'karov, I. S., Yad. Fiz. ISSN 0044-0027, 29(1), Jan. 1979, pp-57-64.
2670. Gul'ko, V. M., Knizhnik, E. I., et al., Skvazhin. Yader.-geofiz. Apparatura s Upravlyaem. Istochnikami Izluch., M. 1978, p-30-5.
2671. Gullholmer, W., and Parker, W., PSPSITF, AERE, Oct. 20 & 21, 1965.
2672. Gullholmer, W., and Parker, W., Chalmers Tekniska Hogskola, Goteborg (Sweden), AERE-R-5097, Paper 17, p-2.
2673. Gunn, G. D., Fox, J. D., et al., Phys. Rev., C, 13(2), Feb. 1976, pp-595-607.
2674. Gunn, G. D., Boyd, R. N., et al., Nucl. Phys., A, 275(2), 10 Jan. 1977, pp-524-532.
2675. Guntherschulze, A., Z. Phys. 100, 1936, p-539.
2676. Gupta, M. K., Indiana Univ., Bloomington (U. S. A.), Thesis (Ph. D), 1975, p-98.
2677. Gupta, M. K., and Walker, G. E., Nucl. Phys., A, 256(3), 19 Jan. 1976, pp-444-460.
2678. Gupta, S. K., Saini, S., et al., Bhabha Atomic Rsch. Centre, Bombay (India), Proc. of Nucl. Phys. and Solid State Phys. Symp., Bangalore, Dec. 27-31, 1973, Vol. 16B, pp-35-38.
2679. Guratzsch, H., Kuehn, B., et al., Nucl. Phys., A, 293(1-2), 12-19 Dec. 1977, pp-109-116.
2680. Gur'ev, V. N., Fiziko-Tekhnicheskij Inst., Nucl. Sci. and Eng. Prob., Voprosy atomnoj nauki i tekhniki, KFTI-75-11, No. 2(14), 1975, p-50.
2681. Gur'ev, V. N., KFTI-75-11, No. 2(14), 1975, pp-48-49.
2682. Gurevich, G. M., Tr. Fiz. Inst., Akad. Nauk SSSR, 41, 1968, pp-146-64.
2683. Gursky, J. C., and Povelites, J. G., INTDS Proc., Los Alamos, U. S. A., Oct. 19-21, 1976.
2684. Gursky, J. C., and O'Rourke, J. A., Nucl. Instr. and Meth. 167(1), 1979, pp-145-9.
2685. Gursky, J. C., and O'Rourke, J. A., INTDS Proc., Garching, FRG, Sept. 11-14, 1978.
2686. Gursky, J. C., and Sherwood,B. A., INTDS Proc., Berkeley, CA, Oct. 19-20, 1977.
2687. Gursky, J. C., and Povelites, J. G., INTDS Proc., Los Alamos, U. S. A., Oct. 19-21, 1976.
2688. Gurney, W. J., Naval Radiological Defense Lab., San Francisco, CA, USNRDL-TR-67-14, Dec. 22, 1966, p-48.

2689. Gurvitz, S. A., Phys. Lett., B, ISSN 0370-2693, 85(1), 30 July 1979, pp-5-8.
2690. Gusakow, M., Ind. At. 13(5-6), 1969, pp-39-46.
2691. Gusdal, M. I., McKee, J. S. C., et al., IEEE Trans. on Nucl. Sci., NS-26(2), April, 1979.
2692. Gusdal, M. I., McKee, J. S. C., et al., TEICCTA, IU, U. S. A., Sept. 18-21, 1978.
2693. Gusev, E. A., Izv. Vyssh. Ucheb. Zaved., Fiz. 10, 1970, pp-129-31.
2694. Gusinow, M. A., Anthes, J. P., et al., Appl. Phys. Lett. 33(9), 1978, pp-800-2.
2695. Gustafson, D. E., Virginia Univ., Charlottesville (U. S. A.), Thesis (Ph. D), 1976, p-72.
2696. Gustafson, D. E., Thornton, S. T., et al., Phys. Rev., C, 13(2), Feb. 1976, pp-691-698.
2697. Gustafson, D. E., and Gomez del Campo, J., Nucl. Phys., A, 262(1), 10 May 1976, pp-96-112.
2698. Gustafsson, C., and Lambert, E., Ann. Phys. (N. Y.), 111(2), April 1978, pp-304-329.
2699. Gosset, J., Gutbrod, H. H., et al., Phys. Rev., C, 16(2), Aug. 1977, pp-629-657.
2700. Gutierrez, C. P., et al., J. Electrochem. Soc. 109, 1962, p-923.
2701. Guyer, H., Meyer, V., et al., Helv. Phys. Acta. 49(2), 18 May 1976, pp-182-183.
2702. Guzhovskij, B. Ya., Pis'ma Zh. Ehksp. Teor. Fiz. 26(5), 5 Sept. 1977, pp-408-412.
2703. Gwin, R., Weston, L. W., et al., Natl. Bur. of Stands. (U. S.), Spec. Publ. 425, Oct. 1975, pp-627-630.
2704. Gyarmati, E., and Nickel, H., Kernforschungsanlage Juelich G.m.b.H. (Germany, F.R.), Jan. 20, 1977, p-3.
2705. Gyles, W., Johnson, R. R., et al., AIP Conf. Proc. ISSN 0094-243X, 54(1), 15 July 1979, pp-522-523.
2706. Haas, B., Centre de Recherches Nucleaires; Strasbourg-1 Univ., 67 (France), CRN-PN-74-44, 1974, p-292.
2707. Haas, B., Chevallier, A., et al., Centre de Recherches Nucleaires, CRN-PN-75-42, 1975, p-7.
2708. Haas, F., Freeman, R. M., et al., Centre de Recherches Nucleaires, CRN-PN-77-15, 1977, p-8.
2709. Haasbroek, F. J., Burdzik, G. F., et al., Council for Scientific and Industrial Rsch., Pretoria (South Africa), 1976, p-13.
2710. Haasbroek, F. J., Burdzik, G. F., et al., Council for Sci. and Ind. Rsch., Pretoria (South Africa), 1976, p-13.
2711. Haase, O., Z. Naturf. a12, 1957, p-941.
2712. Habanec, J., Sebek, Z., et al., Jad. Energ. 18(8), Aug. 1972, pp-276-8.
2713. Hacken, G., Werbin, R., et al., Phys. Rev., C, 17(1), Jan. 1978, pp-43-50.

2714. Hacman, D., Optik 28, 1968, p-115.
2715. Hadjimichael, E., Phys. Lett., B, ISSN 0370-2693, 85(1), 30 July 1979, pp-17-20.
2716. Haeringen, H. van, Nucl. Phys., A, ISSN 0375-9474, 327(1), 17 Sept. 1979, pp-77-98.
2717. Haeusser, O., Alexander, T. K., et al., Nucl. Phys., A, 273(1), 23 Nov. 1976, pp-253-268.
2718. Haeusser, O., Towner, I. S., et al., Nucl. Phys., A, 293 (1-2), 12-19 Dec. 1977, pp-248-268.
2719. Haftel, M. I., Petersen, E. L., et al., Phys. Rev., C, 14 (2), Aug. 1976, pp-419-437.
2720. Haftel, M. I., AIP Conf. Proc. ISSN 0094-243X, 47, 1978, pp-548-549.
2721. Haftel, M. I., and Bassel, R. H., AIP Conf. Proc. ISSN 0094-243X, 47, 1978, pp-550-551.
2722. Hageboe, E., Kjelberg, A., et al., Report 1967, CERN.
2723. Hageboe, E., Kjelberg, A., et al., Ark. Fys. 36(15), 1967, pp-127-32.
2724. Hageboe, E., Proc. Int. Conf. Electromagn. Isotope Separ. Tech. Appl., 7th, 1970, pp-146-157.
2725. Hageboe, E., and Sundell, S., CERN 70-3, 1970, pp-63-80.
2726. Hagemann, U., and Keller, H. J., Zentralinstitut fuer Kernforschung, Annual Report 1975, p-24.
2727. Hagemann, U., ZfK-315, Aug. 1976, pp-14-16.
2728. Hahn, D., Terlecki, G., et al., Nucl. Phys., A, ISSN 0375-9474, 325(1), 13 Aug. 1979, pp-283-304.
2729. Hahn, R. L., Toth, K. S., et al., ORNL-5306, Sept. 1977, pp-60-61.
2730. Haight, R. C., and Grimes, S. M., Meeting on Neutron Data of Structural Materials, Geel, Belgium, 5-8 Dec. 1977, p-16.
2731. Haight, R. C., UCRL-83005, CONF-791058, 1 Oct. 1979, p-12.
2732. Haitsma, Y., and Blok, H. P., Nederlandse Natuurkundige Vereniging, Amsterdam, Spring Meeting of the Dutch Physical Society, March 1978, p-15.
2733. Haggmark, L. G., J. Appl. Phys., 47(1), 1976, pp-357-61.
2734. Halbert, M. L., Dayras, R. A., et al., Phys. Rev., C, 17 (1), Jan. 1978, pp-155-162.
2735. Halbleib, J. A., and Wright, T. P., SAND-79-0385, March 1979, p-42.
2736. Hall, P. J., van, Melssen, J. P. M. G., et al., Nucl. Phys., A, 291(1), 7 Nov. 1977, pp-63-84.
2737. Hall, P. J., van, Melssen, J. P. M. G., et al., Phys. Lett., B, 74(1-2), 27 March 1978, pp-42-44.
2738. Halliday, C. E., RL-79-091, Nov. 1979, p-15.
2739. Halliwell, C., Elias, J. E., et al., Phys. Rev. Lett. 39 (24), 12 Dec. 1977, pp-1499-1502.
2740. Hallock, J. N., Enge, H. A., et al., Nucl. Phys., A, 252(1), 3 Nov. 1975, pp-141-151.

2741. Halpern, G. M., and Kim, H. G., Inertial Confinement Fusion, Dig. Tech. Pap. Top. Meet. (1), TuE 2, 1977, p-5.
2742. Halpern, V., J. Appl. Phys. 40, 1969, p-4627.
2743. Halter, J. M., and Sullivan, R. G., 4th Conf. on Contamination, Washington, D. C., U. S. A., 10-14 Sept. 1978, p-21.
2744. Halverson, J. E., and Johnson, W. H., Jr., Phys. Rev., C, 17(4), April 1978, pp-1414-1416.
2745. Hamm, M. E., Texas Agricultural and Mechanical Univ., College Station (U. S. A.), Thesis (Ph. D), Univ. Microfilms Order No. 77-12,559, 1976, p-117.
2746. Hamm, M., and Nagatani, K., Phys. Rev., C, 17(2), Feb. 1978, pp-586-595.
2747. Hampel, C. A. (ed.), Rare Metals Handbook, 2nd ed., Publishing Corp., London, 1961, p-407.
2748. Hampel, W., Heidelberg Univ. (F.R. Germany), 10 June 1974, p-58.
2749. Hanau, W. B., and Fabian, H., BMVG-FBWT-77-9, Contract T/RF-22/TF/220/21001, 1977, p-37.
2750. Hand, L., Rees, J., et al., "Nucleon Structure", Stanford, CA, Stanford Univ. Press, 1964, pp-364-6.
2751. Hanley, R. P., Habrel, W. A., et al., IEEE Trans. on Nucl. Sci., NS-14(3), June 1967,
2752. Hanley, P. R., Haberl, A. W., et al., from National Particle Accelerator Conf., Washington, D. C., Contract AT(30-1)-3822, Feb. 27, 1967, p-18.
2753. Hanna, S. S., Proc. Int. Symp. Polariz. Phenom. Nucl. React., 4th, 1975, pp-407-22.
2754. Hänni, H. P., Rice Inst., Houston, TX, Report 1959.
2755. Hänni, H. P., Helv. Phys. Acta 33, 1960, pp-987-91.
2756. Hansen, F., Lindahl, A., et al., Proc. Int. EMIS Conf. Low Energy Ion Accel. Mass Sep., 8th, 1973, pp-426-31.
2757. Hansen, H., Hornshoj, P., et al., Nucl. Phys., A, ISSN 0375-9474, 327(1), 17 Sept. 1979, pp-193-206.
2758. Hansen, J., Danish Atomic Energy Comm., Risoe, Rsch. Estab., April 1971, p-15.
2759. Hansen, J. W., and Lundsager, P., Nucl. Instr. and Meth. 160(2), 15 March 1979, pp-203-10.
2760. Hansen, L. F., Grimes, S. M., et al., Nucl. Sci. and Eng. 61(2), Oct. 1976, pp-201-11.
2761. Hansen, O., Maher, J. V., et al., Nucl. Phys., A, 292(1-2), 21-28 Nov. 1977, pp-253-266.
2762. Hansen, P. G., Hogh, J., et al., Nucl. Instr. and Meth. 30, 1964, pp-161-4.
2763. Hansen, E., Jones, E., et al., CERN-65-8, Feb. 25, 1965, p-20.
2764. Hanson, D. L., Yale Univ., New Haven, CT, U. S. A., Thesis, (Ph. D), 1975, p-239.
2765. Hanson, D. L., Stein, N., et al., AIP Conf. Proc. ISSN 0094-243X, 47, 1978, pp-726-727.

2766. Haouat, G., and Seguin, S., CEA-N-1739, July 1974, p-16.
2767. Haouat, G., Seguin, S., et al., CEA-N-1798, June 1975, pp-21-24.
2768. Haouat, G., Lachkar, J., et al., CEA-N-1798, June 1975, pp-25-27.
2769. Haouat, G., Lachkar, J., et al., CEA-N-1798, June 1975, pp-28-29.
2770. Haouat, G., Lachkar, J., et al., CEA-N-1798, June 1975, pp-31-33.
2771. Haouat, G., Lachkar, J., et al., CEA-CONF-3304, 1975, p-19.
2772. Haouat, G., Lachkar, J., CEA-CONF-3305, 1975, p-16.
2773. Haouat, G., Lachkar, J., et al., Natl. Bur. of Stands. (U. S.), Spec. Publ. 425, Oct. 1975, pp-889-892.
2774. Haouat, G., Lachkar, J., et al., Acta Phys. Slovaca 26(1), 1976, pp-51-55.
2775. Haouat, G., Lachkar, J., et al., CEA-CONF-4034, 1977, p-9.
2776. Haouat, G., Lachkar, J., et al., CEA-N-1969, June 1977, pp-102-105.
2777. Haouat, G., Lachkar, J., et al., Nucl. Sci. and Eng. 65(2), Feb. 1978, pp-331-346.
2778. Haque, M. A., Holzer, O. H., et al., Nucl. Instr. and Meth. 47, Jan. 1967, pp-137-40.
2779. Harakeh, M. N., Comfort, J. R., et al., Phys. Lett., B, 62(2), 24 May 1976, pp-155-158.
2780. Harakeh, M. N., Arends, A. R., et al., Nucl. Phys., A, 265 (2), 19 July 1976, pp-189-212.
2781. Harakeh, M. N., Borg, K., van der, et al., Phys. Rev. Lett. 38(13), 28 March 1977, pp-676-679.
2782. Harakeh, M. N., Heyst, B., van, et al., Nucl. Phys., A, ISSN 0375-9474, 327(2), 24 Sept. 1979, pp-373-396.
2783. Harar, S., CEA-CONF-3939, 1977, p-14.
2784. Harar, S., CEA-CONF-3989, 1977, p-18.
2785. Harchol, M., Nucl. Instr. and Meth. 40, Feb. 1966, p-158.
2786. Hardekopf, R. A., Veeser, L. R., et al., Phys. Rev. Lett. 35(24), 15 Dec. 1975, pp-1623-1625.
2787. Hardekopf, R. A., Haglund, R. F., Jr., et al., LASL, LA-7863, July, 1979, p-15.
2788. Hardie, G., Meyer-Schutzmeister, L., et al., Phys. Rev., C, 13(5), May 1976, pp-1874-1883.
2789. Hardt, H. A., DuPont de Nemours (E. I.) and Co., Aiken, SC (U. S. A.), Savannah River Plant, Contract E(07-2)-1, Oct. 1976, p-12.
2790. Hardwick, T. J., and Guentner, W. S., Intern. J. Appl. Radiation and Isotopes 12, Nov. 1961, pp-20-6.
2791. Hardy, D. M., Baker, S. D., et al., Nucl. Instr. and Meth. 98(1), 1972, pp-141-5.
2792. Hariharan, A. V., Knighton, J. B., et al., ANL-7057, Part III, 1965
2793. Harmatz, B., Horen, D. J., et al., Phys. Rev., C, 12(3), Sept. 1975, pp-1083-1086.

2794. Harmon, K. M., BNWL-CC-410, 1965, p-15.
2795. Harmon, K. M., BNWL-CC-459, 25 Jan. 1966, p-24.
2796. Harper, P. V., Skaggs, L. S., et al., Proc. of the 6th Int. Cyc. Conf., Vancouver, Canada, July 18-21, 1972.
2797. Harris, D. C., Beck, J. N., et al., Nucl. Sci. and Eng. 63(4), Aug. 1977, pp-504-507.
2798. Harris, J. W., Cormier, T. M., et al., Phys. Rev. Lett. 38(25), 20 June 1977, pp-1460-1463.
2799. Hartas, A. G., Ioannina Univ. (Greece), Thesis (Ph. D), Univ. Microfilms Order No. 76-10,438, 1975, p-165.
2800. Hartas, A. G., Papadopoulos, C. T., et al., Nucl. Phys., A, 279(3), 4 April 1977, pp-413-429.
2801. Hartin, W. J., and Goodman, L. J., Contract AT(30-1)-2740, Health Phys. 21(2), Aug. 1971, pp-309-15.
2802. Hartley, H., Ponder, A. O., et al., Phil. Mag. XLIII, 1922, p-430.
2803. Hartmann, G., Hubert, D., et al., Deutsches Elektronen-Synchrotron, Hamburg (West Germany), Aug. 1972, p-16.
2804. Hartmann, W., Thalheimer, W., et al., GSI-79-11, Oct. 1979.
2805. Harvey, B. G., Hendrie, D. L., et al., LBL Nuclear Chem., July 1975, pp-74-76.
2806. Harvey, H. W., Hoh, J. C., et al., Isotop. Radiat. Technol. 4, Winter 1966-1967, pp-82-3.
2807. Harvey, J. A., and Hill, N. W., Natl. Bur. of Stands. (U. S.), Spec. Publ. 1, Oct. 1975, pp-244-245.
2808. Harvey, J. A., ORNL-5137, May 1976, pp-1-8.
2809. Harwood, L. H., and Kemper, K. W., Phys. Rev., C, 14(1), July 1976, pp-368-371.
2810. Harwood, L. H., and Kemper, K. W., Phys. Rev., C, 16(3), Sept. 1977, pp-1040-1047.
2811. Harz, U., and Priesmeyer, H. G., et al., Verh. Dtsch. Phys. Ges. 6, 1977, p-1007.
2812. Hasan, H., Nucl. Phys. and Solid State Phys. Symp., Pune, India, 26-30 Dec. 1977, pp-131-134.
2813. Hasegawa, T., Horikawa, N., et al., Nucl. Instr. and Meth. 73, 1969, pp-349-50.
2814. Hashimoto, N., and Kawai, M., Phys. Lett., B, 59(3), 10 Nov. 1975, pp-223-226.
2815. Hashimoto, O., Nagamiya, S., et al., Phys. Lett., B, 62(2), 24 May 1976, pp-233-236.
2816. Hashmi, S. Z. R., Univ. of Texas, Austin, TX, 1967, Thesis, p-74.
2817. Hass, M., Benczer-Koller, N., et al., Phys. Rev., C, 17(3), March 1978, pp-997-1000.
2818. Hassan, A. M., El-Kady, A. A., et al., Arab J. Nucl. Sci. Appl. 10(1), Jan. 1977, pp-77-97.
2819. Hassenzahl, W. V., and Gray, W. H., LA-5833-MS, Contract W-7405-eng-36, Dec. 1974, p-17.

2820. Hassenzahl, W. V., and Gray, W. H., Cryogenics 15(11), Nov. 1975, pp-627-638.
2821. Haste, T. J., and Thomas, B. W., J. Phys., G, 1(9), 1 Dec. 1975, pp-981-994.
2822. Hauser, U., and Kerler, W., Rev. Sci. Instr. 29, 1958, pp-380-2.
2823. Haustein, P. E., Franz, E., et al., Phys. Rev., C, 14(2), Aug. 1976, pp-645-649.
2824. Havens, W. W., Jr., Melkonian, E., et al., Columbia Univ., NY (U. S. A.), Dept. of Mech. and Nucl. Eng., 1977, p-46.
2825. Havloujian, J., Ravalli, G., et al., Aust. J. Phys. 28(3), June 1975, pp-247-250.
2826. Haxton, W. C., Phys. Lett., B, 76(2), 22 May 1978, pp-165-169.
2827. Hayes, R. E., and Roberts, A. R. V., J. Sci. Instr. 39, 1962, p-428.
2828. Hayman, P. J., from Conf. on Polarized Targets and Sources, Saclay, France, Dec. 5-9, 1966, pp-373-6.
2829. Haywood, F. F., Burson, Z. G., et al., ORNL-TM-3397, Contract W-7405-eng-26, June 1972, p-47.
2830. Haywood, F. F., Banta, H. E., et al., IEEE Trans. Nucl. Sci., NS-18(3), June 1971, pp-802-803.
2831. Haywood, F. F., Banta, H. E., et al., ORNL-4720, Oct. 1971, pp-112-15.
2832. Haywood, F. F., Ward, D. R., et al., ORNL-4584, July 31, 1970, pp-187-97.
2833. Heagney, J. M., and Heagney, J. S., INTDS Proc., Garching, FRG, Sept. 11-14, 1978.
2834. Heagney, J. M., and Heagney, J. S., INTDS Proc., Berkeley, U. S. A., Oct. 19-20, 1977.
2835. Heagney, J. M., and Heagney, J. S., LA-6850-C, June 1977, pp-92-99.
2836. Heagney, J. M., and Heagney, J. S., INTDS Proc., Argonne, U. S. A., Sept. 30-Oct. 2, 1975.
2837. Heagney, J. M., Report 1975, Univ. Wash., Seattle.
2838. Heagney, J. M., Nucl. Instr. and Meth. 102(3), 1972, pp-451-5.
2839. Heagney, J. M., 3rd Int. Symp. on Rsch. Mat. for Nucl. Meas., Gatlinburg, TN, U. S. A., 5-8 Oct. 1971, pp-101-7.
2840. Heagney, J. M., Washington Univ., Seattle, Oct. 1971, pp-101-7.
2841. Heaton, H. T., II, Grundl, J. A., et al., Natl. Bur. of Stands. (U. S.), Spec. Publ. 1, Oct. 1975, pp-266-69.
2842. Heaton, H. T., II, Gilliam, D. M., et al., ANL-76-90, 1976, pp-333-352.
2843. Heavens, O. S., Thin Film Phys. (London: Methuen), 1970.

2844. Heavens, O. S., J. Sci. Instr. 36, 1959, p-95.
2845. Hechtl, E., Nucl. Instr. and Meth. 139(1), 1976, pp-79-81.
2846. Hedemann, M. A., Nucl. Instr. and Meth. 141, 1977, pp-377-9.
2847. Heeringa, W., Rijksuniversiteit Groningen (Netherlands), Afdeling Experimentele Natuurkunde, 20 June 1977, p-93.
2848. Hefter, E. F., Boschitz, E. T., et al., Nucl. Phys., A, 275(1), 4 Jan. 1977, pp-212-228.
2849. Hegedues, F., Chakraborty, S., et al., Helv. Phys. Acta 50(2), 25 March 1977, p-194.
2850. Hegewisch, S., Technische Univ. Muenchen (Germany, F. R.), Fachbereich Physik. Diss. (D. Sc.), 22 March 1976, p-95.
2851. Heggie, J. C. P., and Switkowski, Z. E., Nucl. Instr. and Meth. 147(2), 1 Dec. 1977, pp-425-429.
2852. Hegland, P. M., Brown, R. E., et al., Phys. Rev. Lett. 39(1), 4 July 1977, pp-9-12.
2853. Heimbach, C. R., Lehman, D. R., et al., Phys. Rev., C, 16 (6), Dec. 1977, pp-2135-2150.
2854. Heimlich, F. H., Roessle, E., et al., Proc. of the Int. Symp. on Interaction Studies in Nuclei held in Mainz, F. R. Germany, 17-20 Feb. 1975, pp-439-442.
2855. Heimlich, F. H., Huber, G., et al., Nucl. Phys., A, 267(3), 30 Aug. 1976, pp-493-502.
2856. Heinemeier, J., Nucl. Instr. and Meth. 148(1), Jan. 1, 1978, pp-65-75.
2857. Heinrich, J. T., Nucl. Instr. and Meth. 31, Dec. 11, 1964, pp-337-8.
2858. Heits, B., Friederichs, H., et al., Phys. Rev., C, 15(5), May 1977, pp-1742-1757.
2859. Heller, K., Skubic, P., et al., Phys. Rev., D, 16(9), 1 Nov. 1977, pp-2737-2745.
2860. Hellmeister, H. P., Kaup, U., et al., Phys. Lett., B, ISSN 0370-2693, 85(1), 30 July 1979, pp-34-37.
2861. Helus, F., Maier-Borst, W., et al., 7th Int. Conf. on Cyc. and Their Applications, Zurich, Switzerland, 19-22 Aug. 1975,
2862. Helus, F., Radiochem. Radioanal. Letters 3, 1970, p-45.
2863. Helus, F., Sahm, U., et al., Radiochem. Radioanal. Lett. 39(1), 1979, pp-9-17.
2864. Hemstreet, J. M., Thesis, Baton Rouge, LA, Louisiana State Univ., 1963, p-57.
2865. Hemingway, J. W., Nucl. Instr. and Meth. 50, 1967, pp-61-70.
2866. Henderson, D. B., LA-4863, Contract W-7405-eng-36, June 1972, p-10.
2867. Henderson, T. M., Simms, R. J., et al., Adv. Cryog. Eng. 23, 1978, pp-682-9.
2868. Henderson, T. M., Solomon, D. E., et al., Laser Interact. Relat. Plasma Phenom. 4A, 1977, pp-305-16.
2869. Hendricks, C. D., UCRL-81588(Rev. 1), 8 May 1979, p-9.
2870. Hendricks, C. D., J. Nucl. Mater. 1979, pp-85-86(A).

2871. Hendricks, C. D., and Johnson, W. L., IEEE-IAS Conf. on "Electrostatics", Cleveland, OH, U. S. A., 1-4 Oct. 1979, p-14.
2872. Hendricks, C. D., U. S. Dept. of Energy, U. S. 4,190,016 (Cl. 118-724; G21B1/00, 26 Feb. 1980, Appl. 20,121, 13 March 1979, p-8.
2873. Hendricks, C. D., Inertial Confinement Fusion, Dig. Tech. Pap. Top. Meet. (1), ThE12, 1977, p-4.
2874. Hendricks, C. D., Report 1977, UCRL-80209, p-5.
2875. Hendry, A. W., Arndt, R. A., et al., ANL, Contract W-31-109-eng-38, Feb. 1978, p-38.
2876. Hendry, G. O., Proc. of the 6th Int. Cyc. Conf., Vancouver, Canada, July 18-21, 1972.
2877. Hendry, A. W., Arndt, R. A., et al., ANL, Report 1977.
2878. Henning, C. A. O., and Vermaak, J. S., Phil. Mag. 22, 1970, p-269.
2879. Henning, W., Eisen, Y., et al., Phys. Rev., C, 17(6), June 1978, pp-2245-2247.
2880. Henrich, E., and Wolf, K. G., Proc. of Conf. held at St. Catherine's Coll., Oxford, 22-23 Sept. 1969.
2881. Henry, E. A., Nucl. Data Sheets, 17(2), Feb. 1976, pp-287-328.
2882. Herbert, S., Ind. Diamond Rev., Feb. 1972, pp-54-7.
2883. Herbillon, G., Massin, J. P., et al., Sci. Appl., Coll. Publ. 74, 1978, pp-137-150.
2884. Hereward, H. G., Ranft, J., et al., CERN-65-1, Jan. 6, 1965, p-70.
2885. Hering, W. R., Puchta, H., et al., Phys. Rev., C, 14(4), Oct. 1976, pp-1451-1454.
2886. Herlach, F., McBroom, R., et al., IEEE Trans. Nucl. Sci., NS-18(3), June 1971, pp-809-13.
2887. Herman, M., Marcinkowski, A., et al., Nucl. Phys., A, 297(2), 13 March 1978, pp-335-346.
2888. Hermans, J. A. J., Engelbertink, G. A. P., et al., Europ. Physical Soc., Geneva (Switz.), Europ. Conf. on Nucl. Phys. with Heavy Ions, 6-10 Sept. 1976, p-181.
2889. Hermes, E. A., and Pruys, H. S., Nucl. Instr. and Meth. 113(3), 1973, pp-459-64.
2890. Hermsdorf, D., Kiessig, G., et al., Kernenergie 20(6), June 1977, pp-166-174.
2891. Herold, T. R., Contract AT(07-2)-1, Nucl. Instr. and Meth. 81, 1970, pp-49-55.
2892. Herr, H., Kadansky, V., Nucl. Instr. and Meth. 121(1), 1974, pp-1-3.
2893. Herrera, J. C., Kolata, J. J., et al., Phys. Rev. Lett. 40(3), 16 Jan. 1978, pp-158-161.
2894. Herschbach, K., Rev. Sci. Instr. 37, Feb. 1966, pp-171-2.

2895. Herscovitz, V. E., Maris, T. A. J., et al., AIP Conf. Proc. 33, 1976, pp-334-335.
2896. Heusch, B., Haas, F., et al., Strasbourg-1 Univ., 67 (France), Centre de Recherches Nucleaires, 1976, p-11.
2897. Heydegger, H. R., Turkevich, A. L., et al., Phys. Rev., C, 14(4), Oct. 1976, pp-1506-1514.
2898. Hickey, G. T., Zeller, A. F., et al., Aust. Inst. of Nucl. Sci. and Eng., Lucas Heights, 6th AINSE Nucl. Phys. Conf. 9-11 Feb. 1976, p-57.
2899. Hickey, G. T., Crawley, G. M., et al., ANU-P-642, July 1976, p-10.
2900. Hickman, R. S., Kenney, R. W., et al., Rev. Sci. Instr. 30, 1959, pp-983-5.
2901. Hicks, R. S., Auer, I. P., et al., Nucl. Phys., A, 278(2), 7 March 1977, pp-261-284.
2902. Hiebert, J. C., Graves, R. G., et al., Phys. Rev. Lett. 37(5), 2 Aug. 1976, pp-276-278.
2903. High, M. D., and Cujec, B., Nucl. Phys., A, 259(3), 22 March 1976, pp-513-522.
2904. High, M. D., Cujec, B., et al., Europ. Physical Soc. Geneva (Switz.), Europ. Conf. on Nucl. Phys. with Heavy Ions, 6-10 Sept. 1976, p-132.
2905. High, M. D., Cujec, B., et al., Europ. Physical Soc. Geneva (Switz.), Europ. Conf. on Nucl. Phys. with Heavy Ions, 6-10 Sept. 1976, p-133.
2906. High, M. D., and Cujec, B., Nucl. Phys., A, 278(1), 28 Feb. 1977, pp-149-162.
2907. High, M. D., and Cujec, B., Nucl. Phys., A, 282(1), 16 May 1977, pp-181-188.
2908. Highfill, R. R., and Wieland, W. B., IEEE Trans. on Nucl. Sci., NS-26(2), April 1979.
2909. Hildebrandt, D., Manns, R., et al., Radiat. Eff. ISSN 0033-7579, 33(4), Sept. 1977, pp-251-252.
2910. Hildebrandt, K. D., Lynen, U., et al., Verh. Dtsch. Phys. Ges. 6, 1977, p-996.
2911. Hill, D. A., Hasher, B. A., et al., Phys. Lett. 23, Oct. 3, 1966, pp-63-4.
2912. Hill, D. A., Ketterson, J. B., et al., Phys. Rev. Lett. 23, Sept. 1, 1969, pp-460-2.
2913. Hill, D., ANL/HEP-7034, from Symp. on Polarization at High Energy, Argonne, IL, 15 April 1970, pp-147-52.
2914. Hill, D., Masaike, A., et al., from Int. Conf. on Instrumentation for High Energy Physics, Dubna, USSR, 8 Sept. 1970, pp-595-9.
2915. Hill, D. A., ANL/HEP-7208(Vol. 3), 1971, pp-1084-93.
2916. Hill, D. A., Proc. - Summer Study High-Energy Phys. Polarized Beams, ANL/HEP 75-02, XV, 1974, p-23.
2917. Hill, D., AIP Conf. Proc. 35, 1976, pp-494-507.

2918. Hill, D., Moretti, A., et al., Report, 1976, ANL-HEP-TR-76-27, p-113.
2919. Hill, D., Miller, R. C., et al., Nucl. Instr. and Meth. 150(2), 1 April 1978, pp-331-2.
2920. Hill, J. C., Shirk, D. G., et al., Phys. Rev., C, 12(6), Dec. 1975, pp-1978-1982.
2921. Hill, J. J., and Hill, D. A., Nucl. Instr. and Meth. 116(2), 1974, pp-269-74.
2922. Hill, K. J., and Nelson, R. S., Nucl. Instr. and Meth. 38, 1965, p-15.
2923. Hille, M., Hille, P., et al., Nucl. Phys., A, 252(2), 10 Nov. 1975, pp-496-508.
2924. Hille, P., Rudolph, K., et al., Nucl. Phys., A, 266(1), 2 Aug. 1976, pp-253-261.
2925. Hillebrand, S., Puumalainen, P., et al., Nucl. Instr. and Meth. 111(3), 1973, pp-539-540.
2926. Hillier, M., Lomer, P. D., et al., Proc. Conf. Accel. Targets Des. Prod. Neutrons, 3rd, 1968, pp-125-45.
2927. Hillier, M., Lomer, P. D., et al., Proc. of the 3rd Conf. on Accel. Targets Des. for the Prod. of Neutrons, Liege (Belgium), Sept. 18-19, 1967.
2928. Hillis, D. L., Bingham, C. R., et al., Phys. Rev., C, 12(1), July 1975, pp-260-278.
2929. Hillis, D. L., Gross, E. E., et al., Phys. Rev. Lett. 36(6), 9 Feb. 1976, pp-304-306.
2930. Hillis, D. L., Tenn. Univ., Knoxville (U. S. A.), Thesis (Ph. D), 1976, Univ. Microfilms Order No. 76-17,725, p-282.
2931. Hilton, P. A., Cryogenics, ISSN 0011-2275, 17(9), Sept. 1977, pp-532-533.
2932. Hins, A. G., TISRMNM; ORNL, Oct. 5-8, 1971.
2933. Hinterberger, F., Rossen, P. V., et al., Nucl. Phys., A, 253(1), 17 Nov. 1975, pp-125-144.
2934. Hinterberger, F., Rossen, P. V., et al., Nucl. Phys., A, 259(3), 22 March 1976, pp-385-398.
2935. Hinterberger, F., Rossen, P. V., et al., Nucl. Phys., A, 263(3), 14 June 1976, pp-460-476.
2936. Hiramatsu, S., Isagawa, S., et al., Jpn. J. Appl. Phys. 19(1), 1980, pp-161-7.
2937. Hiramatsu, S., Isagawa, S., et al., Natl. Lab. High Energy Phys., Ohomachi, Japan, KEK-PREPRINT-78-8, 1978, p-16.
2938. Hiraoka, E., Furuta, J., et al., Annu. Rep. Radiat. Cent. Osaka Prefect, 13, 1972, pp-86-93.
2939. Hiraoka, E., Furuta, J., et al., Annu. Rep. Radiat. Cent. Osaka Prefect, 13, March 1973.
2940. Hirata, M., Phys. Rev. Lett. 40(11), 13 March 1978, pp-704-706.
2941. Hirata, M., Koch, J. H., et al., Ann. Phys. (NY), ISSN 0003-4916, 120(1), July 1979, pp-205-248.

2942. Hirayama, H., Nucl. Instr. and Meth. 147(3), 15 Dec. 1977, pp-563-569.
2943. Hirsch, P. B., Howie, A., et al., Electron Microscopy of Thin Crystals (London: Butterworth), Chap. 2, 1965.
2944. Hirsh, M. N., Eisner, P. N., et al., Rev. Sci. Instr. 39, Oct. 1968, pp-1547-55.
2945. Hirth, J. P., and Pound, G. M., "Condensation and Evaporation Progress in Mat. Sci.", MacMillan, NY, 1963.
2946. Hlavac, S., Kristiak, J., et al., Acta Phys. Slovaca 26(1), 1976, pp-64-67.
2947. Hockenbury, R. W., Knox, H. R., et al., Natl. Bur. of Stands. (U. S.), Spec. Publ. 425, Oct. 1975, pp-905-907.
2948. Hockenbury, R. W., Yip, W., et al., BNL-NCS-22500, March 1977, pp-242-245.
2949. Hoefert, M., Europ. Organ. for Nucl. Rsch., Geneva (Switz.), CONF-690540, pp-76-86.
2950. Hoenig, M. O., IEEE Trans. Nucl. Sci., NS-16, June 1969, pp-627-30.
2951. Hoenig, M. O., Sonini, H. E., et al., Cryogenics 9, Oct. 1969, pp-349-53.
2952. Hoffmann, D., Sobotta, G., et al., Nucl. Instr. and Meth. 118(2), 15 June 1974, pp-321-326.
2953. Hoffman, D. C., Wilhelmy, J. B., et al., LA-UR-77-2901, CONF-771225-2, 1977, p-35.
2954. Hoffman, G. W., Blanpied, G. S., et al., Phys. Rev. Lett. 40(19), 8 May 1978, pp-1256-1259.
2955. Hoffman, G. W., Blanpied, G. S., et al., Phys. Lett., B, ISSN 0031-9163, 76(4), 19 June 1978, pp-383-387.
2956. Hoffmann, H., Ber. Kernforschungsanlage Juelich (Conf.), 1980, Jul-Conf-34, pp-127-53.
2957. Hogan, W. J., Jankowicz, R., et al., IEEE Trans. Nucl. Sci., NS-12(4), Aug. 1965, pp-251-2.
2958. Hogerton, J. F., and Graff, R. C., Reactor Handbook - Mat., Tech. Info. Serv., U.S.A.E.C., declassified 1955, p-133.
2959. Hogstrom, K. R., Phys. Rev., C, 14(2), Aug. 1976, pp-753-54.
2960. Hoke, G. R., and Newman, E., U.S.A.E.C., ORNL-3021, 1961, p-7.
2961. Holland, L., and Ojha, S. M., Thin Solid Films 48, 1978, pp-L21-3.
2962. Holland, L., and Ojha, S. M., Thin Solid Films 40, 1977, pp-L31-2.
2963. Holland, L., J. Vac. Sci. Technol. 14(1), 1977, pp-5-15.
2964. Holland, L., and Ojha, S. M., Thin Solid Films 38, 1976, pp-L17-9.
2965. Holland, L., Thin Film Microelectronics (London: Chapman and Hall), 1965.

2966. Holland, L., "Vacuum Deposition of Thin Films", Chapman et Hall, London, 1963, p-146.
2967. Holland, L., "Vacuum Deposition of Thin Films", Chapman et Hall, London, 1961.
2968. Holland, L., "Vacuum Deposition of Thin Films", Wiley, NY, 1956.
2969. Hollister, N., Nucleonics 22(6), June 1964, pp-68-9.
2970. Holmgren, H. D., et al., Rev. Sci. Instr. 25, 1954, p-1026.
2971. Holmqvist, B., and Wiedling, T., Aktiebolaget Atomenergi, Stockholm, Dec. 1963, p-13.
2972. Holstein, B. R., Phys. Rev., D, 13(9), 1 May 1976, pp-2499-2501.
2973. Holt, R. J., Smith, A. B., et al., Natl. Bur. of Stands. (U. S.), Spec. Publ. 1, Oct. 1975, pp-246-249.
2974. Holt, R. J., and Jackson, H. E., Natl. Bur. of Stands. (U. S.), Spec. Publ. 425, Oct. 1975, pp-784-787.
2975. Holt, R. J., and Jackson, H. E., Phys. Rev. Lett. 36(5), 2 Feb. 1976, pp-244-248.
2976. Holt, R. J., Laszewski, R. M., et al., BNL-NCS-22500, March 1977, pp-35-36.
2977. Holt, R. J., Zeidman, B., et al., Phys. Lett., B, 69(1), 18 July 1977, pp-55-57.
2978. Holt, R. J., Jackson, H. E., et al., ANL-78-66, 1 April 1977-31 March 1978, pp-116-118.
2979. Holubarsch, W., and Goennenwein, F., Univ., Tuebingen, Ger., ORO-4856-26, 1975, pp-591-592.
2980. Honig, A., LBL-500, from 2nd Int. Conf. on Polarized Targets, Berkeley, CA, 30 Aug. 1971, pp-99-102.
2981. Honig, A., Phys. Rev. Lett. 19, Oct. 30, 1967, pp-1009-10.
2982. Honig, R. E., and Kramer, D. A., RCA Review 30(2), 1969.
2983. Honig, R. E., and Cramer, D. A., RCA Laboratories, Princeton, NJ, Fall 1968.
2984. Honjo, G., Shinozaki, S., et al., Appl. Phys. Lett. 9, 1966, p-23.
2985. Honzatko, J., and Kajfosz, J., Czech. J. Phys. 22(1), 1972, pp-38-41.
2986. Hoogenboom, A. M., Rev. Sci. Instr. 32, Dec. 1961, pp-1395-7.
2987. Hooper, H. R., Davidson, J. M., et al., Phys. Rev., C, 15(5), May 1977, pp-1665-1670.
2988. Hooton, B. W., Nucl. Instr. and Meth. 27, 1964, p-338.
2989. Hora, H., Kane, E. L., et al., Nucl. Instr. and Meth. 150(3), 1978, pp-589-92.
2990. Horen, D. J., and Harvey, J. A., BNL-NCS-22500, March 1977, p-205.
2991. Horen, D. J., Harvey, J. A., et al., Phys. Lett., B, 67(3), 11 April 1977, pp-268-270.
2992. Horen, D. J., Harvey, J. A., et al., Phys. Rev. Lett. 38 (23), 6 June 1977, pp-1344-1348.

2993. Horen, D. J., Harvey, J. A., et al., ORNL-5306, Sept. 1977, pp-112-113.
2994. Horen, D. J., Mueller, D., et al., ORNL-5306, Sept. 1977, p-114.
2995. Horlitz, G., Wolff, S., et al., Nucl. Instr. and Meth. 68, Feb. 15, 1969, pp-213-21.
2996. Horlock, P. L., Thakur, M. L., et al., Postgrad. Med. J. 51(601), Nov. 1975, pp-751-754.
2997. Horn, F. L., Report, 1972, BNL-16850, p-14.
2998. Horn, F. L., BNL-16850, from 20th Remote Systems Technol. Conf., Idaho Falls, ID, 17 Sept. 1972, p-14.
2999. Horn, S. J., Calif. Rsch. and Develop. Co., Livermore, CA, Feb. 18, 1954, Decl. Feb. 26, 1957, p-26.
3000. Hornstra, F., Jr., IEEE Trans. Nucl. Sci., NS-23(1), 1976, pp-198-201.
3001. Hornstra, F., Jr., Knott, M., et al., IEEE Trans. Nucl. Sci., NS-16, June 1969, p-823.
3002. Hornyak, W. F., Chang, C. C., et al., Univ. of Maryland, College Park, 1975, pp-354-355, ORO-4856-26.
3003. Hornyak, W. F., and Wall, N. S., ORO-4856-26, 1975, pp-356-357.
3004. Horowitz, Y. S., and Bell, R. E., Nucl. Instr. and Meth. 75, 1969, pp-5-12.
3005. Horsch, F., and Michaelis, W., Institut fuer Angewandte Kernphysik, May 1970, p-10.
3006. Hortig, G., Mokler, P., et al., Z. Phys. 210, 1968, pp-312-13.
3007. Hosek, J., and Truhlik, E., Institute of Nucl. Phys., 250 68 Rez near Prague, Czech.
3008. Hoshi, N., Hyuga, H., et al., Nucl. Phys., A, ISSN 0375-9474, 324(2-3), 30 July-6 Aug. 1979, pp-234-252.
3009. Hoshida, M., Miyahara, H., et al., Int. J. Appl. Radiat. & Isot. (GB), 28(7), July 1977, pp-633-40.
3010. Hoshino, T., and Arima, A., Phys. Rev. Lett. 37(5), 2 Aug. 1976, pp-266-269.
3011. Houde, A. L., Vacuum Microbalance Techniques 3 (NY: Plenum), 1963, p-109.
3012. Houghton, W. J., and Cohen, E. R., Contract AT-11-1-GEN-8, April 16, 1952, decl. with deletions, Feb. 26, 1957, p-95, North Am. Aviation, Inc., Downey, CA.
3013. Hourst, J. B., Roche, M., et al., Nucl. Instr. and Meth. 145(1), Aug. 15, 1977, pp-19-24.
3014. Houska, C. R., TISRMNM; ORNL, Oct. 5-8, 1971.
3015. Howard, J. E., Appl. Opt. 16(10), 1977, pp-2764-73.
3016. Howard, F. T., (ed.), ORNL-1795, Nov. 18, 1954, decl. Aug. 12, 1959, Contract W-7405-eng-26, p-29.
3017. Howe, F. A., PSPSITF; AERE, Oct. 20-21, 1965.
3018. Howe, F. A., AERE-R-5097, Paper 24, p-5.
3019. Howerton, R. J., and Luce, J. S., UCRL-16857, 4 Aug. 1975, p-27.

3020. Hruby, J. J., Proc. of a Conf. held at St. Catherine's Coll., Oxford, 22-23 Sept. 1969.
3021. Hruby, J. J., from Int. Conf. on the Use of Cyc. in Chem., Metallurgy, and Biol., Oxford, England (London; Butterworth and Co., Ltd.), 1970, pp-149-58.
3022. Hseuh, H., and Macias, E. S., Phys. Rev., C, 14(1), July 1976, pp-345-346.
3023. Hsu, T. H., Tseng, P. K., et al., Nucl. Instr. and Meth. 106(3), 1973, pp-513-17.
3024. Hsu, T. H., Lin, E. K., et al., Annual Report Inst. Phys., Acad. Sin., 1972, pp-25-33.
3025. Hua, C. Y., Jiang, X. Q., et al., INTDS Proc., Garching, FRG, Sept. 11-14, 1978.
3026. Huang, S. L., Hsu, C. C., et al., Nucl. Phys., A, 288(1), 19 Sept. 1977, pp-141-151.
3027. Hubbell, H. J., J. Appl. Phys. 30(7), July 1959, pp-981-4.
3028. Huber, P., and Schopper, H., Proc. of the 2nd Int. Symp. on Polarization Phenom. of Nucleons, Karlsruhe, Sept. 6-10, 1965, p-537.
3029. Hudis, J., and Katcoff, S., Phys. Rev., C, 13(5), May 1976, pp-1961-1965.
3030. Hudson, E. D., Lord, R. S., et al., Proc. of the 6th Int. Cyc. Conf., Vancouver, Canada, July 18-21, 1972.
3031. Hudson, G. M., Kemper, K. W., et al., Phys. Rev., C, 12(2), Aug. 1975, pp-474-480.
3032. Hudson, G. M., Florida State Univ., Tallahassee (U. S. A.), Thesis (Ph. D), Univ. Microfilms Order No. 77-13,317, 1976, p-170.
3033. Huebel, H., Kleinrahm, A., et al., Nucl. Phys., A, 294 (1-2), 2-9 Jan. 1978, pp-177-190.
3034. Huefner, J., Pirner, H. J., et al., Phys. Lett., B, 59(3), 10 Nov. 1975, pp-215-217.
3035. Huesken, H., Nucl. Phys., A, 291(1), 7 Nov. 1977, pp-206-220.
3036. Huizenga, J. R., Birkelund, J. R., et al., ANL/PHY-76-2 (Vol. 1), May 1976, pp-1-30.
3037. Huizenga, J. R., Birkelund, J. R., et al., Phys. Rev. Lett. 37(14), 4 Oct. 1976, pp-885-888.
3038. Huizenga, J. R., Schroeder, W. U., et al., Rochester Univ., NY (U. S. A.), COO-3496-63, 1977, p-20.
3039. Hulek, Z., Cesk. Casopis Fys. 14, 1964, pp-467-79.
3040. Hulet, E. K., Wild, J. F., et al., Phys. Rev. Lett. 26, 1971, p-523.
3041. Hulstman, L., Vis, R. D., et al., Nucl. Instr. and Meth. 120(3), 15 Sept. 1974, pp-537-538.
3042. Hulstman, L., Blok, H. P., et al., Nucl. Phys., A, 251(2), 20 Oct. 1975, pp-269-288.
3043. Humblet, J., Dyer, P., et al., Nucl. Phys., A, 271(1), 26 Oct. 1976, pp-210-220.

3044. Hussain, M., Rahman, F., et al., Nucl. Sci. Appl., Ser. B; 5, Oct. 1969, pp-45-6.
3045. Hussein, H. M., and Zohni, O., Nucl. Phys., A, 267(2), 23 Aug. 1976, pp-303-316.
3046. Hutcheon, D. A., and Gill, D. R., Nucl. Phys., A, 248(3), 1 Sept. 1975, pp-397-405.
3047. Hutcheon, R. M., Schriber, S. O., et al., Med. Phys. 6(3), 1979, pp-211-15.
3048. Huttlin, G. A., Aymar, J. A., et al., Nucl. Phys., A, 263 (3), 14 June 1976, pp-445-459.
3049. Huw-Shing, Tzeng, Jenn Ying, Liu, et al., Nucl. Instr. and Meth. 150(2), 1 April 1978, pp-143-4.
3050. Hwang, C. F., and Sanders, T. M., Jr., Pub. in Int. Symp. on Polarization, Helvetica Phys. Acta, Suppl. 6, 1960, pp-122-33.
3051. Hwang, C. F., and Sanders, T. M., Jr., Helv. Phys. Acta, Suppl. 6, 1961, pp-122-33.
3052. Hwang, C. F., Hasher, B. A., et al., Nucl. Instr. and Meth. 51, May 1967, pp-254-6.
3053. Hwang, C., BNL-20415, Dec. 1975, pp-341-343.
3054. Hwang, W. P., and Primakoff, H., Phys. Rev., C, 18(1), July 1978, pp-414-444.
3055. Hwang, W. P., and Primakoff, H., Phys. Rev., C, 18(1), July 1978, pp-445-461.
3056. Ialongo, G., Aerospace Corp., El Segundo, CA, Contract F04701-69-C-0066, 30 Dec. 1969, p-44.
3057. Ibanez, L., and Sanchez-Gomez, J. L., Z. Phys., A, 279(3), Nov. 1976, pp-339-343.
3058. Iferov, G. A., Pokhil, G. P., et al., Zh. Eksp. Teoret. Fiz., Pis'ma Redakt. 5, April 15, 1967, pp-250-3.
3059. Igo, G., Flynn, E. R., et al., Phys. Rev. C, 3, 1971, p-349.
3060. Ikossi, P. G., McDonald, A. M., et al., Nucl. Phys., A, 297(1), 6 March 1978, pp-1-9.
3061. Imai, K., Hatanaka, K., et al., Nucl. Phys., A, ISSN 0375-9474, 325(2-3), 20-27 Aug. 1979, pp-397-407.
3062. Imanishi, N., Fujiwara, I., et al., Nucl. Phys., A, 263(1), 31 May 1976, pp-141-149.
3063. Imasaki, K., Miyamoto, S., et al., J. Phys. Soc. Jpn. 40 (5), May 1976, pp-1529-1530.
3064. Imasaki, K., Miyamoto, S., et al., Phys. Rev. Lett. 43(26), 1979, pp-1937-40.
3065. Imazato, J., Z. Phys., A, 277(2), May 1976, pp-117-127.
3066. Imbusch, A., Rev. Sci. Instr. 38, July 1967, pp-974-5.
3067. Inamura, T., Kearns, F., et al., Nucl. Phys., A, 270(1), 12 Oct. 1976, pp-255-268.
3068. Inamura, T., Ishihara, M., et al., Phys. Lett., B, 68(1), 9 May 1977, pp-51-54.

3069. Indra, R. N., Kulkarni, B. A., Proc. Chem. Symp., 1st, 2, 1970, pp-121-8.
3070. Ingermarsson, A., and Fagerstroem, B., Phys. Scr. 13(4), April 1976, pp-208-212.
3071. Ippolitov, V. T., and Pozdnyakov, A. V., Conf. on Nucl. Spectro. and Nucl. Structure, Riga, USSR, 27-30 March 1979, p-572.
3072. Ipson, S. S., McLean, K. C., et al., Nucl. Phys., A, 253 (1), 17 Nov. 1975, pp-189-215.
3073. Irshad, M., Asai, J., et al., Nucl. Phys., A, 265(2), 19 July 1976, pp-349-364.
3074. Irvine, J. W., Jr., J. Phys. and Chem. 46, 1942, pp-910-14.
3075. Isagawa, S., Ishimoto, S., et al., Nucl. Instr. and Meth. 154(2), 1978, pp-213-18.
3076. Isbasescu, M., and Panaitescu, I., Rev. Roum. Phys. 13, 1968, pp-711-21.
3077. Ishihara, M., Numao, T., et al., ANL/PHY-76-2(Vol. 2), May 1976, pp-617-623.
3078. Ishimoto, S., Isagawa, S., et al., Nucl. Instr. and Meth. 171(2), 1980, pp-269-74.
3079. Ishimoto, S., Isagaa, S., et al., Natl. Lab. High Energy Phys., KEK (Jpn.), 1979, p-20.
3080. Ishkhanov, B. S., Kapitonov, I. M., et al., Nucl. Phys., A, 283(2), 13 June 1977, pp-307-325.
3081. Ismail, H. A., Hallock, J. N., et al., Phys. Rev., C, 12(2), Aug. 1975, pp-708-709.
3082. Issinskii, I. B., and Miznikov, K. P., JINR-UCRL-Trans-538, 1960, p-13.
3083. Istvan, U., Kozlem. 15(3), 1973, pp-215-216.
3084. Ivanenko, A. I., Report 1978, JINR-13-11667, p-13.
3085. Ivanov, K. N., and Petrzhak, K. A., Int. A.E.A., Vienna (Austria), Int. Nucl. Data Comm., Nucl. Phys. Rsch. in the USSR, Feb. 1977, p-19.
3086. Iverson, M. S., and Rost, E., Phys. Rev., C, 14(5), Nov. 1976, pp-1733-1738.
3087. Iverson, M. S., Colo. Univ., Boulder (U. S. A.), Thesis (Ph. D), 1976, p-125.
3088. Iversen, S., Obst, A., et al., Phys. Rev. Lett. 40(1), 2 Jan. 1978, pp-17-20.
3089. Iwasaki, Y., Phys. Rev. Lett. 36(24), 14 June 1976, pp-1464-1466.
3090. Iyengar, T. S., Soman, S. D., et al., from Tritium Symp., Las Vegas, NE, U. S. A., 30 Aug. 1971, pp-764-772.
3091. Jachcinski, C. M., Cormier, T. M., et al., Phys. Rev., C, 17(3), March 1978, pp-1263-1265.
3092. Jackson, D. F., and Rhoades-Brown, M., Nucl. Phys., A, 266 (1), 2 Aug. 1976, pp-61-71.
3093. Jackson, D. F., and Rhoades-Brown, M., Nucl. Phys., A, 286 (2), 15 Aug. 1977, pp-354-370.

3094. Jackson, H. E., Tabor, S. L., et al., Phys. Rev. Lett. 39 (25), 19 Dec. 1977, pp-1601-1604.
3095. Jackson, H. E., Meyer-Schuetzmeister, L., et al., ANL-77-60, 1977, pp-24-25.
3096. Jackson, H. E., Kaufman, S. B., et al., ANL-77-60, 1977, pp-25-26.
3097. Jackson, H. E., Kovar, D. G., et al., Phys. Rev. Lett. 35 (17), 27 Oct. 1975, pp-1170-1172.
3098. Jackson, H. E., Kaufman, S. B., et al., ANL-78-66, 1 April 1977-31 March 1978, pp-24-26.
3099. Jackson, J. D., Conf. on Polarized Targets and Sources, Saclay, France, Dec. 5-9, 1966, pp-3-26.
3100. Jackson, S. V., Walters, W. B., et al., Phys. Rev., C, 13 (2), Feb. 1976, pp-803-830.
3101. Jacob, M., Conf. on Polarized Targets and Sources, Saclay, France, Dec. 5-9, 1966, pp-235-59.
3102. Jacob, M., CERN-67-24(Vol. 4), Paper 3, p-15.
3103. Jacob, N. P., Jr., and Markowitz, S. S., Phys. Rev., C, 13(2), Feb. 1976, pp-754-766.
3104. Jacobs, E., Thierens, H., et al., Phys. Rev., C, 14(5), Nov. 1976, pp-1874-1877.
3105. Jacobs, J. R., Creten, L. W., et al., Nucl. Instr. and Meth. 61(2), May 1, 1968, pp-157-72.
3106. Jacobson, R. L., 27th Annual Conf. on Physical Electronics, Cambridge, MA, March 20-22, 1967, pp-26-8.
3107. Jaeckle, R., Pilkuhn, H., et al., Phys. Lett., B, 76(2), 22 May 1978, pp-177-181.
3108. Jahn, H., Broeders, C. H. M., et al., Natl. Bur. of Stands. (U. S.), Spec. Publ. 425(1), Oct. 1975, pp-350-353.
3109. Jahn, R., Stahel, D. P., et al., Phys. Lett., B, 65(4), 6 Dec. 1976, pp-339-342.
3110. Jahn, R., Stahel, D. P., et al., Phys. Rev., C, 18(1), July 1978, pp-9-22.
3111. Jahn, R., Wozniak, G. J., et al., Phys. Rev. Lett. 37(13), 27 Sept. 1976, pp-812-816.
3112. Jain, M., Texas Agricultural and Mechanical Univ., College Station (U. S. A.), LAMPF, ORO-4449-8, CONF-751223-3, 1975, p-5.
3113. Jaklovsky, J., INTDS Newsletter 1979.
3114. Jaklovsky, J., New England Nuclear, U. S. A., Internal publication, 1979.
3115. Jaklovsky, J., New England Nuclear, U. S. A., Internal Report, 1979.
3116. Jaklovsky, J., Technical Center, Aluminum Co. of America, U. S. A., Internal Reports, 1975.
3117. Jaklovsky, J., World Conf. of the Int. Nucl. Target Develop. Soc., Boston, U. S. A., Oct. 1-3, 1979.
3118. Jaklovsky, J., New England Nuclear, U. S. A., Internal publication, 1980.

3119. Jakobsson, B., Lindkvist, B., et al., Fysiska Institutionen, LUIP-7706, May 1977, p-3.
3120. Jakobson, M. J., Burleson, G. R., et al., Phys. Rev. Lett. 38(21), 23 May 1977, pp-1201-1204.
3121. James, D. R., and Fletcher, N. R., Phys. Rev., C, 17(6), June 1978, pp-2248-2252.
3122. James, D. R., Morgan, G. R., et al., Nucl. Phys., A, 274 (1-2), 7 Dec. 1976, pp-177-182.
3123. James, G. D., Dabbs, J. W. T., et al., Phys. Rev., C, 15(6), June 1977, pp-2083-2097.
3124. Jan, G. J., Pering, C. C., et al., Phys. Lett., B, ISSN 0370-2693, 85(1), 30 July 1979, pp-25-28.
3125. Janczyszyn, J., and Loska, L., Radiochem. Radioanal. Lett. 3, 24 April 1970, pp-343-8.
3126. Janes, G. S., "Techniques of High Energy Physics", David M. Ritson, ed., NY Interscience Publishers, Inc., 1961, pp-465-85.
3127. Jankowicz, R., Rev. Sci. Instr. 39(6), 1968, pp-848-53.
3128. Janssens, R., El Masri, Y., et al., Nucl. Phys., A, 282(3), 20 June 1977, pp-493-520.
3129. Jarczyk, L., Okolowicz, J., et al., Phys. Rev., C, 17(6), June 1978, pp-2106-2112.
3130. Jarczyk, L., Kamys, B., et al., AIP Conf. Proc. ISSN 0094-243X, 47, 1978, pp-680-681.
3131. Jarczyk, L., Kamys, B., et al., Nucl. Phys., A, ISSN 0375-9474, 325(2-3), 20-27 Aug. 1979, pp-510-524.
3132. Jarmie, N., Jett, J. H., et al., Report, LA-6611, 1977, p-24.
3133. Jarmie, N., Morrison, L. J., et al., Nucl. Instr. and Meth. 116(3), 1974, pp-451-452.
3134. Jarmie, N., Rev. Sci. Instr. 37, Dec. 1966, LA-DC-8106, pp-1670-1.
3135. Jarmie, N., Jett, J. H., et al., LA-6611, Feb. 1977, p-24.
3136. Jarmie, N., Phys. Rev., C, 16(5), Nov. 1977,
3137. Jarmie, and Jett, BNL-NCS-22500, March 1977, pp-124, 129.
3138. Jarrett, J. H., 3rd Int. Symp. on Rsch. Mat. for Nucl. Measurements, Gatlinburg, TN, 5 Oct. 1971, pp-147-59.
3139. Jarvis, O. N., and Shah, M., Nucl. Instr. and Meth. 93(1), 1971, pp-157-61.
3140. Jary, J., Lagrange, C., et al., CEA-N-2084, June 1979, p-98.
3141. Jastrzebski, J., Kaczarowski, R., et al., J. Phys. (Paris), 37(12), Dec. 1976, pp-1383-1385.
3142. Jauho, P., and Pirila, P., Phys. Status Solidi 42(2), 1970, pp-757-66.
3143. Javorsky, C. A., and Catlett, D. S., Microstruct. Sci. 5, 1977, pp-259-72.
3144. Jeffries, C. D., Univ. of Calif., Berkeley, 1972, pp-217-262.

3145. Jeffries, C. D., 3rd Int. Symp. on Polarization Phenomena in Nucl. Reactions, Madison, WI, Aug.-Sept. 1970, pp-351-72.

3146. Jeffries, C. D., Conf. on Polar. Targets and Sources, Saclay, France, Dec. 5-9, 1966, pp-147-67.

3147. Jeffries, C. D., 2nd Int. Symp. on Polar. Phenom. of Nucleons, Huber, P., and Schopper, H. (eds.), Basel and Stuttgart, Birkhaeuser Verlag, 1966, pp-105-12.

3148. Jenefsky, R. F., Joseph, C., et al., Nucl. Phys., A, 290 (2), 31 Oct. 1977, pp-407-434.

3149. Jenkins, I. L., and Wain, A. G., Int. Conf. on the Use of Cyc. in Chem., Metall., and Biol., Oxford, England, 1970, pp-168-82.

3150. Jenkins, T. M., and Nelson, W. R., Health Phys. 17, Aug. 1969, pp-305-12.

3151. Jenkins, T. M., and Nelson, W. R., Trans Amer. Nucl. Soc. 11, June 1968, pp-406-7.

3152. Jenny, B., Grueebler, W., et al., Verh. Dtsch. Phys. Ges. 6, 1977, p-822.

3153. Jensen, B., and Nelson, J. W., Nucl. Instr. and Meth. 93(1), 1971, pp-137-40.

3154. Jensen, C. M., Lanier, R. G., et al., Phys. Rev., C, 15(6), June 1977, pp-1972-1976.

3155. Jensen, M., Tilley, D. R., et al., Phys. Rev. Lett. ISSN 0031-9007, 43(9), 27 Aug. 1979, pp-609-611.

3156. Jeremie, H., Deguise, J. C., et al., Nucl. Instr. and Meth. 120(2), 1 Sept. 1974, pp-355-356.

3157. Jessen, P. L., Proc. Conf. Accel. Targets Des. Prod. Neutrons, 3rd, 1968, pp-147-63.

3158. Jessen, P. L., Kaman Nuclear, Colorado Springs, CO, KN-67-655(R), p-14.

3159. Jessen, P. L., Trans. Amer. Nucl. Soc. 10, Nov. 1967, p-452.

3160. Jessenberger, J., and Hink, W., Z. Phys., A, 275(4), Dec. 1975, pp-331-337.

3161. Jobst, J. E., Burson, Z. G., et al., 18th Annu. Amer. Nucl. Soc. Conf., Las Vegas, NE, 18 June 1972, p-12.

3162. Joehsson, C., Hoffmann, H., et al., Physik Kondensierten Materie 3, 1965, pp-193-9.

3163. Johansen, P. J., Siemens, P. J., et al., Volume 3, ISBN 0 444 85257 3, Amsterdam, Netherlands, North-Holland, 1979, pp-1129-1155.

3164. Johansson, K., and Norlin, L. O., Uppsala Univ. (Sweden), Fysiska Institutionen, April 1974, p-40.

3165. Johnsen, S. W., Brady, F. P., et al., Phys. Rev. Lett. 38(20), 16 May 1977, pp-1123-1125.

3166. Johnson, A. B., Jr., Kabele, T. J., et al., Report BNWL-2097, 1976, p-142.

3167. Johnson, A. B., Jr., Kabele, T. J., et al., Report BNWL-2097, 1977, p-10.

3168. Johnson, C. H., Galonsky, A., et al., Phys. Rev. Lett. 39 (25), 19 Dec. 1977, pp-1604-1607.
3169. Johnson, C. H., Bair, J. K., et al., Phys. Rev., C, 15(1), Jan. 1977, pp-196-216.
3170. Johnson, K. F., LA-6561-T, Oct. 1976, p-223.
3171. Johnson, M. B., Baer, H., et al., LA-7892C, July 1979, pp-343-351.
3172. Johnson, M. B., and Cooper, M. D., AIP Conf. Proc. 33, 1976, pp-276-277.
3173. Johnson, M. L., Romero, J. L., et al., Nucl. Instr. and Meth. 169(1), 1980, pp-179-84.
3174. Johnson, P. O., et al., Nucl. Instr. and Meth. 106, 1973, p-83.
3175. Johnson, R. G., Irish, J. D., et al., Can. J. Phys. 3(15), 1 Aug. 1975, pp-1434-1442.
3176. Johnson, R. H., Dorning, J. J., et al., Natl. Bur. of Stands. (U. S.), Spec. Publ. 425(Vol. 1), Oct. 1975, pp-169-172.
3177. Johnson, R. R., Phys. Quantum Electron. 3, 1976, pp-175-89.
3178. Johnson, R. R., Masterson, T. G., et al., Nucl. Phys., A, 296(3), 27 Feb. 1978, pp-444-460.
3179. Johnson, R. R., Masterson, T., et al., Phys. Rev. Lett. ISSN 0031-9007, 43(12), 17 Sept. 1979, pp-844-847.
3180. Johnson, W. L., Letts, S. A., et al., UCRL-83000, 19 Sept. 1979, p-4.
3181. Johnson, W. L., Letts, S. A., et al., Report 1979, UCRL-83000, p-4.
3182. Johnson, W. S., and Gibbons, J. F., Stanford Univ. Bookstore, 1969.
3183. Johnsson, B., Nilsson, M., et al., Lund Univ. (Sweden), Fysiska Institutionen, 19 Oct. 1976, p-18.
3184. Johnsson, B., Jaerund, A., et al., Z. Phys., A, 276(4), April 1976, p-410.
3185. Johnsson, B., Nilsson, M., et al., Nucl. Phys., A, 278(3), 14 March 1977, pp-365-371.
3186. Johnston, A., Aust. Inst. of Nucl. Sci. and Eng., Lucas Heights, 7th AINSE Nucl. Phys. Conf., 1978, p-47.
3187. Jolivette, P. L., Goss, J. D., et al., Phys. Rev., C, 13(1), Jan. 1976, p-439.
3188. Jolly, R. K., and White, H. B., Jr., Nucl. Instr. and Meth. 97(1), 1971, pp-103-5.
3189. Jolly, R. K., Kane, J. R., et al., Nucl. Instr. and Meth. 151(1-2), 1 May 1978, pp-183-188.
3190. Jolly, R. K., Kane, J. R., et al., College of William and Mary, Williamsburg, VA (U. S. A.), 1978.
3191. Joly, S., and Adam, A., CEA-N-1875, April 1976, pp-104-106.
3192. Jones, C. M., Johnson, J. W., et al., Nucl. Instr. and Meth. 68, Feb. 1, 1969, pp-77-87.
3193. Jones, L. H., and Swanson, B. I., J. Chem. and Phys. 63 (12), 15 Dec. 1975, pp-5401-5410.

3194. Jones, P. M. S., Edmondson, W., et al., J. Nucl. Mater. 23, Aug. 1967, pp-309-12.
3195. Jongsma, H. W., and Verheul, H., Nucl. Instr. and Meth. 72, 1969, pp-51-5.
3196. Jorgenson, T., LASL Report LADC-209, 1944.
3197. Jouanigot, J., Cabe, J., et al., N.I.M. Vol.47(1), 1967, p-105.
3198. Jovanovich, J. V., Smith, C. A., et al., Phys. Rev. Lett. 37(10), 6 Sept. 1976, pp-631-633.
3199. Joye, A. M. R., Australian Nat. Univ., Canberra, Thesis (Ph. D), May 1977, p-v.
3200. Joyeux, J., and Van Audenhove, J., TISRMNM, ORNL, Oct. 5-8, 1971.
3201. Jung, P., Viehweg, J., et al., Nucl. Instr. and Meth. ISSN 0029-554X, 154(2), 15 Aug. 1978, pp-207-212.
3202. Jung, P., Viehweg, J., et al., Nucl. Instr. and Meth. ISSN 0029-554X, 154(2), 15 Aug. 1978, pp-207-212.
3203. Jurney, E. T., and Motz, H. T., LA-DC-8335, Contract W-7405-eng-36, Nov. 2, 1966, p-14.
3204. Kabachenko, A. P., Kuznetsov, I. V., et al., Yad. Fiz. 25 (2), Feb. 1977, pp-249-254.
3205. Kabir, S. M., Nucl. Instr. and Meth. 109(3), 1973, pp-533-36.
3206. Kaellne, J., Anderson, A. N., et al., AIP Conf. Proc. ISSN 0094-243X, 47, 1978, pp-602-603.
3207. Kaellne, J., Stetz, A. W., et al., Phys. Lett., B, 74(3), 10 April 1978, pp-170-172.
3208. Kaellne, J., Thiessen, H. A., et al., Phys. Rev. Lett. 40 (6), 6 Feb. 1978, pp-378-381.
3209. Kaellne, J., Sundberg, O., et al., Nucl. Phys., A, 274 (3-4), 21 Dec. 1976, pp-509-518.
3210. Kaeppeler, F., Institut fuer Angewandte Kernphysik, April 1973, p-81.
3211. Kagawa, Y., Biochim. Biophys. Acta 131(3), 1967, pp-586-8.
3212. Kahane, S., Bar-Noy, T., et al., Nucl. Phys., A, 280(1), 11 April 1977, pp-180-188.
3213. Kailas, S., Saini, S. S., et al., Proc. of the Nucl. Phys. and Solid State Phys. Symp., Bombay, Dec. 27-31, 1974, Vol. 17B, pp-78-80.
3214. Kailas, S., Mehta, M. K., et al., Phys. Rev., C, ISSN 0556-2813, 20(4), Oct. 1979, pp-1272-1278.
3215. Kaipov, D. K., Kosyak, Yu. G., et al., Izv. Akad. Nauk SSSR, Ser. Fiz. ISSN 0367-6765, 43(1), Jan. 1979, pp-37-44.
3216. Kaiser, T., and Von Gunten, H. R., J. Inorg. and Nucl. Chem., ISSN 0022-1902, 40(3), 1978, pp-377-379.
3217. Kaiser, H. F., IEEE Trans. Nucl. Sci., NS-12(3), June 1965, pp-519-26.
3218. Kalbreier, W., Knezovic, A., et al., IEEE Trans. Nucl. Sci., NS-24(3), 1977, pp-1568-70.

3219. Kalebin, S. M., IAEA-186, 2, 1976, pp-121-133.
3220. Kalinsky, D., Melnik, D., et al., Weizmann Inst. of Sci., Rehovoth (Israel), 1977, p-9.
3221. Kalish, R., Borchers, R. R., et al., Nucl. Phys. A147, 1970, p-161.
3222. Kallipke, G., and Schmitt, L., Verh. Dtsch. Phys. Ges. 6, 1977, p-821.
3223. Kalpakchieva, R., Oganessian, Yu. Ts., et al., Phys. Lett., B, 69(3), 15 Aug. 1977, pp-287-289.
3224. Kambara, T., Takai, M., et al., J. Phys. Soc. Jpn. 44(3), March 1978, pp-704-711.
3225. Kamermans, R., Jongsma, H. W., et al., Nucl. Phys., A, 266 (2), 9 Aug. 1976, pp-346-364.
3226. Kamermans, R., Smits, J. W., et al., Nederlandse Natuurkundige Vereniging, Utrecht, Sectie Kernfysica, Autumn Meeting, Najaarsvergadering, 15 Oct. 1976, p-10.
3227. Kamermans, R., van Driel, J., et al., Phys. Rev., C, 17(5), May 1978, pp-1555-1558.
3228. Kamys, B., Bobrovska, A., et al., Izv. Akad. Nauk SSSR, Ser. Fiz. 39(1), Jan. 1975, pp-75-86.
3229. Kanada, H., Kaneko, T., et al., AIP Conf. Proc. ISSN 0094-243X, 47, 1978, pp-498-499.
3230. Kanda, Y., Nakamura, A., et al., Radioisotopes 28(12), 1979, pp-757-9.
3231. Kankowsky, R., Fritz, J. C., et al., Nucl. Phys., A, 263 (1), 31 May 1976, pp-29-46.
3232. Kantelo, M. V., and Hogan, J. J., Phys. Rev., C, 14(1), July 1976, pp-64-74.
3233. Kanter, H., Ann. Phys. 20, 1957, p-144.
3234. Kaplan, D. M., Fisk, R. J., et al., Phys. Rev. Lett. 40(7), 13 Feb. 1978, pp-435-438.
3235. Kaplan, M., Carnegie-Mellon Univ., Pittsburgh, PA (U. S. A.), July 1975, p-73.
3236. Kaptein, E. J., Blok, H. P., et al., Nucl. Phys., A, 260 (1), 29 March 1976, pp-141-162.
3237. Kapustsik, A., Madeya, M., et al., Leningrad. Nauka 1978, p-163.
3238. Karam, R. A., Stanford, G., et al., Annual Meeting of Amer. Nucl. Soc., Boston, MA, June 13-17, 1971.
3239. Karandashov, V. G., Roslyakov, V. I., et al., Prib. Tekh. Eksp. 5, 1979, pp-231-2.
3240. Karasawa, T., Seki, H., et al., Proc. of Conf. held at St. Catherine's Coll., Oxford, 22-23 Sept. 1969.
3241. Karasek, F. J., ANL, Illinois, AERE-R-5097, Paper 22, p-4.
3242. Karasek, F. J., Mat. Sci. Div., ANL, Argonne, IL, 1980.
3243. Karasek, F. J., INTDS Newsletter, July, 1979.
3244. Karasek, F. J., INTDS Proc., Boston, U. S. A., Oct. 1-3, 1979.

3245. Karasek, F. J., INTDS Proc., Garching, FRG, Sept. 11-14, 1978.
3246. Karasek, F. J., Nucl. Instr. and Meth. 167(1), 1979, pp-165-6.
3247. Karasek, F. J., Nucl. Instr. and Meth. 102, 1972, pp-457-8.
3248. Karasek, F. J., 3rd Int. Symp. on Rsch. Mat. for Nucl. Measurements, Gatlinburg, TN, 5 Oct. 1971, p-4.
3249. Karasek, F. J., PSPSITF, AERE, Oct. 20-21, 1965.
3250. Karasek, F. J., Nucl. Sci. and Eng. 17(3), 1963, pp-15-19.
3251. Karatzas, P. T., Couchell, G. P., et al., Nucl. Sci. and Eng. 67(1), July 1978, pp-34-53.
3252. Karban, O., Basak, A. K., et al., Nucl. Phys., A, 269(2), 5 Oct. 1976, pp-312-326.
3253. Karban, O., Burcham, W. E., et al., Nucl. Phys., A, 292 (1-2), 21-28 Nov. 1977, pp-1-19.
3254. Karban, O., Basak, A. K., et al., Nucl. Phys., A, 266(2), 9 Aug. 1976, pp-413-423.
3255. Kardontchik, A., and Kalish, R., Tech.-Israel, Haifa, 1974.
3256. Kardontchik, A., and Kalish, R., Nucl. Instr. and Meth. 131(3),, 28 Dec. 1975, pp-399-402.
3257. Karev, V. N., Klyucharev, A. P., et al., Fizika Metallicheskikh Plenok, Kiev, Naukova Dumka, 1969, pp-92-6.
3258. Karev, V. N., and Medyanik, V. N., Fizika Metallicheskikh Plenok, Kiev, Naukova Dumka, 1969, pp-97-108.
3259. Karev, V. N., Klyucharev, A. P., et al., Fizika Metallicheskikh Plenok, Kiev, Naukova Dumka, 1969, pp-87-91.
3260. Karev, V. N., and Klyucharev, A. P., Ukr. Fiz. Zh. 10, Sept. 1965, pp-907-10.
3261. Kari, K., and Cierjacks, S., Verh. Dtsch. Phys. Ges. 6, 1977, p-970.
3262. Kari, K., and Cierjacks, S., Inst. fuer Angewandte Kernphysik, Annual Report, Oct. 1979 p-27.
3263. Karim, M., and Overley, J. C., Natl. Bur. of Stands. (U. S.), Spec. Publ. 425, Oct. 1975, pp-788-791.
3264. Karl, G., and Tadic, D., Phys. Rev., C, ISSN 0556-2813, 20(5), Nov. 1979, pp-1959-1961.
3265. Karmohapatro, S. B., Conf. on the Phys. of Electromagnetic Separation Methods, Orsay, France, July 1962, p-5.
3266. Karol, P. J., Phys. Rev. Lett. 36(6), 9 Feb. 1976, pp-338-339.
3267. Karolyi, J., and Szalay, A., Nucl. Instr. and Meth. 36, Oct. 1965, p-353.
3268. Karzmark, C. J., and Pering, N. C., Phys. Med. Biol. 18 (3), May 1973, pp-321-354.
3269. Kas, J., and Novak, D., Nucl. Instr. and Meth. 99(1), 1972, pp-359-63.
3270. Kashy, E., Perry, R. R., et al., Nucl. Instr. and Meth. 4, 1959, pp-167-70.

3271. Kaspaul, A. F., and Kaspaul, E. E., Trans. 10th Nat. Vac. Symp., Macmillan, NY, 1963.
3272. Kato, S., Kubono, S., et al., Phys. Lett., B, 62(2), 24 May 1976, pp-153-154.
3273. Kato, S., Nucl. Instr. and Meth. 75, 1969, pp-293-6.
3274. Katori, K., Ishigami, T., et al., ANL/PHY-76-2(2), May 1976, pp-625-635.
3275. Katori, K., Ooi, T., et al., AIP Conf. Proc. ISSN 0094-243X, 47, 1978, pp-592-593.
3276. Katori, K., Furuno, K., et al., Phys. Rev. Lett. 40(23), 5 June 1978, pp-1489-1493.
3277. Katsaurov, L. N., Kratk. Soobshch. Fiz. 10, Oct. 1971, pp-56-63.
3278. Katsaurov, L. N., Pribory i Tekhnika Eksp. (USSR), 4, March 1971, pp-35-36.
3279. Katz, R., and Lee, R., Rev. Sci. Instr. 25, 1956, pp-58-62.
3280. Katzer, M. F., Calif. Rsch. and Develop. Co., Livermore, CA, LWS-24589, CRD-T2B-98, Decl. April 5, 1957, p-13.
3281. Kaufman, S. B., Weisfield, M. W., et al., Phys. Rev., C, 13(1), Jan. 1976, pp-253-256.
3282. Kawakami, H., Koike, M., et al., Nucl. Phys., A, 262(1), 10 May 1976, pp-52-60.
3283. Kay, E., Adv. Electron Phys. 17, 1962, pp-245-322.
3284. Kayser, J., Kaetzmer, O., et al., Kernenergie, ISSN 0023-0642, 21(9), Sept. 1978, pp-300-302.
3285. Kazanskii, G. S., Mikhailov, A. I., et al., Instr. and Exp. Tech. 5, Sept.-Oct. 1962, pp-887-92.
3286. Kazanskii, G. S., Mikhailov, A. I., et al., JINR-P-849, 1961, p-17.
3287. Kazanskii, G. S., Mikhailov, A. I., et al., JINR-P-849, 1961, p-41.
3288. Keeling, M. J., Nucl. Instr. and Meth. 119(2), 15 July 1974, p-401.
3289. Keinonen, J., Rascher, R., et al., Phys. Rev., C, 14(1), July 1976, pp-160-170.
3290. Keinonen, J., and Anttila, A., Nucl. Instr. and Meth. 160 (2), 1979, pp-211-15.
3291. Kekelis, G. J., Florida State Univ., Tallahassee (U. S. A.), Thesis (Ph. D), 1975, p-246.
3292. Kekelis, G., Zisman, M. S., et al., LBL-8151, 1978, pp-12-13.
3293. Keller, N. K., Lee, J. K. P., et al., Can. J. Phys. 47, 15 March 1969, pp-611-15.
3294. Keller, R., and Mueller, H. H., Nucl. Instr. and Meth. 119(2), 15 July 1974, pp-321-322.
3295. Kelley, G. M., and Dropesky, B. J., LASL, LA-6850-C, 1977, pp-38-47.
3296. Kelley, G. M., and Dropesky, B. J., INTDS Proc., Los Alamos, U. S. A., Oct. 19-21, 1976.

3297. Kelly, G. M., and Dropesky, B. J., INTDS Proc., Argonne, U. S. A., Sept. 30-Oct. 2, 1975.
3298. Kellner, E., and Maier-Komor, P., Universitaet Muenchen, Aug. 1978.
3299. Kellner, E., and Maier-Komor, P., INTDS Proc., Berkeley, Oct. 19-20, 1977.
3300. Kelsey, C. A., Chenevert, G. M., et al., Eur. J. Cancer 10 (4), 1974, pp-257-8.
3301. Kelsey, C. A., DeLuca, P. M., et al., Symp. Neutron Dosim. Biol. Med., 2nd, (Pt. 2), 1974, pp-971-7.
3302. Kelsey, C. A., Spalek, G. C., et al., 1st Symp. on Neutron Dosim. in Biol. and Med., Neuherberg/Munich, Germany, 15 May 1972.
3303. Kelsey, C. A., Boone, M. L. M., et al., Meeting on Neutron Sources and Applications, Augusta, Ga., 18 April 1971.
3304. Kelsey, C. A., Boone, M. L. M., et al., Radiol. 98, March 1971, pp-686-8.
3305. Kelly, J. C., J. Sci. Instr. 36, 1959, p-89.
3306. Kemnitz, P., Funke, L., et al., Zentralinstitut fuer Kernforschung, Rossendorf bei Dresden, Annual Report 1976, pp-50-52.
3307. Kemp, K., and Jensen, F. P., Nucl. Instr. and Meth. 142 (1-2), 1-15 April 1977, pp-101-103.
3308. Kemper, A., Nederlandse Organisatie voor Toegepast Natuurwetenschappelijk Onderzoek, Delft, Centraal Laboratorium, 1967, p-43.
3309. Kemper, K. W., Zeller, A. F., et al., Aust. Nat. Univ., Canberra, Dec. 1977, p-10.
3310. Kemper, K. W., Moore, G. E., et al., Phys. Rev., C, 15(5), May 1977, pp-1726-1731.
3311. Kemper, K. W., Haynes, D. S., Nucl. Instr. and Meth. 88, 1970, pp-289-93.
3312. Kendall, S., Need, J. L., et al., INTDS Proc., Boston, U. S. A., Oct. 1-3, 1979.
3313. Kenna, B. T., and Conrad, F. J., Health Phys. 12, April 1966, pp-564-6.
3314. Kennedy, D. L., Linnard, B. J., et al., Aust. Inst. of Nucl. Sci. and Eng., Lucas Heights, 7th AINSE Nucl. Phys. Conf., 1978, p-34.
3315. Kenny, M. J., Allen, B. J., et al., Aust. J. Phys. 30(5), Nov. 1977, pp-605-616.
3316. Kenny, M. J., and Allen, B. J., Aust. J. Phys. 30(5), Nov. 1977, pp-591-597.
3317. Kenny, M. J., Allen, B. J., et al., Aust. Atomic Energy Comm. Rsch. Estab., Lucas Heights, Jan. 1977, p-24.
3318. Kenny, M. J., Allen, B. J., et al., Nucl. Phys., A, 270(1), 12 Oct. 1976, pp-164-174.
3319. Kerekatte, S. S., Sekharan, K. K., et al., Indian J. Pure Appl. Phys. 5, March 1967, pp-107-8.

3320. Kernohan, A., Drake, T. E., et al., Phys. Rev., C, 16(1), July 1977, pp-239-242.
3321. Kessel, Q. C., Rev. Sci. Instr. 40, Jan. 1969, pp-68-70.
3322. Kessler, G., Adv. in Cryo. Eng. 15, Proc. of Conf., Univ. of Calif., Los Angeles, June 16-18, 1969, pp-443-6.
3323. Kessler, G., DESY-67/44, Deutsches Elektronon-Synchrotron, Hamburg, Nov. 1967, p-17.
3324. Ketel, T. J., Vervaet, E. A. Z. M., et al., Nederlandse Natuurkundige Vereniging, Utrecht, Sectie Kernfysica, Autumn Meeting, 15 Oct. 1976, p-9.
3325. Kettner, K. U., Lorenz-Wirzba, H., et al., Verh. Dtsch. Phys. Ges. 6, 1977, p-999.
3326. Kettner, K. U., Lorentz-Wirzba, H., et al., Europ. Physical Soc., Geneva (Switz.), Europ. Conf. on Nucl. Phys. with Heavy Ions, 6-10 Sept. 1976, p-131.
3327. Key, J. F., Idaho Nat. Eng. Lab., Idaho Falls, U. S. A., 1979, p-8.
3328. Keyworth, G. A., LA-UR-76-1418, CONF-760715-1, 1976, p-17.
3329. Keyworth, G. A., Moore, M. S., et al., LA-UR-76-1318, CONF-760647-5, 1976, p-28.
3330. Khalin, N. F., Golognya, V. Ya., et al., Prib. Tekh. Eksp. 16(5), 1973, pp-29-31.
3331. Khalin, N. F., Golovnya, V. Ya., et al., Prib. Tekh. Eksp. 14(2), March-April 1971, pp-38-40.
3332. Khandelwal, G. S., Pritchard, W. M., et al., Nucl. Sci. and Eng. 60(4), Aug. 1976, pp-481-486.
3333. Kharchenko, V. F., and Levashev, V. P., Phys. Lett., B, 60(4), 2 Feb. 1976, pp-317-320.
3334. Khodyachikh, A. F., Arkatov, Yu. M., et al., Summaries of Reports of 26th Conf. on Nucl. Spectroscopy and Nucl. Structure, Baku, 3-6 Feb. 1976, p-446.
3335. Khokhlov, Uy. K., Sov. At. Energy, U. S. A., 37(3), Sept. 1974, pp-961-2.
3336. Khomyakov, G. K., Matyash, P. P., et al., Instr. Exp. Tech. 2, March-April 1970, pp-361-3.
3337. Khozyainov, M. S., and Seredin, Yu. V., Moscow, Vsesoyuznii Nauchno-Issledovatel'skii Inst. Yaderno Geofizikii i Geokhimii, 1973, pp-139-153.
3338. Khurshudian, L. S., and Aganiants, A. O., Akademiya Nauk Armyanskoi SSR, Erevan, Institut Fiziki, 1970, p-11.
3339. Khurshudyan, L. S., and Agan'yants, A. O., Inst. Exp. Tech. 14(4), July-Aug. 1971, pp-1010-12.
3340. Khvastunov, V. M., Verezovoj, V. P., et al., Yad. Fiz. 25 (5), May 1977, pp-921-925.
3341. Kidder, R. E., UCRL-83342, CONF-791109-1, Sept. 1979, p-18.
3342. Kiebele, U., Baumgartner, E., et al., Helv. Phys. Acta, ISSN 0018-0238, 51(5-6), 31 July 1979, pp-726-742.
3343. Kienle, F., Hamburg Univ. (Germany, F. R.), Fachbereich Physik. Diss. (D. Sc.), 1975, p-63.

3344. Kikochi, T., Konishi, E., et al., Hoken Botsori 14(3), 1979, pp-171-5.
3345. Kilian, K., AIP Conf. Proc. 33, 1976, pp-497-506.
3346. Kim, B. T., Greiner, A., et al., Phys. Rev., C, ISSN 0556-2813, 20(4), Oct. 1979, pp-1396-1407.
3347. Kim, J. C., Depommier, P., et al., Nucl. Instr. and Meth. 143(2), 1 June 1977, pp-371-377.
3348. Kim, Jung-Do, Japan Atomic Energy Rsch. Inst., Tokyo, March 1976, p-18.
3349. Kim, J., Davis, R. C., et al., Inst. of Electrical and Electronics Eng., Inc., 1977, pp-1593-1594.
3350. Kim, J., and Morgan, O. B., ANL/CTR, 75, (ANL/CTR-75-4, Proc. Int. Conf. Radiat. Test Facil. CTR Surf. Mater. Program, 1975), pp-250-8.
3351. Kim, K., Smoot, B. J., et al., Appl. Phys. Lett. 34(4), 1979, pp-282-3.
3352. Kimura, I., Hayashi, S. A., et al., J. Nucl. Sci. Technol. 8(3), March 1971, pp-173-5.
3353. Kindel, J. M., and Lindman, E. L., Nucl. Fusion 19(5), 1979, pp-597-606.
3354. King, A. G., DiNaro, S., et al., Trans. Am. Nucl. Soc. 34, 1980, p-181.
3355. King, C. H., Shahabuddin, M. A. M., et al., Nucl. Phys., A, 270(2), 19 Oct. 1976, pp-399-412.
3356. King, H. T., and Slater, D. C., Nucl. Phys., A, 283(3), 20 June 1977, pp-365-393.
3357. King, L. J., Bigelow, J. E., et al., ORNL, Oak Ridge, TN, U. S. A., Report 1979, CONF-790949-1, p-10.
3358. King, L. J., Bigelow, J. E., et al., ORNL-5415, Contract W-7405-eng-26, Aug. 1978, p-30.
3359. King, L. J., and Collins, E. D., ORNL-TM-2434, Contract W-7405-eng-26, Dec. 1969, p-43.
3360. King, N. M., and Wilson, E. J. N., RHEL-2000, pp-183-9.
3361. King, N. S. P., Romero, J. L., et al., Phys. Lett., B, 69 (2), 1 Aug. 1977, pp-151-153.
3362. King, T. R., Proc. of a Conf. held at St. Catherine's Coll., Oxford, 22-23 Sept. 1969.
3363. Kinzer, J. E., Atomics Int., Canoga Park, CA, CONF-660511, pp-3.8.1-21.
3364. Kirchner, R., Klepper, O., et al., Verh. Dtsch. Phys. Ges. 6, 1977, pp-886-887.
3365. Kirk, J., J. Sci. Instr. 38, Nov. 1961, pp-439-41.
3366. Kirkby, P., and Link, W. T., Can. J. Phys. 44, 1966.
3367. Kirouac, G. J., and Eiland, H. M., Natl. Bur. of Stands. (U. S.), Spec. Publ. 425, Oct. 1975, pp-776-779.
3368. Kirsten, T. A., and Schaeffer, O. A., State Univ. of New York, Stony Brook, Contract AT(30-1)3629, pp-75-157.
3369. Kishikawa, T., and Shinomiya, C., Radiochem. Radioanal. Lett. 7(1), 15 June 1971, pp-15-21.

3370. Kiss, A., Koltay, E., et al., Nucl. Phys., A, 282(1), 16 May 1977, pp-44-52.
3371. Kiss, A., Aspelund, O., et al., Nucl. Phys., A, 262(1), 10 May 1976, pp-1-18.
3372. Kissener, H. R., and Eramzhyan, R. A., AIP Conf. Proc. ISSN 0094-243X, 54(1), 15 July 1979, pp-424-425.
3373. Kisslinger, L. S., and Wu, C. S., AIP Conf. Proc. 33, 1976, pp-198-201.
3374. Kisslinger, L. S., and Saharia, A., AIP Conf. Proc. 33, 1976, pp-184-185.
3375. Kisslinger, L. S., and Wang, W. L., Ann. Phys. (N. Y.), 99(2), Aug. 1976, pp-374-407.
3376. Kistner, O. C., Sunyar, A. W., et al., Phys. Rev., C, 17 (4), April 1978, pp-1417-1427.
3377. Kitching, P., AIP Conf. Proc. ISSN 0094-243X, 47, 1978, pp-638-639.
3378. Kitching, P., Miller, C. A., et al., AIP Conf. Proc. 36, 1977, CONF-760567, pp-182-186.
3379. Kivikas, T., Lund Univ. (Sweden), LUSY-7101, Nov. 1971, p-36.
3380. Kivikas, T., Lund Univ. (Sweden), LUSY-7007, Sept. 1970, p-75.
3381. Kivits, H. P. M., DeRooij, F. A. J., et al., Nucl. Instr. and Meth. 164, 1979, pp-225-229.
3382. Klapdor, H. V., Reiss, H., et al., Max-Planck-Institut fuer Kernphysik, Heidelburg, ORO-4856-26, 1975, pp-425-426.
3383. Klapdor, H. V., Reiss, H., et al., Nucl. Phys., A, 262(1), 10 May 1976, pp-157-188.
3384. Kleinheinz, P., Stefanini, A. M., et al., Nucl. Phys., A, 283(2), 13 June 1977, pp-189-222.
3385. Kline, F. J., and Hayward, E., Phys. Rev., C, 17(5), May 1978, pp-1531-1534.
3386. Klingenbeck, K., Phys. Lett., B, ISSN 0370-2693, 85(1), 30 July 1979, pp-21-24.
3387. Klinger, W., and Hauer, N., Madison, WI, Univ. of Wisconsin Press, Polarization Phenomena in Nucl. Reactions, 1971.
3388. Kluge, W., Matthaey, H., et al., Nucl. Phys., A, ISSN 0029-5582, 302(1), 12 June 1978, pp-93-124.
3389. Klyuchnikov, A. A., Kupryashkin, V. T., et al., Summaries of Reports of 26th Conf. on Nucl. Spectroscopy and Nucl. Structure, Baku, 3-6 Feb. 1976, p-127.
3390. Kneissl, U., Leister, K. H., et al., Nucl. Phys., A, 264 (1), 21 June 1976, pp-30-44.
3391. Kneissl, U., Leister, K. H., et al., Nucl. Phys., A, 272 (1), 9 Nov. 1976, pp-125-132.

3392. Kniedler, M. J., and Silverman, J., Symp. on Utilization of Large Radiation Sources and Accelerators in Ind. Processing, Munich, Germany, 1968, p-27.
3393. Knighton, J. B., and Steunenberg, R. K., ANL-7057, Part I, 1965.
3394. Knitter, H. H., IAEA-208(Vol. 1), 1978, pp-183-195.
3395. Knoepfle, K. T., Mairle, G., et al., Kneis, W. (ed.), Kernforschung, Karlsruhe, Annual Report KFK-2868 Oct. 1979, pp-43-45.
3396. Knoepfle, K. T., Wagner, G. J., et al., Phys. Lett., B, 74(3), 10 April 1978, pp-191-194.
3397. Knoll, P., Zentralinstitut fuer Kernphysik, Rossendorf bei Dresden, ZFK-RCH-1, ORNL-tr-2400, p-86.
3398. Knox, H. D., White, R. M., et al., Ohio Univ., Athens (U. S. A.), March 1977, p-28.
3399. Knox, W. J., Phys. Rev. 81, 1951, pp-693-7.
3400. Knudsen, J. G., and Katz, D., Fluid Dynamics and Heat Transfer, McGraw-Hill, New York, 1958.
3401. Knuepfer, W., and Huber, M. G., Proc. of the June Workshop in Intermediate Energy Electromagnetic Interactions with Nuclei, held at MIT, June 13-24, 1977, pp-368-71.
3402. Knuepfer, W., and Huber, M. G., Proc. of the June Workshop in Intermediate Energy Electromagnetic Interactions with Nuclei, held at MIT, June 13-24, 1977, pp-363-67.
3403. Knutson, L. D., Hichwa, B. P., et al., Phys. Rev. Lett. 35(23), 8 Dec. 1975, pp-1570-1573.
3404. Knutson, L. D., and Haeberli, W., Phys. Rev., C, 12(5), Nov. 1975, pp-1469-1477.
3405. Knutson, L. D., Ann. Phys. (N. Y.), 106(1), July 1977, pp-1-25.
3406. Knutson, L. D., Colby, P. C., et al., Phys. Lett., B, ISSN 0370-2693, 85(2-3), 13 Aug. 1979. pp-209-211.
3407. Koang, D. H., et al., Inst. des Sciences Nucleaires, ISN-75-45, 1975, p-24.
3408. Kobayashi, S., Nisimura, K., et al., Tokyo Univ., Inst. for Nuclear Study, Jan. 22, 1962, p-7.
3409. Kobisk, E. H., and Adair, H. L., Nucl. Instr. and Meth. 167(1), 1979, pp-153-60.
3410. Kobisk, E. H., INTDS Proc., Boston, U. S. A., Oct. 1-3, 1979.
3411. Kobisk, E. H., Quinby, T. C., et al., INTDS Proc., Boston, U. S. A., Oct. 1-3, 1979.
3412. Kobisk, E. H., and Adair, H. L., INTDS Proc., Garching, FRG., Sept. 11-14, 1978.
3413. Kobisk, E. H., ORNL Report 1978, CONF-780943-2
3414. Kobisk, E. H., ORNL Proc. Conf. Appl. Small Accel., 3rd, 1975, p-2.
3415. Kobisk, E. H., 3rd Conf. on the use of Small Accel., Denton, TX, U. S. A., 21 Oct. 1974, p-23.

3416. Kobisk, E. H., Nucl. Instr. and Meth. 102, 1972, pp-1-610.
3417. Kobisk, E. H., TISRMNM, Gatlinburg, TN, Oct. 5-8, 1971.
3418. Kobisk, E. H., and Grisham, W. B., Mater. Res. Bull. 4, 1969, p-651.
3419. Kobisk, E. H., INTDS, Munchen-Garching, F. R. Germany, 11-14 Sept. 1978.
3420. Kobisk, E. H., Isotop. Radiat. Technol. 7, Fall 1969, pp-1-19.
3421. Kobisk, E. H., ORNL-4308, Contract W-7405-eng-26, Oct. 1968, p-37.
3422. Kobisk, E. H., Nucleonics 24(8), Aug. 1966, pp-122-4.
3423. Kobisk, E. H., ORNL-P-1900, Contract W-7405-eng-26, Dec. 28, 1965, p-10.
3424. Kobisk, E. H., PSPSITF, AERE, Oct. 20 & 21, 1965.
3425. Kobisk, E. H., ORNL, AEC Access. Nos. 65 (AERE-R-5097, Paper 26), p-3.
3426. Kobisk, E. H., ORNL-3829, Contract W-7405-eng-26, Sept. 1965, p-40.
3427. Kobisk, E. H., Trans. Am. Nucl. Soc. 7, Nov. 1964, pp-347-8.
3428. Kobisk, E. H., ORNL-TM-718, Contract W-7405-eng-26, Jan. 16, 1964, p-14.
3429. Kobisk, E. H., ORNL-3400, Contract W-7405-eng-26, Jan. 15, 1963, p-14.
3430. Kobisk, E. H., ORNL-TM-1047, pp-109-12.
3431. Kobisk, E. H., ORNL, AERE-R-5097, Paper 26, p-3.
3432. Koch, J. H., and Woloshyn, R. M., AIP Conf. Proc. 33, 1976, pp-610-611.
3433. Koch, R., and Thies, H. H., Nucl. Phys., A, 272(2), 16 Nov. 1976, pp-296-302.
3434. Koch, S., and Jugelt, P., Isotopenpraxis 2, June 1966, pp-251-2.
3435. Kocher, D. C., and Haeberli, Nucl. Phys., A, 252(2), 10 Nov. 1975, pp-381-415.
3436. Kocher, D. C., Bertrand, F. E., et al., Phys. Rev., C, 14(4), Oct. 1976, pp-1392-1411.
3437. Kochevanov, V. A., Pribory i Tekhn. Eksperim. 2, March-April 1965, pp-26-30.
3438. Koene, B. K. S., and Chrien, R. E., Phys. Rev., C, 16(2), Aug. 1977, pp-588-596.
3439. Koene, B. K. S., Epstein, M. B., et al., AIP Conf. Proc. ISSN 0094-243X, 47, 1978, pp-636-637.
3440. Koenig, V., Grueebler, W., et al., Helv. Phys. Acta 51(1), 18 May 1978, p-92.
3441. Koenig, V., Grueebler, W., et al., AIP Conf. Proc. ISSN 0094-243X, 47, 1978, pp-740-741.
3442. Koerner, H. J., Eisen, Y., et al., Verh. Dtsch. Phys. Ges. 10(7), 1975, pp-723-724.

3443. Koersner, I., Glantz, L., et al., Tandemaccelerator-laboratoriet, Uppsala, 1976, p-37.
3444. Koester, L., Knopf, K., et al., Z. Phys., A, 292(1), Aug. 1979, pp-95-103.
3445. Kohler, W., Schmidt-Boecking, H., et al., Nucl. Phys., A, 262(1), 10 May 1976, pp-113-124.
3446. Kohler, W., and Bethge, K., Verh. Dtsch. Phys. Ges. 10(7), 1975, p-765.
3447. Kohlmeyer, B., Pfeffer, W., et al., Nucl. Phys., A, 292 (1-2), 21-28 Nov. 1977, pp-288-300.
3448. Kohnle, A., and Goennewein, F., Verh. Dtsch. Phys. Ges. 6, 1977, p-972.
3449. Kohout, G., Buschmann, J., et al., Nucl. Instr. and Meth. 68, Feb. 15, 1969, pp-325-8.
3450. Kolalis, R. P., Sadkovskij, V. S., et al., Leningrad Nauka, 1978, p-185.
3451. Kolata, J. J., and Oothoudt, M., Phys. Rev., C, 15(6), June 1977, pp-1947-1958.
3452. Kolata, J. J., Fuller, R. C., et al., Phys. Rev., C, 16(2), Aug. 1977, pp-891-894.
3453. Kolata, J. J., Freeman, R. M., et al., Centre de Recherches Nucleaires, 1976, p-10.
3454. Kolata, J. J., Freeman, R. M., et al., Phys. Lett., B, 65(4), 6 Dec. 1976, pp-333-336.
3455. Kolesov, I. V., Lobanov, Yu. Y., et al., JINR-P13-5353, 1970, p-22.
3456. Kollarits, R. V., and Geller, K. N., Drexel Univ., Philadelphia, ORO-4856-26, 1975, pp-536-537.
3457. Kollarits, R. V., Drexel Univ., Philadelphia, PA (U. S. A.), Thesis (Ph. D), 1976, p-264.
3458. Kollath, K. J., and Lucas, C. B., Z. Phys., A 292(3), 1979, pp-215-218.
3459. Koller, L. R., Sci. Found. of Vacuum Technique, New York, London, John Wiley & Sons Inc., 1962.
3460. Koltun, D. S., and Myhrer, F., Rochester Univ., NY (U. S. A.), Dept. of Phys. and Astron., 1975, p-11.
3461. Koltypin, E. A., Instr. Exp. Tech. 4, 1960, pp-654-5.
3462. Kolybasov, V. M., AIP Conf. Proc. 33, 1976, pp-394-402.
3463. Komarov, V. I., Kosarev, G. E., et al., Phys. Lett., B, 69(1), 18 July 1977, pp-37-40.
3464. Komarov, V. L., Kosarev, G. E., et al., JINR-E-1-9460, 1976, p-8.
3465. Komarov, V. V., Popova, A. M., Izv. Akad. Nauk SSSR, Ser. Fiz. ISSN 0367-6765, 42(4), April 1978, pp-868-873.
3466. Komarov, V. V., and Popova, A. M., Izv. Akad. Nauk SSSR, Ser. Fiz. ISSN 0367-6765, 43(1), Jan. 1979, pp-151-54.
3467. Komarov, V. I., Kosarev, G. E., et al., Annual Report 1978, Gemeinsamer Jahresbericht, ZfK-385, June 1979, p-9.

3468. Komarov, V. I., Kosarev, G. E., et al., JINR-E-1-12393, 1979, p-8.
3469. Komochkov, M. M., JINR-P16-7335, 1973, p-20.
3470. Komura, K., and Tanaka, S., J. Inorg. and Nucl. Chem. 38 (12), 1976, pp-2157-2160.
3471. Komura, K., Mitsugashira, T., et al., Bull. Inst. Chem. Res., Kyoto Univ. 47(2), 1969, pp-79-82.
3472. Kondo, K., Lambrecht, R. M., et al., BNL Report 1976.
3473. Kondo, K., Lambrecht, R. M., et al., Int. J. Appl. Radiat. & Isot. (GB), 28(9), Sept. 1977, pp-765-71
3474. Kondo, K., Lambrecht, R. M., et al., Brookhaven National Lab. Report 1975.
3475. Kondo, Y., Abe, Y., et al., AIP Conf. Proc. ISSN 0094-243X, 47, 1978, pp-504-505.
3476. Kondo, Y., Matsuse, T., et al., Kyoto Univ., ORO-4856-26, 1975, pp-532-533.
3477. Kondratiouk, L., Lombard, R. M., et al., CEA-CONF-4102, 1977, p-1.
3478. Kondratiko, Ya. M., Korinets, V. N., et al., Sov. At. Energy 34(1), Jan. 1973, pp-69-70.
3479. Kondratyuk, L. A., Lev, F. M., et al., Yad. Fiz. ISSN 0044-0027, 29(4), April 1979, pp-1081-1090.
3480. Kondratyuk, L. A., Shevchenko, L. V., et al., Seminar on Nucl. Theory in ITEP, Sept.-Dec. 1978, ITEF-68(1979).
3481. Konijn, J., Panman, J. K., et al., AIP Conf. Proc. ISSN 0094-243X, 54(1), 15 July 1979, pp-434-435.
3482. Konjachin, N. A., Krasnov, N. N., et al., 7th Int. Conf. on Cyc. and their Application, Zurich, Switz., August 19-22, 1975.
3483. Konobeevskij, E. S., Musaelyan, R. M., et al., Neutron Physics, Pt. 1, Proc. of the 4th All-union Conf. on Neutron Physics, 1977, pp-265-268.
3484. Kononov, V. N., Yurlov, B. D., et al., Yad. Fiz. 27(1), 1978, pp-10-17.
3485. Kononov, V. N., Poletaev, E. D., et al., At. Ehnerg. 43(4), Oct. 1977, pp-303-305.
3486. Kononov, V. N., Poletaev, E. D., et al., Yad. Fiz. 27(2), Feb. 1978, pp-298-300.
3487. Kononov, V. N., Yurlov, B. D., et al., Yad. Fiz. 26(5), 1977, pp-947-955.
3488. Kononov, V. N., and Metlev, A. A., Gosudarstvennyj Komitet po Ispol'zovaniyu Atomnoj Ehnergii SSSR, Obninsk, FEI-338, 1972, p-5.
3489. Konyakhin, N. A., Krasnov, N. N., et al., Prib. Tekh. Eksp. 5, Sept.-Oct. 1969, pp-34-7.
3490. Konyakhin, N. A., Krasnov, N. N., et al., Instr. Exp. Tech. 5, Sept.-Oct. 1969, pp-1119-25.

3491. Van Konynenburg, R. A., Barschall, H. H., et al., Int. Conf. Radiat. Test Facil. CTR Surf. Mater. Program, 1975, pp-171-82.
3492. Koonin, S., Maruhn-Rezwani, V., et al., Europ. Phys. Soc., Geneva (Switz.), Europ. Conf. on Nucl. Phys. with Heavy Ions, 6-10 Sept. 1976, p-145.
3493. Koonin, S., Maruhn-Rezwani, V., et al., ANL/PHY-76-2(2), May 1976, pp-637-643.
3494. Koopman, R. P., UCRL-52275, 30 June 1977, p-210.
3495. Kostin, V. Ya., Kopanets, E. G., et al., Ukr. Fiz. Zh. 20(11), Nov. 1975, pp-1787-1794.
3496. Kopecky, J., Vennink, R., et al., Rijksuniversiteit Utrecht (Netherlands), Robert van de Graaff Lab., Annual Report 1977, March 1978, p-17.
3497. Kopf, U., Kraft, G., et al., GSI-79-11, Oct. 1979
3498. Koralewski, J., Rurarz, E., et al., Nukleonika 14, 1969, pp-1-10.
3499. Korbel, K., and Mruk, W., Nukleonika 12, 1967, pp-313-18.
3500. Korenman, G. Ya., and Popov, V. P., Izv. Akad. Nauk SSSR, Ser. Fiz. ISSN 0367-6765, 42(4), April 1978, pp-861-867.
3501. Korkisch, J., and Orlandini, K. A., Anal. Chem. 40, June 1968, pp-1127-30.
3502. Korn, H. P., Weinzierl, P., et al., Acta Phys. Austriaca 41(3-4), 1975, pp-335-340.
3503. Korotkin, Yu. S, Report 1973, JINR-P6-7400, p-20, from Nucl. Sci. Abstr. 29(4), 1974, p-8251.
3504. Korotkin, Yu. S., Radiokhimiya 16(3), 1974, pp-377-82.
3505. Korotkin, Yu. S., Internal Report, JINR, 1973.
3506. Korotkin, Yu. S., Report 1973, JINR-P6-7400, p-20.
3507. Korzh, I. A., Mishchenko, V. A., et al., Yad. Fiz. 26(6), 1977, pp-1151-1157.
3508. Kosanke, K. L., Edmiston, M. D., et al., Nucl. Instr. and Meth. 124(2), 1 March 1975, pp-365-74.
3509. Kosinov, G. A., Lagutin, I. G., et al., Leningrad Nauka 1978, p-196.
3510. Koslowsky, V., and Parsons, J. R., Prog. Report, July-Sept. 1976.
3511. Kostritsa, A. A., and Nejmotin, E. I., At. Ehnerg. 40(3), March 1976, pp-244-245.
3512. Kotajima, K., and Beringer, R., Rev. Sci. Instr. 41, May 1970, pp-632-5.
3513. Kotel'nikova, G. V., Kolpachev, A. G., et al., Fiziko-Ehnergeticheskij Inst., 1976, p-18.
3514. Kotov, V. I., and Sabsovich, L. L., Pribory i Tekh. Eksperimenta 6, 1957, pp-19-21.
3515. Kovalev, A. I., Prokof'ev, A. N., et al., Int. Conf. on Instrumentation for High Energy Physics, Dubna, USSR, 8 Sept. 1970, pp-585-7.

3516. Kovalev, A. I., Komar, A. P., et al., Sov. Phys.-Tech. Phys. 15(10), April 1971, pp-1750-3.

3517. Kovalev, V. P., Kharin, V. P., et al., At. Energ. 32(2), Feb. 1972, pp-173-5.

3518. Kovalev, V. P., Kharin, V. P., et al., Med. Radiol. 15(5), May 1970, pp-49-54.

3519. Kovar, D. G., Geesaman, D. F., et al., Phys. Rev., C, ISSN 0556-2813, 20(4), Oct. 1979, pp-1305-1331.

3520. Kovar, D. G., Eisen, Y., et al., ANL/PHY-76-2(2), May 1976, pp-645-653.

3521. Kownacki, J., Sujkowski, Z., et al., Contributed papers of the Int. Symp. on High-Spin States and Nucl. Structure held at Dresden, 19-24 Sept. 1977, pp-29-30.

3522. Kozma, L., Acta Chim. Acad. Sci. Hung. 102(3), 1979, pp-267-75.

3523. Kozyr', Yu. E., Prokopets, G. A., et al., Ukr. Fiz. Zh. 20(12), Dec. 1975, pp-2061-2063.

3524. Kozyr', Yu. E., Plyujko, V. A., et al., Ukr. Fiz. Zh. 23(3), March 1978, pp-373-376.

3525. Kraemer, S., Technische Hochschule Darmstadt, Fachbereich Anorganische Chemie und Kernchemie, Diss. (D. Sc.), 16 June 1975, p-170.

3526. Kraemmer, P., Drenckhahn, W., et al., Verh. Dtsch. Phys. Ges. 10(7), 1975, p-755.

3527. Kraeutle, H., Nucl. Instr. and Meth. 137(3), 1976, pp-553-7.

3528. Kraft, O. E., Naumov, Yu. V., et al., Summaries of Reports of 26th Conf. on Nucl. Spectro. and Nucl. Structure, Baku, 3-6 Feb. 1976, p-56.

3529. Kraft, O. E., Naumov, Yu. V., et al., Leningrad Nauka, 1978, p-12.

3530. Kraft, O. E., Naumov, Yu. V., et al., Izv. Akad. Nauk SSSR, Ser. Fiz. ISSN 0367-6765, 42(4), April 1978, pp-759-64.

3531. Kramers, J. D., Earth Planet. Sci. Lett. 34(3), April 1977, pp-419-431.

3532. Krane, K. S., Sharma, T. C., et al., Phys. Rev., C, ISSN 0556-2813, 20(5), Nov. 1979, pp-1873-1877.

3533. Krane, K. S., Phys. Rev., C, 17(6), June 1978, pp-2213-18.

3534. Krasnov, N. N., and Sevastjanov, Yu. G., Int. J. Appl. Radiat. Isot., ISSN 0020-708X, 30(12), Dec. 1979, pp-783-784.

3535. Krasnov, N. N., Dmitriyev, P. P., et al., Inst. of Phys. and Power Eng., Obninsk, USSR

3536. Krasnov, N. N., Dmitriyev, P. P., et al., Int. Conf. on the Use of Cyc. in Chem., Metall., and Biol., Oxford, Eng., see CONF-690924, 1970, pp-159-67.

3537. Kratz, J. V., Bruechle, W., et al., Europ. Phys. Soc., Geneva (Switz.), Europ. Conf. on Nucl. Phys. with Heavy Ions, 6-10 Sept. 1976, p-175.

3538. Kratz, J. V., Bruechle, W., et al., Nucl. Phys., A, ISSN 0375-9474, 332(3/4), Dec. 1979, pp-477-500.
3539. Kratz, J. V., Bruechle, W., et al., Internal Report, Gesellschaft fuer Schwerionenforschung m. b. H., Darmstadt.
3540. Kratz, J. V., Schaedel, M., et al., Lecture meeting on Principles and Applications of Nucl.-, radio-, and radiation chem., Lindau, Germany, F. R., 16-19 Oct. 1978.
3541. Krause, H. H., Arnold, W., et al., Z. Phys., A 293(4), 1979, pp-343-349.
3542. Kreiner, A. J., Bermudez, G. G., et al., Phys. Lett., B, ISSN 0370-2693, 83(1), 23 April 1979, pp-31-33.
3543. Kreische, W., Niedrig, H., et al., Phys. Rev., C, 17(6), June 1978, pp-2006-2010.
3544. Kreische, W., Maar, H. U., et al., Nucl. Instr. and Meth. 150(2), 1 April 1978, p-325.
3545. Kreische, W., Maar, H. U., et al., Internal Report 1977.
3546. Kreische, W., Maar, H. U., et al., Internal Report 1976.
3547. Kreische, W., Maar, H. U., et al., Internal Report 1975.
3548. Krestovnikov, A. N., Panteleev, G. V., et al., Internal Report 1976.
3549. Krestovnikov, A. N., Panteleev, G. V., et al., Zh. Fiz. Khim. 49(9), 1975, pp-2442-5.
3550. Krieger, C. I., INTDS Newsletter, July 1979.
3551. Krimmel, E., Dullnig, H., et al., Hochvakuumkammer zum Bestrahlen von Targets, German (F. R.) patent document 2403349/A/, Int. Cl. H01J 37-10, 7 Aug. 1975, p-17.
3552. Krimmel, E., Dullnig, H., et al., Berlin Report 1976.
3553. Krisch, A. D., Proc. of the 4th Int. Symp. on Polar. Phenom. in Nucl. React., 25-29 Aug. 1975, pp-41-7.
3554. Krivan, V., Katedra Radiochemie a Radiacnej Chemie Slovenskej Vysokej Skoly Technickej, Bratislava, Chem. Zvesti 19, 1965, pp-699-702.
3555. Krivonosov, G. A., Ekhichev, O. I., et al., Summ. of Reports of 26th Conf. on Nucl. Spectro. and Nucl. Struct., Baku, 3-6 Feb. 1976, p-152.
3556. Krivonosov, G. A., Ekhichev, O. I., et al., Summ. of Reports of 26th Conf. on Nucl. Spectro. and Nucl. Struct., Baku, 3-6 Feb. 1976, p-153.
3557. Kroll, W., Z. Anorg. Chem. 234, 1937, 42.
3558. Krug, H., Bonn Univ. (Germany, F. R.), Mathematisch-Naturwissenschaftliche Fakultaet, Diss. (D. Sc.), 24 July 1974, p-205.
3559. Kruger, O. L., and Pardue, W. M., Battelle Columbus Labs., Ohio, Dec. 1971, p-86.
3560. Kruger, S. T., Rice Univ., Houston, TX, Diss. Abstr. Int. B, 32(4), 1971, pp-2336-7.

3561. Ksenzov, V. G., Yad. Fiz. ISSN 0044-0027, 29(2), Feb. 1979, pp-340-342.
3562. Ksenzov, V. G., Yad. Fiz. ISSN 0044-0027, 28(5), 1978, pp-1249-1257.
3563. Ku, T. H., and Karol, P. J., Phys. Rev., C, 16(5), Nov. 1977, pp-1984-1995.
3564. Ku, T. H. H., Carnegie-Mellon Univ., Pittsburgh, PA (U. S. A.), Thesis (Ph. D), 1975, p-111.
3565. Ku, T. H., and Karol, P. J., Nucl. Instr. and Meth. 121 (3), 1 Nov. 1974, pp-537-540.
3566. Ku, T. H., and Karol, P. J., Internal Report 1973.
3567. Kuan, H. M., and Shirk, D. G., Phys. Rev., C, 13(2), Feb. 1976, pp-883-886.
3568. Kuang-Ch'ang, H., Wu Li Hsueh Pao 20(10), 1964, pp-1056-7.
3569. Kuang-Ch'ang, H., Internal Report 1963. See J. in No. 3568.
3570. Kubo, K., Nagatani, K., et al., Phys. Rev., C, 15(5), May 1977, pp-1758-1773.
3571. Kubo, K., Nair, K. G., et al., Phys. Rev. Lett. 37(4), 26 July 1976, pp-222-225.
3572. Kubono, S., Kato, S., et al., Tokyo Univ., Tanashi (Japan), Nov. 1975.
3573. Kubota, M., Amano, H., et al., JAERI-13545, 1969, p-12.
3574. Kucharski, M., Inst. of Nucl. Rsch., Warsaw (Poland), INR-1551/1/C/B, 1975, p-17.
3575. Kuchnir, F. T., Waterman, F. M., et al., Proc. of 3rd Symp. on Neutron Dosimetry in Biol. and Med., EUR 5848 DE/EN/FR, Brussels-Luxemburg, 1978, p-369.
3576. Kuchnir, F. T., Waterman, F. M., et al., IEEE 76 CH 1175 9 NPS, 1977, p-513.
3577. Kuchnir, F. T., Waterman, F. M., et al., 4th Ann. Conf. on the Use of Small Accel., Denton, TX (U. S. A.), 25 Oct. 1976, p-4.
3578. Kuchnir, F. T., Skaggs, L. S., et al., Proc. of the 6th Int. Cyc. Conf., Vancouver, CA, July 18-21, 1972.
3579. Kuchnir, F. T., Skaggs, L. S., et al., AIP Conf. Proc. 9, 1972, p-638.
3580. Kudo, Y., and Honda, T., Osaka City Univ., ORO-4856-26, 1975, pp-321-322.
3581. Kudo, Y., Honda, T., et al., Osaka City Univ., ORO-4856-26, 1975, pp-419-420.
3582. Kuehn, P. R., O'Donnell, F. R., et al., Nucl. Instr. and Meth. 102, 1972, pp-403-7.
3583. Kuehn, P. R., O'Donnell, F. R., et al., 3rd Int. Symp. on Rsch Mat. for Nucl. Measurements, Gatlinburg, TN, 5 Oct. 1971, pp-39-46.
3584. Kuehn, W., et al., Koeln Univ. (Germany, F. R.), Spiering, H. (ed.), Erlangen-Nuernberg Univ., Erlangen (Germany, F. R.), Physikalisches Inst. Annual Report 1976, Jahresbericht 1976, pp-26-28.

3585. Kuenhold, A. K., Duggan, L. J., et al., Sci. and Ind. Appl. of Small Accel., 4th Conf., 1976.
3586. Kugler, E., Chalmers Tekniska Hogskola, Goteborg (Sweden), Institutionen for Fysik, Feb. 1970, p-15.
3587. Kuhn, W., Martin, H., et al., Z. Phys. Chem. Abt. 213, 1941, p-B50.
3588. Kuhn, W., Martin, H., et al., Z. Phys. Chem. Abt. 93, 1933, p-B21.
3589. Kretschmer, W., Loeh, H., et al., Kneis, W., (ed.), et al., Inst. fuer Angewandte Kernphysik, Ann. Rep., Oct. 1979.
3590. Krieger, T. J., et al., Magurno, B. A. (ed.), BNL-NCS-50446, April 1975, pp-92-107.
3591. Kuiper, H., Diekamp, J., et al., Nucl. Instr. and Meth. 98 (1), 1972, pp-151-5.
3592. Kuiper, H., Z. Phys. 232, 1970, pp-325-41.
3593. Kukavadze, G. M., Trebukhovskii, V. V., et al., Instr. Exp. Tech. 4, July-Aug. 1970, pp-993-5.
3594. Kuks, I. M., Selitskij, Ya. A., et al., IAEA, Vienna (Austria), Int. Nucl. Data Comm., Feb. 1976.
3595. Kulkarni, R. G., Phys. Scr. 13(4), April 1976, pp-213-16.
3596. Kullberg, R., Oskarsson, A., et al., Lund Univ. (Sweden), Fysiska Inst., LUIP-7706, May 1977, p-3.
3597. Kumabe, I., Nucl. Instr. and Meth. 46, 1967, pp-177-8.
3598. Kumar, A., Soni, S. K., et al., Izv. Akad. Nauk SSSR, Ser. Fiz. 40(10), Oct. 1976, p-2150.
3599. Kume, S., Meeting on Nuclei with Short life, Osaka (Japan), 30 June 1969.
3600. Kumpf, H., Moeller, K., et al., Zentralinstitut fuer Kernforschung, Rossendorf bei Dresden, June 1977, p-58.
3601. Kunsch, B., Eder, O. J., J. Phys., E, 6(6), June 1973, pp-576-578.
3602. Kusno, D., Univ. Oregon, Eugene, U. S. A., 1979, p-159.
3603. Kusuhara, M., Nucl. Instr. and Meth. 83, 1970, p-328.
3604. Kutschera, K., Kuiper, H., et al., Nucl. Phys. A183(3), 1972, pp-593-600.
3605. Kutschera, W., Brown, B. A., et al., Phys. Rev., C, 12(3), Sept. 1975, pp-813-820.
3606. Kutschera, W., Brown, B. A., et al., Verh. Dtsch. Phys. Ges. 10(7), 1975, p-702.
3607. Kuz'menko, P. P., Golovinskij, L. P., et al., Izv. Akad. Nauk SSSR, Neorg. Mater. 11(8), Aug. 1975, pp-1365-68.
3608. Kuz'minykh, V. A., and Vorob'ev, S. A., Nucl. Instr. and Meth. 167(3), 1979, pp-483-8.
3609. Kuz'menko, V. S., Mitrofanova, A. V., et al., Pis'ma Zh. Ehksp. Teor. Fiz. 23(3), 5 Feb. 1976, pp-174-176.
3610. Kwiatkowski, K., Maryland Univ., Baltimore (U. S. A.), Thesis (Ph. D), 1976, p-131.
3611. Kwinta, J., Nucl. Instr. and Meth. 167(1), 1979, pp-65-70.

3612. Kwinta, J., INTDS Proc., Garching, FRG., Sept. 11-14, 1978.
3613. Kwinta, J., and Amoudry, F., INTDS Proc., Chalk River, Canada, Oct. 1-3, 1975.
3614. Kwinta, J., and Amoudry, F., ACNTDS, CRNL, Oct. 1-3, 1974.
3615. Lachkar, J., Haouat, G., et al., CEA-CONF-3303, 1975, p-14.
3616. Lachkar, J. C., Sigaud, J., et al., CEA-R-4715, Nov. 1975, p-51.
3617. Lachkar, J. C., Natl. Bur. of Stand. (U. S.), Spec. Publ. 493, Oct. 1977, pp-93-100.
3618. Lachkar, J. C., CEA-CONF-3940, 1977, p-8.
3619. Lachkar, J., Haouat, G., et al., Natl. Bur. of Stands. (U. S.), Spec. Publ. 425, Oct. 1975, pp-897-900.
3620. Ladage, A., Dittmann, P., et al., Mater. Soveshch. Besfil'movym Iskrovym Strimernym Kameram, 1969, pp-41-50.
3621. Lafourcade, L., David, M. J., et al., J. Microsc. Spectrosc. Electron. 2(6), Dec. 1977, pp-279-284.
3622. Laget, J. M., 7th Int. Conf. on High Energy Physics and Nuclear Structure, Zurich (Switz.), 29 Aug.-2 Sept. 1977.
3623. Laget, J. M., CEA-CONF-4390, 1978, p-20.
3624. Lagunas-Solar, M. C., Jungerman, J. A., et al., Int. J. Appl. Radiat. Isot. ISSN 0020-708X, 29(3), March 1978, pp-159-165.
3625. Lagunas-Solar, M. C., and Jungerman, J. A., Int. J. Appl. Radiat. Isot. 30(1), 1 Jan. 1979, pp-25-32.
3626. Lajtai, A., Kecskemeti, J., et al., IAEA, Vienna (Austria), Int. Nucl. Data Comm., Feb. 1978, p-6.
3627. Lal, D., et al., J. Appl. Phys. 40, 1969, p-3257.
3628. Lamb, E., ORNL, DP-1066(Vol. 1), pp-II.3-14.
3629. Lambrecht, R. M., and Wolf, P. A., Radiat. Rsch. 52, 1972.
3630. Lambrecht, R. M., Ritter, E., et al., Int. J. Appl. Radiat. Isot. 28(6), 1977, pp-567-71.
3631. Lamehi-Rachti, M., Levi, C., et al., CEA-N-1861 (nd), 1 Oct. 1974-1 Sept. 1975, pp-21-24.
3632. Lamehi-Rachti, M., Levi, C., et al., CEA-N-1861 (nd), 1 Oct. 1974-1 Sept. 1975, pp-32-35.
3633. Lamehi-Rachti, M., Levi, C., et al., Europ. Phys. Soc. Geneva (Switz.), Europ. Conf. on Nucl. Phys. with Heavy Ions, 6-10 Sept. 1976, p-51.
3634. Lamehi-Rachti, M., Levi, C., et al., Europ. Phys. Soc. Geneva (Switz.), Europ. Conf. on Nucl. Phys. with Heavy Ions, 6-10 Sept. 1976, p-52.
3635. Lammer, G., and Lammer, M., IAEA-208(Vol. 1), 1978, pp-301-318.
3636. Lamontagne, C. R., Frois, B., et al., Laval Univ., Quebec City, Quebec, LBL-8151, July 1, 1977-June 30, 1978, pp-62-63.

3637. Lancman, H., Sparks, R. J., et al., Nucl. Phys., A; 257(1), 26 Jan. 1976, pp-29-36.
3638. Landau, R. H., Baer, H., et al., LA-7892C, July 1979, pp-150-166.
3639. Landaud, G., Devaux, A., et al., College de France, 75-Paris, Lab. de Physique Atomique et Moleculaire, June 1977, p-17.
3640. Landowne, S., Dasso, C. H., et al., Phys. Lett., B, 70(3), 10 Oct. 1977, pp-292-296.
3641. Lane, R. O., White, R. M., et al., Int. Conf. on the Interactions of Neutrons with Nuclei, Vol. II, CONF-760715-P2, 1976, pp-1042-1057.
3642. Lane, R. O., Ohio Univ., Athens (U. S. A.), COO-2490-3, 1976, p-59.
3643. Lane, R. O., COO-2490-10, 1979, p-73, Ohio Univ.
3644. Lanford, W. A., Phys. Rev., C, 16(3), Sept. 1977, pp-988-1000.
3645. Lang, F. M., et al., CEA 545, CEA, 1956.
3646. Lang, J., Mueller, R., et al., Europ. Phys. Soc., Geneva (Switz.), Europ. Conf. on Nucl. Phys. with Heavy Ions, 6-10 Sept. 1976, p-134.
3647. Lang, J., Mueller, R., et al., Europ. Phys. Soc., Geneva (Switz.), Europ. Conf. on Nucl. Phys. with Heavy Ions, 6-10 Sept. 1976, p-39.
3648. Lang, J., Mueller, R., et al., Europ. Phys. Soc., Geneva (Switz.), Europ. Conf. on Nucl. Phys. with Heavy Ions, 6-10 Sept. 1976, p-48.
3649. Lang, J., Mueller, R., et al., AIP Conf. Proc. ISSN 0094-243X, 47, 1978, pp-682-683.
3650. Langanke, K., Friedrich, H., et al., AIP Conf. Proc. ISSN 0094-243X, 47, 1978, pp-704-705.
3651. Lange, A., Nucl. Instr. and Meth. 98(3), 1972, pp-605-6.
3652. Lange, J., Muenzel, H., et al., Radiochim. Acta 11, 1969, pp-121-3.
3653. Langevin, M., Barreto, J., et al., Europ. Phys. Soc., Geneva (Switz.), Europ. Conf. on Nucl. Phys. with Heavy Ions, 6-10 Sept. 1976, p-108.
3654. Langevin, M., Barreto, J., et al., Phys. Rev., C, 14(1), July 1976, pp-152-159.
3655. Langevin, M., Barreto, J., et al., Internal Report, Paris-11 Univ., 91-Orsay (France), Inst. de Phys. Nucl.
3656. Langley, R. A., and Brice, D. K., Nucl. Instr. and Meth. 149(1-3), 15 Feb.-1 March 1978, pp-195-197.
3657. Langley, R. A., SAND-75-0331, Report 1975, p-33.
3658. Langsdorf, A., Jr., ANL, EUR-1815.e(Rev.), pp-254-7.
3659. Lantto, V., J. Phys. D. 6(9), 1973, pp-1058-66.
3660. Lapidus, L. I., JINR-P2-3217, 1967, p-17.

3661. Lapikas, L., Box, G., et al., Nucl. Phys., A, 253(2), 24 Nov. 1975, pp-324-354.
3662. Lark, N. L., et al., Nucl. Phys. A139, 1969, p-481.
3663. Larsen, J. T., UCRL-80875, CONF-780461-2, Report 1978, p-48.
3664. Larson, D. C., Harvey, J. A., et al., ORNL, TN (U. S. A.), CONF-760622-12, 1976, p-6.
3665. Laszewski, R. M., Ill. Univ., Urbana (U. S. A.), Thesis (Ph. D), 1975, p-127.
3666. Laszewski, R. M., Holt, R. J., et al., BNL-NCS-22500, March 1977, pp-33-34.
3667. Latrous, H., CEA-R-3065, 1967, p-54.
3668. Lattuada, M., Vinciguerra, D., et al., Lett. Nuovo Cim. 21(14), 8 April 1978, pp-497-501.
3669. Latuszynski, A., Zuber, K., et al., JINR-E6-7780, Report 1974, p-24.
3670. Latzel, G., Paetz, H., et al., Verh. Dtsch. Phys. Ges. 6, 1977, p-961.
3671. Laubenstein, R. A., North Amer. Aviation, Inc., Downey, CA, Contract AT-11-1-GEN-8, Aug. 20, 1951, p-30.
3672. Laughlin, J. S., Tilbury, R. S., et al., Sloan-Kettering Inst. for Cancer Rsch., NY, Contract AT(30-1)-910, Sept. 30, 1969, p-25.
3673. Lauer, K. F., Nucl. Instr. and Meth. 102, 1972, p-589.
3674. Lauer, K. F., AEC Symp. Ser. 23, 1971, pp-488-99.
3675. Lauer, K. F., TISRMNM, ORNL, Oct. 5-8, 1971.
3676. Lauer, K. F., AERE-R-5097, Paper 27, 1965, p-11.
3677. Lauer, K. F., PSPSITF, AERE, Oct. 20-21, 1965.
3678. Lauer, K. F., and Verdingh, V., Nucl. Instr. and Meth. 21, 1963, p-161.
3679. Laumer, H., Davids, C. N., et al., Nucl. Instr. and Meth. 120(3), 15 Sept. 1974, pp-535-536.
3680. Laurent, H., Massolo, C. P., et al., Nucl. Phys., A, 284 (3), 11 July 1977, pp-501-512.
3681. Lauritzen, T. A., and Comprelli, F. A., Nucl. Appl. Technol. 3(6), 1967, pp-390-1.
3682. Lavi, N., and Nir-El, Y., J. Inorg. and Nucl. Chem. 38(12), 1976, pp-2133-2134.
3683. Lavrukhina, A. K., Ustinova, G. K., et al., Sov. At. Energy 34(1), Jan. 1973, pp-29-35.
3684. Lawson, D. L., Nucleonics 10(11), Nov. 1952, pp-61-5.
3685. Lawson, R. P. W., Freeman, J. H., et al., Nucl. Instr. and Meth. 131(3), 1976, pp-567-8.
3686. Lazar, I., Stud. Cercet. Fiz. 26(1), 1974, pp-115-131.
3687. Lazzarini, A. J., Cosman, E. R., et al., Phys. Rev. Lett. 40(22), 29 May 1978, pp-1426-1429.
3688. Leaver, K. D., and Chapman, B. N., Thin Films (Wykeham Publ. Ltd., London, 1971).

3689. Lebowitz, E., and Greene, M. W., Int. J. Appl. Radiat. Isotop. 21, Oct. 1970, pp-625-7.
3690. Lechpammer, T., and Babarovic, B., Fizika 4(23), 1971.
3691. Lecoguie, R., Leconte, Ph., et al., Nucl. Instr. and Meth. 143(1), 15 May 1977, pp-13-16.
3692. Ledger, A. M., Contract NAS5-23581, Aug. 1977, p-43.
3693. Lee, H. C., and Earle, E. D., Nucl. Instr. and Meth. 131 (1), 1 Dec. 1975, pp-199-202.
3694. Lee, H. C., Khanna, F. C., et al., Phys. Lett., B, 65(3), 22 Nov. 1976, pp-201-204.
3695. Lee, I. Y., Aleonard, M. M., et al., Phys. Rev. Lett. 38 (25), 20 June 1977, pp-1454-1457.
3696. Lee, S. M., Berg, G., et al., Verh. Dtsch. Phys. Ges. 10 (7), 1975, p-768.
3697. Lee, S. M., and Kueln, W., Europ. Phys. Soc., Geneva (Switz.), Europ. Conf. on Nucl. Phys. with Heavy Ions, 6-10 Sept. 1976, p-122.
3698. Lee, S. M., Motobayashi, T., et al., Europ. Phys. Soc., Geneva (Switz.), Europ. Conf. on Nucl. Phys. with Heavy Ions, 6-10 Sept. 1976, p-123.
3699. Lee, S. M., Adloff, J. C., et al., AIP Conf. Proc. ISSN 0094-423X, 47, 1978, pp-596-597.
3700. Lee, T. S. H., and Pittel, S., Nucl. Phys., A, 256(3), 19 Jan. 1976, pp-509-520.
3701. De Leeuw, J. H., Haasz, A. A., et al., Int. Conf. Radiat. Test Facil. CTR Surf. Mater. Program, 1975, pp-315-34.
3702. Lefort, M., Proc. of Conf. held at St. Catherine's Coll., Oxford, 22-23 Sept. 1969.
3703. Legrain, R., Cassagnou, Y., et al., 2nd Int. Conf. on Meson-Nuclear Phys., Houston, TX, U. S. A., 5-6 March 1979, p-2.
3704. Lehar, F., and Winternitz, P., Cesk. Cas. Fys. 17, 1967, pp-158-211.
3705. Lehar, F., and Winternitz, P., Fortschr. Phys. 15, 1967, pp-495-536.
3706. Leite, R. J., INTDS Newsletter, July 1979.
3707. Leitz, W., Mahnke, H. E., et al., Nucl. Phys., A, 258(1), 16 Feb. 1976, pp-103-108.
3708. Leleux, P., Boxman, M., et al., Nucl. Instr. and Meth. ISSN 0029-554X, 152(2-3), 15 June 1978, pp-495-504.
3709. Lemaire, M., and Low, K. S., Phys. Rev., C, 16(1), July 1977, pp-183-191.
3710. Lemaire, M. C., Nagamiya, S., et al., Phys. Lett., B, ISSN 0370-2693, 85(1), 30 July 1979, pp-38-42.
3711. Lemere, M., Tang, Y. C., et al., Nucl. Phys., A, 266(1), 2 Aug. 1976, pp-1-15.
3712. Lemley, J. R., and Keyworth, G. A., 3rd Int. Symp. on Polar. Phenom. in Nucl. React., Madison, WI, Aug.-Sept. 1970, pp-887-9.

3713. Lemmel, H. D., Natl. Bur. of Stands. (U. S.), Spec. Publ. 1, Oct. 1975, pp-286-292.
3714. Lenske, H., Bonn Univ. (Germany, F. R.), July 1979, p-140.
3715. Lenz, F., AIP Conf. Proc. 33, 1976, pp-403-417.
3716. DeLeo, R., D'Erasmo, G., et al., Nucl. Phys., A, 254(1), 1 Dec. 1975, pp-156-168.
3717. DeLeo, R., D'Erasmo, G., et al., Phys. Rev., C, ISSN 0556-2813, 20(4), Oct. 1979, pp-1244-1250.
3718. Lepine, A., Volant, C., et al., Nucl. Phys., A, 289(1), 10 Oct. 1977, pp-187-204.
3719. Lepretre, A., Beil, H., et al., Nucl. Phys., A, 258(2), 23 Feb. 1976, pp-350-364.
3720. Lerner, J., ANL/PHY/MSD-76-1, 1975, pp-60-67.
3721. Lerner, J., INTDS Proc., Berkeley, U. S. A., Oct. 19-20, 1977.
3722. Lerner, J., INTDS Proc., Argonne, U. S. A., Sept. 30-Oct. 2, 1975.
3723. Lerner, J., Nucl. Instr. and Meth. 102, 1972, pp-373-8.
3724. Lerner, J., TISRMNM, ORNL, Oct. 5-8, 1971.
3725. Leroy, J. L., Szabo, I., et al., Neutron standard reference data (IAEA, Vienna, 1972).
3726. Lesiecki, H., Mack, G., et al., Verh. Dtsch. Phys. Ges. 10(7), 1975, pp-866-867.
3727. Lesniak, H., and Lesniak, L., Acta Phys. Pol., Ser. B, 5(5), 1974, pp-703-724.
3728. Lesniak, H., Lesniak, L., et al., Nucl. Phys., A, 267(3), 30 Aug. 1976, pp-503-531.
3729. Letkhov, V. S., Science 180, 1973, p-451.
3730. Letts, S. A., Johnson, W., et al., Report 1979, UCID-18324, p-10.
3731. Leugers, B., Cierjacks, S., et al., ANL-76-90, 1976, pp-246-257.
3732. Leugers, B., et al., Kneis, W., et al., (eds.), KFK-2868, Kernforschungszentrum Karlsruhe G.m.b.H., Inst. fuer Angewandte Kernphysik, Ann. Rep., Oct. 1979, pp-2-4.
3733. Leugers, B., Annual Report KFK-2868, Oct. 1979, p-10-11.
3734. Leung, M. K., PTB-ND-5, Dec. 1973, p-31.
3735. Lev, A., and Beres, W. P., Phys. Rev., C, 13(6), June 1976, pp-2585-2587.
3736. Lev, A., and Beres, W. P., Phys. Rev., C, 14(1), July 1976, pp-354-356.
3737. Levelut, A. M., Lambert, M., et al., UCRL-Trans-10092, from Compt. Rend. 255: 319-21, 1962, p-8.
3738. Levelut, A. M., UCRL-Trans-10090, J. Phys. 24: 560-4, 1963, p-16.
3739. Levermore, C. D., Caflisch, R. E., et al., UCRL-79934, CONF-771136-31, Report 1977, p-13.
3740. Levin, E., and Eisenberg, J. M., Nucl. Phys., A, 292(3), 5 Dec. 1977, pp-459-476.

3741. Levin, I. A., and Khokhlova, I. M., Gosudarstvennyj Nauchno-Issledovatel'skij i Proektnyj Inst. Azotnoj Promyshlennosti i Produktov Organicheskogo Sinteza, Moscow (USSR), INIS-mf-3317, 1976, pp-25-26.
3742. Levine, G. S., Squier, D. M., et al., Proc. of the Int. Cong. on Protection against Accel. and Space Radiation, 1971, pp-798-813.
3743. Levintov, I. I., Kanauets, V. P., et al., Akademiya Nauk SSSR, NP-14345, 1964, p-8.
3744. Lewis, A. B., Mississippi Univ., Dept. of Physics and Astronomy, Contract AT(40-1)-2891, June 12, 1967, p-12.
3745. Lewis, A. B., Contract AT(40-1)-2891, June 1, 1966, p-32.
3746. Lewis, B., Surface Sci. 21, 1970, pp-273, 289.
3747. Lewis, D. A., Morsch, H. P., et al., Minn. Univ., Minneapolis (U. S. A.), John H. Williams Lab. of Nucl. Phys., Annual Report 1975, pp-187-198.
3748. Lewis, D. A., and Petersen, J. F., ANL/PHY-76-2(2), May 1976, pp-655-661.
3749. Lewis, R. A., et al., Ann. Meeting of Am. Nucl. Soc., Boston, MA, June 13-17, 1971.
3750. Lewis, R. N., Van Loon, L. S., et al., Rev. Sci. Instr. 40, May 1969, pp-698-702.
3751. Lewis, V. E., Nucl. Instr. and Meth. 64, 1968, pp-293-6.
3752. Liaw, J. R., and Spinrad, B. I., Trans. Am. Nucl. Soc. 22, Nov. 1975, pp-672-673.
3753. DiLiberto, S., Meddi, F., et al., Nucl. Phys., A, 296(3), 27 Feb. 1978, pp-519-532.
3754. Liboff, R. L., Nucl. Eng. Des. ISSN 0029-5493, 54(1), Sept. 1979, pp-119-123.
3755. Lichtenstadt, J., Marinov, A., et al., Israel Phys. Soc., Jerusalem, 1976 Annual Meeting, p-12.
3756. Lidsky, L. M., and Colombant, D., Trans. Nucl. Sci., NS-14, June 1967, pp-945-9.
3757. Lieb, B. J., Lankford, W. F., et al., Phys. Rev., C, 14(4), Oct. 1976, pp-1515-1533.
3758. Lieber, A., Nucl. Instr. and Meth. 26, Feb. 1964, pp-51-4.
3759. Liese M. H., and Bertsche, U., 7th Int. Conf. on Cyc. and Their Application, Zurich, Switz., Aug. 19-22, 1975.
3760. Lightbody, J. W., Jr., Penner, S., et al., Phys. Rev., C, 14(3), Sept. 1976, pp-952-964.
3761. Likar, A., and Sever, F., Nucl. Phys., A, 295(3), 6 Feb. 1978, pp-405-423.
3762. Liljenzin, J. O., LBL-1912, Contract W-7405-eng-48, June 1973, p-60.
3763. Liljestrand, R., Blanpied, G., et al., Nucl. Instr. and Meth. 138(3), 1 Nov. 1976, pp-471-477.
3764. Lilley, J. S., Franey, M., et al., John H. Williams Lab., Minnesota Univ., Minneapolis (U. S. A.), Annual Report 1975, pp-122-128.

3765. Lillie, A. B., and Connor, J. P., Rev. Sci. Instr. 22, 1951, pp-210-11.
3766. Lin, C. S., Hou, W. S., et al., Nucl. Phys., A, 275(1), 4 Jan. 1977, pp-93-99.
3767. Lin, Erh-K'ang, Kiang, G. C., et al., Chin. J. Phys. 6(2), 1968, pp-67-73.
3768. Lin, L., Sherman, A., et al., Phys. Rev. Lett. 37(6), 9 Aug. 1976, pp-327-330.
3769. Lind, J. M., Princeton Univ., NJ (U. S. A.), Thesis (Ph. D), 1975, p-108.
3770. Lind, J. M., Garvey, G. T., et al., Nucl. Phys., A, 276(1), 17 Jan. 1977, pp-25-60.
3771. Lind, V. G., McAdams, R. E., et al., AIP Conf. Proc. ISSN 0094-243X, 47, 1978, pp-646-647.
3772. Lind, V. G., McAdams, R. E., et al., AIP Conf. Proc. ISSN 0094-243X, 54(1), 15 July 1979, pp-284-285.
3773. Lindberg, J. F., Grimson, J., et al., Proc. Conf. Remote Syst. Technol. 26th, 1979, pp-65-75.
3774. Lindberg, J., Grimson, J., et al., Conf. on Rem. Syst. Tech. 23rd, Proc. ANS Winter Meet., San Francisco, CA, Nov. 1975, pp-24-31.
3775. Lindeke, K., Specht, S., et al., Phys. Rev., C, 12(5), Nov. 1975, pp-1507-1510.
3776. Lindgren, L. J., and Sandell, A., Lund Univ. (Sweden), Fysiska Inst., March 1977, p-20.
3777. Lindgren, R. A., Bendel, W. L., et al., Phys. Rev., C, 14(5), Nov. 1976, pp-1789-1799.
3778. Lindl, J. D., and Bangerter, R. O., UCRL-77042, Report 1975, p-36.
3779. Lindquist, L. O., and Scarbrough, E. C., LA-6936-MS, Nov. 1977, p-27.
3780. Lindquist, L. O., and Scarbrough, E. C., LA-7221-MS, May 1978, p-14.
3781. Lindquist, L. O., and Scarbrough, E. C., Internal Report, LASL, 1977.
3782. Lindsey, W. J., Roggenkamp, P. L., et al., Nucl. Technol. 13(1), 1972, pp-78-82.
3783. Lindsky, M. L., IEEE Trans. on Nucl. Sci., June 1967.
3784. Liou, H. I., and Chrien, R. E., Nucl. Sci. and Eng. 62(3), March 1977, pp-463-478.
3785. Lipas, P. O., Haapakoski, P., et al., Jyvaeskylae Univ. (Finland), Dept. of Physics, JU-RR-75-1, May 1975, p-40.
3786. Lipas, P. O., Haapakoski, P., et al., Phys. Scr. 13(6), June 1976, pp-339-350.
3787. Lipparini, E., Orlandini, G., et al., Phys. Rev., C, 16(2), Aug. 1977, pp-812-815.
3788. Lipták, J. and Ryba, M., Ceskoslov. casopis fys. 1961, pp-149-53.

3789. Lishenko, L. G., Nazarova, T. S., et al., Prib. Tekh. Eksp. 1, Jan.-Feb. 1968, pp-19-20.
3790. Lishenko, L. G., Nazarova, T. S., et al., Prib. Tekh. Eksp. 4, July-Aug. 1968, pp-37-8.
3791. Lishenko, L. G., Nazarova, T. S., et al., Ukr. Fiz. Zh. 12, March 1967, pp-515-16.
3792. Liskien, H., Int. Conf. on the Inter. of Neutrons with Nuclei, CONF-760715-P2, 1976, pp-1110-1124.
3793. Lisowski, P. W., Rhea, T. C., et al., Nucl. Phys., A, 259(1), 8 March 1976, pp-61-74.
3794. Lister, C. J., et al., J. Phys. G: Nucl. Phys. 2, 1976, p-577.
3795. Lister, C. J., et al., J. Phys. G: Nucl. Phys. 3, 1977, p-L75.
3796. Lister, C. J., et al., J. Phys. G: Nucl. Phys. 4, 1978, p-907.
3797. Litherland, A. E., Ollerhead, R. W., et al., Can. J. Phys. 45, 1967, p-1901.
3798. Littauer, R., Rev. Sci. Instr. 29, 1958 Feb., pp-178-9.
3799. Little, R. C., and Block, R. C., BNL-NCS-22500, March 1977, pp-243, 246, 253.
3800. Little, W. A., Stanford Univ., CA (U. S. A.), Thesis (Ph. D), 1976, p-128.
3801. Litz, L. M., and Batzel, R. E., Calif. Rsch. and Develop. Co., Livermore Rsch. Lab., Livermore, CA, CRD-T2C-87, April 2, 1957, p-18.
3802. Liu, L. C., and Shakin, C. M., Phys. Rev., C, 14(5), Nov. 1976, pp-1885-1892.
3803. Liu, L. C., and Shakin, C. M., Phys. Rev., C, 16(1), July 1977, pp-333-339.
3804. Liu, L. C., Phys. Rev., C, 17(5), May 1978, pp-1787-1798.
3805. Liu, Q. K. K., von Oertzen, W., et al., AIP Conf. Proc. ISSN 0094-243X, 47, 1978, pp-732-733.
3806. Liukkonen, E., Metag, V., et al., Nucl. Instr. and Meth. 125(1), 1975, pp-113-17.
3807. Liuti, G., Dondes, S., et al., J. Chem. Phys. 44, 1966, p-4052.
3808. Livingston, A. E., Berry, M. G., et al., INTDS Proc., Berkeley, U. S. A., Oct. 19-20, 1977.
3809. Livingston, A. E., Berry, M. G., et al., Nucl. Instr. and Meth. 198, 1978, p-125.
3810. Livingston, J. T., and Angerman, C. L., DuPont de Nemours (E. I.) and Co., Aiken, SC, Savannah River Lab., Contract AT(07-2)-1, June 9, 1966, p-14.
3811. Livingston, M. S., Rev. Sci. Instr. 22, 1951, pp-428-9.
3812. Lo, J. W., Hungerford, E. V., et al., Phys. Rev., C, ISSN 0556-2813, 20(4), Oct. 1979, pp-1479-1489.
3813. Locher, M. P., and Myhrer, F., AIP Conf. Proc. 33, 1976, pp-340-341.

3814. Locher, M. P., Amado, R. D., et al., Phys. Rev., C, 17(1), Jan. 1978, pp-403-405.
3815. Lock, J. A., Nucl. Phys., A, 298(2), 3 April 1978, pp-253-268.
3816. Lodin, G., Nilsson, L., et al., Phys. Scr. 14(1-2), July 1976, pp-27-30.
3817. Lodin, G., Nilsson, L., et al., Tandemaccelerator-laboratoriet, Uppsala (Sweden), TLU-41/75, 1975, p-17.
3818. Loeffler, F., Deutsches Elektronen-Synchrotron, Hamburg, Feb. 1965, p-40.
3819. Loepfe, E., Hoszar, I., et al., J. Labelled Compd. Radio-pharm. 13(2), 1977, p-241.
3820. Logan, C. M., Booth, R., et al., Nucl. Instr. and Meth. 145(1), 1977, pp-77-9.
3821. Logan, C. M., Anderson, J. D., et al., Int. Conf. Radiat. Test Facil. CTR Surf. Mater. Program. 1975, pp-410-20.
3822. Logan, C. M., UCRL-51634, Contract W-7405-eng-48, 8 Aug. 1974, p-13.
3823. Lohs, K. P., and Gal, A., Nucl. Phys., A, 292(3), 5 Dec. 1977, pp-375-412.
3824. Lombard, R., CEA-N-1861 (nd), 1 Oct. 1974-1 Sept. 1975, pp-189-190.
3825. Lombard, R., CEA-N-1861 (nd), 1 Oct. 1974-1 Sept. 1975, pp-211-213.
3826. Lomon, E. L., AIP Conf. Proc. ISSN 0094-243X, 54(1), 15 July 1979, pp-318-331.
3827. Londergan, J. T., Nixon, G. D., et al., Phys. Lett., B, 65(5), 20 Dec. 1976, pp-427-431.
3828. London, H., The Separation of Isotopes,George Newnes Ltd., London, 1961.
3829. Longo, G., and Saporetti, F., Natl. Bur. of Stands. (U. S.), Spec. Publ. 1, Oct. 1975, pp-346-349.
3830. Longo, G., and Saporetti, F., Phys. Lett., B, 65(1), 25 Oct. 1976, pp-15-18.
3831. Loos, J. S., and Dawson, S. L., Am. J. Phys. 46(5), May 1978, pp-560-564.
3832. Loose-Wagenbach, I., and Clausnitzer, G., Nucl. Instr. and Meth. 150(2), 1 April 1978, pp-345-7.
3833. Lorant, M., Electroplating Metal Finishing 16(1), 1963, p-15.
3834. Lostis, P., These de Doctorat ès Sciences Physiques (Université Paris), 1958.
3835. Lotts, A. L., ORNL-4370, pp-193-7.
3836. Lotts, A. L., ORNL-4170, pp-184-91.
3837. Lotts, A. L., ORNL-4870, Oct. 1973, pp-84-86.
3838. Lotts, A. L., ORNL-4470, pp-163-7.
3839. Lougheed, R., and Hulet, E. K., INTDS Proc., Chalk River, Canada, Oct. 1-3, 1974.

3840. Lougheed, R. W., Wild, J. F., et al., BNL-NCS-22550, March 1977, pp-99, 101, 102-104.
3841. Louvel, M., Caen Univ., 14 (France), Lab. de Physique Corpusculaire, These (D. es S.), 1977, p-145.
3842. Lovas, I., Rogge, M., et al., Verh. Dtsch. Phys. Ges. 10 (7), 1975, p-751.
3843. Lovchikova, G. N., Kotel'nikova, G. V., et al., Yad. Fiz. ISSN 0044-0027, 28(5), 1978, pp-1144-1147.
3844. Love, L. O., INTDS Proc., Chalk River, Canada, Oct. 1-3, 1974.
3845. Love, L. O., and Bell, W. A., ORNL-3606, Contract W-7405-eng-26, May 1964, p-53.
3846. Love, L. O., Gwinn, H. R., et al., Nucl. Instr. and Meth. 38, 1965, pp-87-90.
3847. Love, L. O., ORNL-TM-658, Contract W-7405-eng-26, Nov. 14, 1963, p-45.
3848. Love, W. G., Scott, A., et al., Phys. Lett., B, 73(3), 27 Feb. 1978, pp-277-280.
3849. Love, W. G., Phys. Rev., C, 17(5), May 1978, pp-1876-78.
3850. Lovell, S., and Rollinson, E., J. Phys. E, (London), 1 Oct. 1968, pp-1032-5.
3851. Lovett, B., Mishina, M., et al., BNL-19582, 28 Dec. 1974, p-7.
3852. Lovhoiden, G., Andersen, P. H., et al., Nucl. Phys., A, ISSN 0375-9474, 327(1), 17 Sept. 1979, pp-64-76.
3853. Lovoi, P. A., New Mexico Univ., Albuquerque (U. S. A.), Thesis (Ph. D), 1975, p-108.
3854. Low, K. S., Tamura, T., et al., Phys. Rev., C, 13(6), June 1976, pp-2579-2581.
3855. Lowe, A. T., and Fries, R. J., Inertial Confinement Fusion, Dig. Tech. Pap. Top. Meet. (1), 1977, ThE5, p-4.
3856. Lowenheim, A. L., Mod. Electroplating (Wiley and Sons, NY), 1963.
3857. Lowry, M., Schweitzer, J. S., et al., Nucl. Phys., A, 259 (1), 8 March 1976, pp-122-128.
3858. Lozowski, W. R., and Rife, T. M., INTDS Proc., Boston, U. S. A., Oct. 1-3, 1979.
3859. Lozowski, W. R., INTDS Proc., Berkeley, U. S. A., Oct. 19-20, 1977.
3860. Lozowski, W. R., and Rife, T. M., INTDS Newsletter, Aug. 1980.
3861. Lu, P. C., Univ. of IL., Urbana, U. S. A., Univ. Microfilms Order No. 78-04,078, 1977, p-86.
3862. Lucas, R., Nifenecker, H., et al., Europ. Phys. Soc., Geneva (Switz.), Europ. Conf. on Nucl. Phys. with Heavy Ions, 6-10 Sept. 1976, p-138.
3863. Ludewigt, B., Loehner, H., et al., Z. Phys., A, 292(1), Aug. 1979, pp-35-41.

3864. Ludwig, E. J., Clegg, T. B., et al., Phys. Rev. Lett. 40 (7), 13 Feb. 1978, pp-441-444.
3865. Ludziejewski, J., Rurarz, E., et al., Nukleonika 14, 1969, pp-247-51.
3866. Ludziejewski, J., Rurarz, E., et al., INR Rep. No. 1020/ IA/PL, 1968, p-5.
3867. Luedecke, H., and Schroeter, H., Nucl. Instr. and Meth. 97(3), 1971, pp-613-14.
3868. Luers, B., Felvinci, J. P., et al., BNL-NCS-22500, March 1977, pp-70, 75-77.
3869. Lui, Y. W., Karban, O., et al., Nucl. Phys., A, 297(2), 13 March 1978, pp-189-205.
3870. Lukac, P., Radiochem. Radioanal. Lett. 3, 12 March 1970, pp-121-7.
3871. Lukas, I. R., Acad. Rep. Populare Romine, Studii Cercetari Fiz. 15, 1964, pp-483-5.
3872. Lukashunas, N. I., and Semenchuk, G. G., Prib. Tekh. Eksp. 3, May-June 1969, pp-41-4.
3873. Lukner, C., Nucl. Instr. and Meth. 167(2), 1979, pp-249-54.
3874. Lumpkin, A. H., KeKelis, G. J., et al., Phys. Rev., C, 13 (6), June 1976, pp-2564-2567.
3875. Lumpkin, A. H., Morgan, G. R., et al., Phys. Rev., C, 16 (1), July, 1977, pp-220-223.
3876. Lumpkin, A. H., Morgan, G. R., et al., Phys. Rev. Lett. 40 (2), 9 Jan. 1978, pp-104-107.
3877. Lundan, A., and Anttila, K., Nucl. Instr. and Meth. 79, 1970, pp-333-40.
3878. Lundgren, F. A., and Lutz, G. J., Trans. Amer. Nucl. Soc. 10, June 1967, pp-89-90.
3879. Lurie, N. A., Harris, L., Jr., et al., IEEE Trans. Nucl. Sci., NS-22(6), Dec. 1975, pp-2573-2575.
3880. Lushchikov, V. I., Taran, Yu. V., et al., Yadern. Fiz. 10, 1969, pp-1178-94.
3881. Luthardt, G., Proc. of 3rd Conf. on Accel. Targ. Designed for the Prod. of Neutrons, Liege (Belgium), Sept. 18-19, 1967, pp-113-23.
3882. Lutz, H. O., Datz, S., Phys. Rev. Lett. 17, 1966, p-285.
3883. Lutz, G., Nuovo Cim. (10), 53A, Jan. 1, 1968, pp-242-54.
3884. Lutz, G., and Schulz, H. D., Deutsches Elektronon-Synchrotron, Hamburg, Sept. 1967, p-34.
3885. Lux, C. R., Purdue Univ., Lafayette, IN (U. S. A.), Thesis (Ph. D), 1975, p-118.
3886. Lux, C. R., and Porile, N. T., Nucl. Phys., A, 277(3), 21 Feb. 1977, pp-413-428.
3887. Lux, C. R., and Porile, N. T., Phys. Rev., C, 18(1), July 1978, pp-148-157.
3888. Lyashenko, V. I., Pontekorvo, D. B., et al., JINR-R-1-9591, 1976, p-16.

3889. Lynch, F. J., Lewis, R. N., et al., Nucl. Instr. and Meth. 159, 1979, p-245.
3890. Lyttkens, J., Bergqvist, I., et al., Phys. Scr. 13(2), Feb. 1976, pp-96-100.
3891. Lyul'ka, V. A., IAE-1241, 1966, p-33.
3892. Maas, J. W., Elsenaar, R. J., et al., Nederlandse Natuurkundige Vereniging, Amsterdam, Spring Meet. 26 and 27 April 1976, p-59.
3893. Maas, J. W., and Endt, P. M., Nederlandse Natuurkundige Vereniging, Utrecht, Sectie Kernfysica, Autumn Meet. 15 Oct. 1976, p-6.
3894. MacArthur, J. D., and Mak, H. B., IEEE Trans. Nucl. Sci. NS-26(1)Pt. 2, Feb. 1979, pp-1220-1222.
3895. Macauley, M. W. S., Singhal, R. P., et al., J. Phys., G, 2(3), March 1976, pp-L35-L38.
3896. MacDonald, N. S., Birdsall, R., et al., UCLA-12-1082, 1976, pp-118-122.
3897. MacDonald, J. A., Hardy, J. C., et al., INTDS Proc., LASL, U. S. A., Oct. 19-21, 1976.
3898. MacDonald, J. R., Univ. of British Columbia, Vancouver, Can., 1964, p-149.
3899. MacDonald, N. S., Int. Symp. on Radiopharm., Atlanta, GA, U. S. A., 12 Feb. 1974, pp-165-173.
3900. Macfarlane, R. D., Oakey, S. N., et al., Proc. of a Conf. held at St. Catherine's Coll., Oxford, 22-23 Sept. 1969.
3901. Macfarlane, R. D., and McHarris, W. C., Nucl. Spectro. and Reactions, Part A, 1974, pp-243-286.
3902. MacKellar, A. D., ANL/PHY-76-2(Vol. 2), May 1976, pp-663-70.
3903. Macklin, R. L., and Gibbons, J. H., Phys. Rev. 109, 1958, p-105.
3904. Macklin, R. L., Halperin, J., et al., Astrophys. J. 217(1), 1 Oct. 1977, pp-222-226.
3905. Macklin, R. L., and Halperin, J., Nucl. Sci. and Eng. 64 (4), Dec. 1977, pp-849-858.
3906. Macklin, R. L., Ingle, R. W., et al., Nucl. Sci. and Eng. ISSN 0029-5639, 71(2), Aug. 1979, pp-205-208.
3907. Macklin, R. L., Winters, R. R., et al., ORNL/TM-7058, Oct. 1979, p-17.
3908. Macphail, M. R., and Summers-Gill, R. G., Nucl. Phys., A, 263(1), 31 May 1976, pp-12-28.
3909. Madey, R., Waterman, F. M., et al., Phys. Rev., C, 14(3), Sept. 1976, pp-801-806.
3910. Madhusudhan, C. P., Treves, S., et al., J. Radioanal. Chem. ISSN 0134-0719, 53(1-2), 1979, pp-299-305.
3911. Madland, D. C., and England, T. R., LA-6430-MS, ENDF-240, July 1976, p-21.
3912. Madueme, G. Ch., and Arita, K., Nucl. Phys., A, 297(2), March 13, 1978, pp-347-364.

3913. Maeck, W. J., Natl. Bur. of Stands. (U. S.), Spec. Publ. (425), Oct. 1975, pp-378-384.
3914. Maeck, W. J., Emel, W. A., et al., Idaho Natl. Eng. Lab., Idaho Falls (U. S. A.), Dec. 1976, p-18.
3915. Maerker, R. E., ORNL-TM-5204, Jan. 1976, p-36.
3916. Maggiore, C., Goldstone, P., et al., IEEE Trans. Nucl. Sci. NS-24, Feb. 1977.
3917. Maguire, C. F., Ramayya, A. V., et al., ORNL-5306, Sept. 1977, pp-39-41.
3918. Magurno, B. A., BNL-NCS-50446, April 1975, pp-207-285.
3919. Magurno, B. A., and Mughabghab, S. F., BNL-NCS-50446, April 1975, pp-42-50.
3920. Magurno, B. A., and Takahashi, H., BNL-NCS-50446, April 1975, pp-70-73.
3921. Mahendra N., and Veeraraghavan, N., Bhabha Atomic Rsch. Ctr., Bombay (India), March 1, 1976, p-6.
3922. Maher, J. V., Peng, J. C., et al., Phys. Rev., C, 14(6), Dec. 1976, pp-2174-2184.
3923. Mahlmeister, J. E., and Bernstein, K., Calif. Rsch. and Develop. Co., Livermore, CA, LWS-24610, Oct. 13, 1952, p-21.
3924. Mahlmeister, J. E., and Katzer, M. F., LWS-24477, July 28, 1952, p-19.
3925. Mahut, B., CEA-N-1733, June 1974, p-13.
3926. Maier, H. J., Nucl. Instr. and Meth. 167(1), 1979, pp-167-70.
3927. Maier, H. J., INTDS Proc., Boston, U. S. A., Oct. 1-3, 1979.
3928. Maier, H. J., INTDS Proc., Garching, F. R., Sept. 11-14, 1978.
3929. Maier, H. J., and Kutschera, W., INTDS Proc., Garching, F. R., Sept. 11-14, 1978.
3930. Maier, H. J., and Maier-Komor, P., INTDS Proc., Garching, F. R., Sept. 11-14, 1978.
3931. Maier, H. J., and Grossmann, R., ANL/PHY/MSD-76-1, 1976, p-167.
3932. Maier, H. J., and Grossman, R., ANL/PHY/MSD-76-1, 1975, pp-167-187.
3933. Maier, H. J., and Grossmann, R., INTDS Proc., ANL, U. S. A., Sept. 30-Oct. 2, 1975.
3934. Maier, H. J., et al., Jahresbericht 1972 des Beschleuniger-laboratoriums der Universität und der Technischen Universität München
3935. Maier-Komor, P., INTDS Proc., ANL, U. S. A., Sept. 30-Oct. 2, 1975.
3936. Maier-Komor, P., INTDS Proc., Chalk River, Canada, Oct. 1-3, 1974.
3937. Maier-Komor, P., INTDS Proc., Los Alamos, U. S. A., Oct. 19-21, 1976.
3938. Maier-Komor, P., and Ranzinger, E., INTDS Proc., Boston, U. S. A., Oct. 1-3, 1979.

3939. Maier-Komor, P., INTDS Proc., Boston, U. S. A., Oct. 1-3, 1979.
3940. Maier-Komor, P., Nucl. Instr. and Meth. 167(1), 1979, pp-125-8.
3941. Maier-Komor, P., Nucl. Instr. and Meth. 167(1), 1979, pp-73-6.
3942. Maier-Komor, P., ACNTDS, CRNL, Oct. 1-3, 1974.
3943. Maier-Komor, P., Nucl. Instr. and Meth. 102, 1972, p-485.
3944. Maier-Komor, P., TISRMNM, ORNL, Oct. 5-8, 1971.
3944a. Maier-Komor, P., INTDS Proc., Garching, FRG., Sept. 11-14, 1978.
3944b. Maier-Komor, P., INTDS Proc., Garching, FRG., Sept. 11-14, 1978.
3945. Maillet, J. P., Dedonder, J. P., et al., Paris-11, Univ., 91-Orsay (France), Inst. de Physique Nucleaire, 1976, p-44.
3946. Maillet, J. P., Dedonder, J. P., et al., Nucl. Phys., A, 271(2), 2 Nov. 1976, pp-253-278.
3947. Maino, F., Muller, W., et al., EAEC, Karlsruhe, 1969, p-22.
3948. Mairle, G., and Wagner, G. J., Verh. Dtsch. Phys. Ges. 10(7), 1975, p-755.
3949. Mairle, G., and Wagner, G. J., Nucl. Phys., A, 253(2), 24 Nov. 1975, pp-253-262.
3950. Maisel, L. I., and Glang, R., Handbook of Thin Film Tech. (McGraw-Hill Inc., NY), 1970.
3951. Maisheev, V. A., Samoilov, A. V., et al., Inst. Fiziki Vysokikh Ehnergij, 1973, p-25.
3952. Maissel, L. I., and Glang, R., Handbook of Thin Film Tech. (McGraw-Hill, NY), 1970.
3953. Maissel, L. I., Physics of Thin Films 3 (Academic Press, NY), 1966, p-31.
3954. Majka, Z., Phys. Lett., B, 76(2), 22 May 1978, pp-161-64.
3955. Majka, Z., Gils, H. J., et al.,Inst. fuer Angewandte Kernphysik, F.R., Germany, July 1979.
3956. Majka, Z., Gils, H. J., et al., Kneis, W., and Nowicki, G. (eds.), Ann. Rep., KFK-2868, Oct. 1979.
3957. Makosky, L. M., and Hojvat, C., Nucl. Instr. and Meth. 74, 1969, pp-342-4.
3958. Makowska-Rzeszutko, M., Dreves, W., et al., Europ. Phys. Soc., Geneva (Switz.), Europ. Conf. on Nucl. Phys. with Heavy Ions, 6-10 Sept. 1976, p-31.
3959. Makowska-Rzeszutko, M., Egelhof, P., et al., Phys. Lett., B, 74(3), 10 April 1978, pp-187-190.
3960. Maksyutenko, B. P., and Shimanskij, A. A., Yad. Fiz. ISSN 0044-0027, 29(1), Jan. 1979, pp-3-9.
3961. Maksyutenko, B. P., Balakshev, Yu. F, et al., IAEA, Vienna (Austria), Int. Nucl. Data Comm., Nucl. Phys. Rsch. in the USSR, Feb. 1976, pp-1-2.

3962. Maksyutenko, B. P., Balakshev, Yu. F., et al., IAEA, Vienna (Austria), Int. Nucl. Data Comm., Nucl. Phys. Rsch. in the USSR, Feb. 1976, pp-2-4.
3963. Malakhov, I. Ya., Deineko, A. S., et al., Izv. Akad. Nauk SSSR, Ser. Fiz. 33, April 1969, pp-676-8.
3964. Malbrough, D. J., Darden, C. W., et al., Phys. Rev., C, 17(4), April 1978, pp-1395-1401.
3965. Maletskos, C. J., and Irvine, J. W., Nucleonics 14(4), 1956, p-84.
3966. Maliska, C., Brasil. Ser. Vet. 8(6), 1973, pp-91-94.
3967. Malmin, R. E., and Paul, P., SUNY, Stony Brook, NY, ORO-4856-26, 1975, pp-540-541.
3968. Malmin, R. E., Kahn, F., et al., Phys. Rev., C, 17(6), June 1978, pp-2097-2105.
3969. Malmin, R. E., Harris, J. W., et al., Phys. Rev., C, 18 (1), July 1978, pp-163-179.
3970. Managoli, V., Robson, D., et al., Nucl. Phys., A, 290(1), 24 Oct. 1977, pp-128-140.
3971. Manchester, K. E., Sibley, C. B., et al., Nucl. Instr. and Meth. 38, 1965, p-169.
3972. Mancusi, M. D., Bair, J. K., et al., Nucl. Instr. and Meth. 68, Feb. 1, 1969, pp-70-6.
3973. Mancusi, M. D., and Jones, C. M., Nucl. Instr. and Meth. 58, Jan. 1968, pp-351-2.
3974. Manero, F., Proc. of 3rd Conf. on Accel. Targets Designed for the Prod. of Neutrons, Liege (Belgium), Sept. 18-19, 1967, pp-213-23.
3975. Mango, S., Runolfsson, Oe., et al., Nucl. Instr. and Meth. 72, 1969, pp-45-50.
3976. Manin, A., and Cholet, D., Proc. of the 3rd Conf. on Accel. Targets Designed for the Prod. of Neutrons, Liege (Belgium), Sept. 18-19, 1967, pp-31-40.
3977. Mann, F. M., and Schenter, R. E., Nucl. Sci. and Eng. 63 (3), July 1977, pp-242-249.
3978. Mano, H., and Honig, A., Nucl. Instr. and Meth. 124(1), 1975, pp-1-10.
3979. Manohar, S. B., Prasanna Venkatesan, P., et al., Proc. of Nucl. Phys. and Solid State Phys. Symp., Calcutta, Dec. 22-26, 1975, Vol. 18B, pp-138-140.
3980. Mantel, J., Damico, C. J., et al., Amer. J. Roent., Radium Ther. Nucl. Med. 112(4), Aug. 1971, pp-707-14.
3981. Manthos, E. J., Adams, R. E., et al., Trans. Amer. Nucl. Soc. 10, Nov. 1967, pp-480-1.
3982. Mantzouranis, G., Phys. Rev., C, 14(5), Nov. 1976, pp-2018-2020.
3983. Marauhashi, A., and Nakamura, T., Health Phys. 33(5), Nov. 1977, pp-479-81.
3984. Marcinkowski, A., INR, Warsaw, Report 1971, p-54.

3985. Marcks, D. F., and Wehrle, R. B., IEEE Trans. Nucl. Sci. NS-14, June 1967, pp-955-9.
3986. Marcks, D. F., and Wehrle, R. B., IEEE Trans. Nucl. Sci. NS-12(3), June 1965, pp-887-94.
3987. Mark, H., Nucl. Instr. and Meth. 28, June 1964, pp-131-8.
3988. Mark, J. W., IEEE Trans. Nucl. Sci. NS-18(3), June 1971, pp-806-8.
3989. Markham, R. G., Austin, S. M., et al., Nucl. Phys., A, 270(2), 19 Oct. 1976, pp-489-500.
3990. Markham, R. G., and Shahabuddin, M. A. M., Phys. Rev., C, 14(6), Dec. 1976, pp-2037-2043.
3991. Markov, A. N., Fradkov, A. B., et al., Kvantovaya Ehlektron. 4(5), May 1977, pp-1132-1134.
3992. Markovskii, E. A., and Krasnoshchekov, M. M., At. Energ. 18, Jan. 1965, pp-72-3.
3993. Marks, T., Baker, M. P., et al., Phys. Rev. Lett. 38(4), 24 Jan. 1977, pp-149-152.
3994. Maroni, C., Massa, I., et al., Phys. Lett., B, 60(4), 2 Feb. 1976, pp-344-346.
3995. Maroni, C., Massa, I., et al., Nucl. Phys., A, 273(2), 30 Nov. 1976, pp-429-444.
3996. Marquardt, N., Buttlar, H. von, et al., Europ. Phys. Soc., Geneva (Switz.), Europ. Conf. on Nucl. Phys. with Heavy Ions, 6-10 Sept. 1976, p-63.
3997. Marquardt, N., Brunot, B., et al., Europ. Phys. Soc., Geneva (Switz.), Europ. Conf. on Nucl. Phys. with Heavy Ions, 6-10 Sept. 1976, p-64.
3998. Marrs, R. E., Adelberger, E. G., et al., Nucl. Phys., A, 277(3), 21 Feb. 1977, pp-429-441.
3999. Marrs, R. E., Adelberger, E. G., et al., Phys. Rev., C, 16(1), July 1977, pp-61-75.
4000. Marrs, R. E., and Pollock, R. E., AIP Conf. Proc. ISSN 0094-243X, 54(1), 15 July 1979, pp-187-188.
4001. Mars, J., French patent document 2397047/E/. Int. Cl. G21g4/00; C23d5/00; C23f7/02; C23f17/00, 5 July 1977, p-9.
4002. Marsh, R. W., Proc. held at St. Catherine's Coll., Oxford, 22-23 Sept. 1969.
4003. Martin, C. T., Raytheon Co., Quincy, MA, 12 July 1977, p-69.
4004. Martin, J. A., Proc. of the 6th Int. Cyc. Conf., Vancouver, Canada, July 18-21, 1972.
4005. Martin, J. A., IEEE Trans. Nucl. Sci. NS-14(2), April 1967, pp-22-4.
4006. Martin, J. A., and Green, L. F., Nucl. Sci. and Eng. 1(2), May 1956, pp-185-90.
4007. Martin, J. A., Livingston, R. S., et al., Nucleonics, March, 1955.

4008. Martin, J. T., and Smith, R. K., LA-6237-MS, Feb. 1976, p-4.
4009. Martin, P. J., Kempter, V., et al., Int. Conf. on the Phys. of Electronic and Atomic Collisions, 10th, Vol. 2, 1977, pp-1294-1295.
4010. Marty, C., Paris-11 Univ., 91-Orsay (France), Inst. de Physique Nucleaire, Nov. 1977, p-5.
4011. Martz, L. M., Sanders, S. J., et al., Phys. Rev., C, ISSN 0556-2813, 20(4), Oct. 1979, pp-1340-1346.
4012. Maruhn, J. A., Cusson, R. Y., et al., ANL/PHY-76-2(Vol. 2), May 1976, pp-671-679.
4013. Maruhn, J. A., and Greiner, W., Phys. Rev., C, 13(6), June 1976, pp-2404-2412.
4014. Maruhn, J. A., and Cusson, R. Y., Nucl. Phys., A, 270(2), 19 Oct. 1976, pp-471-488.
4015. Maruhn, J. A., Cusson, R. Y., et al., Law. Liv. Lab., CONF-770987-2, 1977, p-22.
4016. Maruhn-Reswani, V., Davies, K. T. R., et al., Phys. Lett., B, 67(2), 28 March 1977, pp-134-138.
4017. Marx, D., Nickel, F., et al., Nucl. Instr. and Meth. 163 (1), 1979, pp-15-20.
4018a. Marx, D., Nickel, F., et al., Nucl. Instr. and Meth. 167 (1), 1979, pp-151-2.
4018b. Marx, D., Nickel, F., et al., INTDS Proc., Garching, FRG., Sept. 11-14, 1978.
4019. Masaike, A., ANL-HEP-PR-77-88, Report 1977, p-31.
4020. Masaike, A., LBL-500, Report 1971, pp-295-9.
4021. Masaike, A., Eisenkremer, M., et al., Rev. Sci. Instr. 41, July 1970, pp-1090-1.
4022. Masaike, A., Sugimoto, S., et al., Nucl. Instr. and Meth. 59, Feb. 1968, pp-170-2.
4023. Maschke, A. W., BNL-6426, Contract AT(30-2)-Gen-16, June 6, 1962, p-6.
4024. Maschke, A. W., BNL-24377, CONF-780343-4, Report 1978, p-8.
4025. Mashkarov, Yu. G., and Dejneko, A. S., Prib. Tekh. Ehksp. 1, Jan. 1976, p-35.
4026. Mashkarov, Yu. G., Dejneko, A. S., et al., Izv. Akad. Nauk SSSR, Ser. Fiz. 40(10), Oct. 1976, pp-2189-2193.
4027. Mason, R. J., Fries, R. J., et al., Appl. Phys. Lett. 34 (1), 1979, pp-14-16.
4028. El Masri, Y., Ferte, J. M., et al., Nucl. Phys., A, 271 (1), 26 Oct. 1976, pp-133-161.
4029. Massaad, M. J., Grenoble-1 Univ., 38(France), Inst. des Sciences Nucleaires, These (3e Cycle), 1975, p-107.
4030. Massey, B. J., ORNL-2638, Contract W-7405-eng-26, March 26, 1959, p-5.
4031. Massey, B. J., ORNL, 5th Nat. Symp. on Vac. Tech. Trans., 1958, p-72-5.

4032. Massey, B. J., ORNL-2237, Feb. 21, 1957, p-10.
4033. Mateja, J. F., Notre Dame Univ., IN, (U. S. A.), Thesis (Ph. D), 1976, p-114.
4034. Mateja, J. F., Bieszak, J. A., et al., Phys. Rev., C, 13 (6), June 1976, pp-2269-2287.
4035. Mateja, J. F., Browne, C. P., et al., Phys. Rev., C, 15(5), May 1977, pp-1708-1718.
4036. Mathews, G. J., Sobotka, L. G., et al., LBL-8151, 1978, p-95.
4037. Mathews, G. J., Bigeleisen, P., et al., LBL-8151, 1978, pp-179-180.
4038. Mathews, J., and Buthala, D. A., Rev. Sci. Instr. 34, 1963, p-592.
4039. Mathews, J. W., and Grünbaum, E., Phil. Mag. 11, 1965, p-1233.
4040. Mathiak, E., Eberhard, K. A., et al., Nucl. Phys., A, 259 (1), 8 March 1976, pp-129-156.
4041. Matoba, M., Hyakutake, M., et al., Nucl. Phys., A, ISSN 0375-9474, 325(2-3), 20-27 Aug. 1979, pp-389-396.
4042. Matsuda, S., and Uematsu, T., Nucl. Phys., B, B168(1), 1980, pp-181-8.
4043. Matsuki, S., Yasue, M., et al., Nucl. Instr. and Meth. 94 (2), 1971, p-387.
4044. Matthews, J. L., and Owens, R. O., Nucl. Instr. and Meth. 91, 1971, pp-37-43.
4045. Matthews, J. L., Findlay, D. J. S., et al., Nucl. Phys., A, 267(1), 16 Aug. 1976, pp-51-76.
4046. Matthews, L. D., and Fortner, J. R., Sci. and Ind. Applications of Small Accel., 4th Conf., 1976.
4047. Mattox, D. M., Sandia Corp. Monograph, SC-R-65-582, 1965.
4048. Matusevich, V. A., Chernov, I. P., et al., Summaries of reports of 26th Conf. on Nucl. Spectro. and Nucl. Structure, Baku, 3-6 Feb. 1976, p-203.
4049. Matusevich, V. A., Chernov, I. P., et al., Summ. of reports of 26th Conf. on Nucl. Spectro. and Nucl. Structure, Baku, 3-6 Feb. 1976, p-204.
4050. Mavis, D. G., Dunham, J. S., et al., AIP Conf. Proc. 35, High Energy Phys. Polariz. Beams Targets, 1976, pp-517-518.
4051. Maxman, S. H., Nucl. Instr. and Meth. 50, 1967, pp-53-60.
4052. Maxman, S. H., Rev. Sci. Instr. 35, Nov. 1964, pp-1572-3.
4053. Maxson, R. D., Jolly, K. R., et al., Nucl. Instr. and Meth. 62, 1968, p-276.
4054. Mayer, A., and Catlett, D. S., Inertial Confinement Fusion, Dig. Tech. Pap. Top. Meet. 1, ThE8, 1977, p-3.
4055. Mayer-Boericke, C., and Turek, P., Nucl. Phys., A, 252(2), 10 Nov. 1975, pp-333-342.
4056. Mayfield, R. M., Tope, W. G., et al., ANL-6489, Dec. 1962.

4057. Maynard, M., Palmer, D. C., et al., J. Phys., G, Nucl. Phys., ISSN 0305-4616, 3(12), Dec. 1977, pp-1735-52.

4058. Mazan'ko, B. V., Shevchenko, N. G., et al., Kharkov, Fiziko-Tekhnicheskij Inst., 1977, p-7.

4059. McAdams, R. E., Goulding, C. A., et al., AIP Conf. Proc. ISSN 0094-243X, 47, 1978, pp-640-641.

4060. McAdams, R. E., Otteson, O. H., et al., AIP Conf. Proc. ISSN 0094-243X, 54(1), 15 July 1979, pp-178-179.

4061. McAdams, W. H., Heat Transmission, McGraw-Hill Kogakuska, Ltd., Tokyo, 1954.

4062. McAlpine, R. D., ACNTDS, CRNL, Oct. 1-3, 1974.

4063. McCall, R. C., Nelson, W. R., et al., Proc. of 2nd Int. Conf. on Accel. Dosim. and Experience, SLAC, Stanford, CA, November 1969, pp-684-691.

4064. McCallum, G. J., and Sowerby, B. D., Nucl. Instr. and Meth. 47(1), 1967, pp-45-54.

4065. McCarthy, J., Orphanos, L., et al., Virginia Univ., Charlottesville (U. S. A.), Exp. on the Nucl. Interactions of Pions and Electrons, Progress Rept., Dec. 1, 1976-Nov. 30, 1977, p-1-5.

4066. McCarthy, J., Whitney, R. R., Exp. on the Nucl. Inter. of Pions and Electrons, Progress Rept., Dec. 1, 1976-Nov. 30, 1977, p-6.

4067. McCarthy, J., Minehart, R. C., Exp. on the Nucl. Inter. of Pions and Electrons, Progress Rept., Dec. 1, 1976-Nov. 30, 1977, pp-8-11.

4068. McCarthy, J. S., Whitney, R., et al., Stanford Univ., CA, Contract Nonr-225(67), HEPL-619, Nov. 1969, p-17.

4069. McCaslin, J. B., and Smith, A. R., UCRL-19374, Contract W-7405-eng-48, Oct. 1969, p-13.

4070. McClintock, I. S., and Orr, J. C., Chem. and Phys. of Carbon 11, 1973.

4071. McCormick, R. D., and McCormack, J. D., Nucl. Instr. and Meth. 13, Sept. 1961, pp-147-52.

4072. McCreary, W. J., and Catlett, D. S., LASL, Inertial Confinement Fusion, Dig. Tech. Pap. Top. Meet. (1), ThE7, 1977, p-4.

4073. McCulloch, D., Grusell, E., et al., Nucl. Instr. and Meth. ISSN 0029-554X, 165(2), 1 Oct. 1979, pp-283-287.

4074. McDaniel, E. W., Love, L. O., et al., INTDS Proc., Chalk River, Canada, 1974.

4075. McDaniel, Sinram, et al., BNL-NCS-22500, March 1977, pp-315-317.

4076. McDaniels, D. K., Bergqvist, I., et al., Nucl. Instr. and Meth. 99(1), 1972, pp-77-80.

4077. McDonald, A. B., Alexander, T. K., et al., Nucl. Phys., A, 287(1), 29 Aug. 1977, pp-189-194.

4078. McDonald, A. B., Alexander, T. K., et al., Nucl. Phys., A, 258(1), 16 Feb. 1976, pp-152-156.

4079. McDonald, A. B., Alexander, T. K., et al., Nucl. Phys., A, 273(2), 30 Nov. 1976, pp-451-463.
4080. McDonald, J., and Sjoestrand, N. G., Natl. Bur. of Stands. (U. S.), Spec. Publ. 425, Oct. 1975, pp-810-812.
4081. McElligott, P. E., and Roberts, R. W., J. Appl. Phys. 37, 1966, p-1992.
4082. McEllistrem, M. T., Lachkar, J., et al., Natl. Bur. of Stands. (U. S.), Spec. Publ. 425, Oct. 1975, pp-942-945.
4083. McEllistrem, Coope, et al., BNL-NCS-22500, March 1977, pp-321-323.
4084. McFadden, R. C., and Martin, P. W., Nucl. Instr. and Meth. 113(4), 1973, pp-601-2.
4085. McFee, J. E., Prestwich, W. V., et al., Phys. Rev., C, 13(5), May 1976, pp-1864-1873.
4086. McGeorge, J. C., Shotter, A. C., et al., Nucl. Phys., A, ISSN 0375-9474, 326(1), 3 Sept. 1979, pp-108-118.
4087. McGovern, D. E., Stanford Univ., CA, SAND-74-5393, 1974, p-5.
4088. McGrath, R. L., Cormier, T. M., et al., ANL/PHY-76-2(2), May 1976, pp-681-689.
4089. McGTegart, W. J., The Electrolytic and Chem. Polishing of Metals (London: Pergamon), 1959.
4090. McHarris, W. C., Mich. State Univ., East Lansing, Contract AT(11-1)-1779, Dec. 1, 1968, p-71.
4091. McHenry, H. I., Read, D. T., et al., NBS, Boulder, CO (U. S. A.), Mater. Studies for Magnetic Fusion Energy Applications at Low Temp. II, NBSIR-79-1609, June 1979, pp-297-312.
4092. McKane, R. H., and Honkonen, D. L., IAEA Symp. on Non-destructive Test. in Nucl. Tech., Bucharest, March 1965, p-15.
4093. McKee, J. S. C., UCRL-17675, Contract W-7405-eng-48, Aug. 9, 1967, p-5.
4094. McKeown, R., and Garvey, G. T., Phys. Rev., C, 16(1), July 1977, pp-482-485.
4095. McKeown, R. D., Ph. D. Thesis, Princeton Univ., unpublished.
4096. McNally, J. H., Barnes, J. W., et al., Phys. Rev., C, 9, 1974, p-717.
4097. McNaughton, M. W., King, N. S. P., et al., Nucl. Instr. and Meth. 129(1), 1 Nov. 1975, pp-241-245.
4098. McNaughton, M. W., King, N. S. P., et al., Nucl. Instr. and Meth. 130(2), 15 Dec. 1975, pp-555-557.
4099. Mead, W. C., Report 1977, UCRL-79736, p-25.
4100. Mead, W. C., Lindl, J. D., et al., Report 1977, UCRL-80005, p-22.
4101. Meadows, J., Poenitz, W., et al., ANL/NDM-35, Sept. 1977, p-108.

4102. Meadows, J. W., ANL/NDM-39, March 1978, p-20.
4103. Mechtersheimer, G., Bueche, G., et al., Phys. Lett., B, 73(2), 13 Feb. 1978, pp-115-118.
4104. Meckback, W., Rev. Sci. Instr. 34(2), Feb. 1963, p-188.
4105. Medsker, L. R., Headley, S. C., et al., Phys. Rev., C, 17(1), Jan. 1978, pp-51-55.
4106. Medyanik, V. N., and Kadaner, L. I., Ukr. Fiz. Zh. 13, Jan. 1968, pp-127-9.
4107. Medyanik, V. N., Karev, V. N., et al., Ukr. Fiz. Zh. 11, 1966, pp-560-2.
4108. Medyanik, V. N., Karev, V. N., et al., Ukr. Fiz. Zh. 9, July 1964, pp-798-9.
4109. Meek, R. L., Gibson, W. M., et al., Nucl. Instr. and Meth. 94(3), 1971, pp-435-42.
4110. Meeker, D. J., and Nuckolls, J. H., UCRL-78469, Report 1976, p-27.
4111. Meeker, D. J., Nuckolls, J. H., et al., UCRL-77045, Report 1975, p-24.
4112. Meens, A., INTDS Newsletter, Feb. 1980.
4113. Meens, A., INTDS Newsletter, Jan. 1979.
4114. Meens, A., INTDS Proc., Boston, U. S. A., Oct. 1-3, 1979.
4115a. Meens, A., Nucl. Instr. and Meth. 167(1), 1979, p-173.
4115b. Meens, A., INTDS Proc., Garching, FRG., Sept. 11-14, 1978.
4116. Meens, A., INTDS Proc., Berkeley, U. S. A., Oct. 19-20, 1977.
4117. Meens, A., INTDS Proc., LASL, U. S. A., Oct. 19-21, 1976.
4118. Meens, A., INTDS Proc., LASL, U. S. A., Oct. 19-21, 1976.
4119. Mehr, D. L., and McLaughlin, E. F., Rev. Sci. Instr. 34, 1963, pp-104-5.
4119a. Mehr, D. L., UCRL-9824, Contract W-7405-eng-48, Aug. 17, 1961.
4120. Mehtra, M. K., Kailas, S., et al., Pramana 9(4), Oct. 1977, pp-419-34.
4121. Meier, K. L., Proc. Top. Meet. Technol. Controlled Nucl. Fusion 1978, 3rd (2), pp-1032-40.
4122. Meier, K. L., LA-UR-77-2398, Report 1977, p-6.
4123. Meijer de, R. J., Kamermans, R., et al., Phys. Rev. C, 16(6), Dec. 1977, pp-2442-2444.
4124. Meijer de, R. J., Stahel, D. P., et al., Kernfysisch Versneller Inst., Netherlands, KVI Annual Report 1978, pp-28-30.
4125. Meinel, K., and Debertin, K., Inst. fuer Kernphysik, Universitaet Frankfurt, Aug. 1966, p-24.
4126. Melssen, J. P. M. G., and Poppema, O. J., Nederlandse Natuurkundige Vereniging, Utrecht, Sectie Kern-fysica, Autumn meeting, Oct. 1977, p-12.
4127. Menchaca-Rocha, A., Buenerd, M., et al., LBL-5075, 1975, pp-91-93.
4128. Menchaca-Rocha, A., Buenerd, M., et al., ANL/PHY-76-2(2), May 1976, pp-691-698.

4129. Mende, A., Otto, G., et al., Kernenergie 12, June 1969, pp-209-16.
4130. Mendes, A. T. M., Veta, N., et al., Cienc. Cult. (San Paulo), 27(7), July 1975, p-22.
4131. Mendes, A. T. M., Ueta, N., et al., Rev. Bras. Fis. 6(2), Aug. 1976, pp-139-160.
4132. Menet, J., Lucas, J. J., et al., Europ. Phys. Soc., Geneva (Switz.), Europ. Conf. on Nucl. Phys. with Heavy Ions, 6-10 Sept. 1976, p-124.
4133. Menet, J., Cole, A. J., et al., J. Phys. 38(9), Sept. 1977, pp-1051-59.
4134. Menti, W., Martin, M., et al., Nucl. Instr. and Meth. 31, Dec. 1, 1964, pp-25-8.
4135. Merdinger, J. C., Schulz, N., et al., CRN-PN-75-37, 1975, p-15, Strasbourg-1 Univ., 67 (France), Centre de Recherches Nucleaires.
4136. Merdinger, J. C., Bozek, E., et al., CRN-PN-75-39, 1975, p-7.
4137. LeMere, M., Tang, Y. C., et al., Phys. Rev. 14(1), July 1976, pp-23-27.
4138. Meritet, L., and Proriol, J., Nucl. Phys., A, ISSN 0375-9474, 324(2-3), 30 July-6 Aug. 1979, pp-420-444.
4139. Merkel, M., and Muenzel, H., AED-Conf-78-274-000, Lindau, Germany, F. R., 16-19 Oct. 1978.
4140. Merritt, J. S., ACNTDS, CRNL, Oct. 1-3, 1974.
4141. Merton, T. R., and Hartley, H., Nature 105, 1920, p-104.
4142. Merz, T., Princ. of Nucl. Med., (Wagner, H. N., Jr. (ed.), W. B. Saunders Co., Philadelphia), 1968, pp-722-41.
4143. Mesko, L., and Valek, A., Atomki Kozlem 15(1), 1973, pp-61-63.
4144. Mesquita Bueno de, K. G., Inst. voor Kernphysisch Onderzoek, Amsterdam (Netherlands), Amsterdam Univ., Proefschrift (Dr.), Dec. 1977, p-151.
4145. Metzger, F. R., Phys. Rev., C, 15(6), June 1977, pp-2253-4.
4146. Meyer, F. W., and Anderson, L. W., Univ. of Wisconsin, Madison, WI, U. S. A. 53706.
4147. Meyer, H. O., Nucl. Instr. and Meth. 120(1), 15 Aug. 1974, pp-143-146.
4148. Meyer, H. O., Weitkamp, W. G., et al., Helv. Phys. Acta 48(4), 25 Nov. 1975, p-559.
4149. Meyer, H. O., Weitkamp, W. G., et al., Nucl. Phys., A, 269(2), 5 Oct. 1976, pp-269-280.
4150. Meyer, H. O., and Plattner, G. R., Helv. Phys. Acta 49(5), 29 Oct. 1976, p-788.
4151. Meyer, H. O., and Plattner, G. R., Nucl. Phys., A, 279(1), 21 March 1977, pp-53-69.
4152. Meyer, J., Nahabetian, R. S., et al., AIP Conf. Proc. ISSN 0094-243X, 47, 1978, pp-746-747.

4153. Meyer, M. A., Venter, I., et al., Nucl. Phys., A, 264(1), 21 June 1976, pp-13-29.
4154. Meyer, R. A., Henry, E. A., et al., TEICCTA, IU, U. S. A., Sept. 18-21, 1978.
4155. Meyer, S. L., Thesis, Princeton Univ., Princeton, NJ, 1962, p-133.
4156. Meyer, W. G., Viola, V. E., Jr., et al., Phys. Rev., C, ISSN 0556-2813, 20(5), Nov. 1979, pp-1716-1730.
4157. Meyer-Schuetzmeister, L., Segel, R. E., et al., Phys. Rev., C, 17(1), Jan. 1978, pp-56-65.
4158. Michael, H., Qaim, S. M., et al., Inst. fuer Kernphysik Annual Report 1978, Pt. 3, pp-146-147.
4159. Michaelis, E. G., Proc. of the 6th Int. Cyc. Conf., Vancouver, Canada, July 18-21, 1972.
4160. Michaelis, E. G., and Skarek, P., Nucl. Instr. and Meth. 89, 1970, pp-81-6.
4161. Michalowski, S., Andrews, D., et al., Phys. Rev. Lett. 39(12), 19 Sept. 1977, pp-737-740.
4162. Michel, F., and Vanderpoorten, R., Phys. Rev., C, 16(1), July 1977, pp-142-152.
4163. Michel, F., Inst. fuer Angewandte Kernphysik 1, Karlsruhe, Germany, F. R., 2-4 May 1979, p-200-208.
4164. Michielsen, A., INTDS Newsletter, July 1979.
4165. Michielsen, A., INTDS Newsletter, August, 1980.
4166. Micklinghoff, M., Nucl. Phys., A, 295(2), 30 Jan. 1978, pp-237-255.
4167. Miehe, C., Centre de Recherches Nucleaires, Strasbourg-1 Univ., 67 (France), These (D. es S.), 1975, p-167.
4168. Miescher, E., Helv. Phys. Acta. 14, 1941, p-507.
4169. Migneco, E., Bonsignore, P., et al., Nat. Bur. of Stands. (U. S.), Spec. Publ. 425, Oct. 1975, pp-607-610.
4170. Mihailova, V. Z., and Dermendzhiev, E. G., Nucl. Instr. and Meth. 66, Dec. 1, 1968, pp-25-8.
4171. Mikhalyak, S., Moscow State Univ., 1961, p-12.
4172. Mikulski, J., and Szeglowski, Z., INP-406/C, April 1965, p-4.
4173. Milder, F. L., Janecke, J., et al., Univ. of Michigan, Ann Arbor, U. S. A., ORO-4856-26, 1975, pp-409-410.
4174. Milder, F., Blecher, M., et al., Phys. Lett., B, 72(2), 19 Dec. 1977, pp-159-162.
4175. Mildner, D. F. R., Carpenter, J. M., et al., Rev. Sci. Instr. 45(4), April 1974, pp-572-575.
4176. Millener, D. J., and Alburger, D. E., Phys. Rev., C, ISSN 0556-2813, 20(5), Nov. 1979, pp-1891-1901.
4177. Miller, C. A., Manitoba Univ., Winnipeg (Canada), Dept. of Physics, Thesis (Ph. D), 1974, p-84.
4178. Miller, D. J., RHEL/R-245, 19 Nov. 1971, pp-19-27.
4179. Miller, E. K., Brittingham, J. N., et al., UCRL-52211, Contract W-7405-eng-48, 10 Jan. 1977, p-62.

4180. Miller, G. A., LA-7892C, July 1979, pp-108-118.
4181. Miller, J. R., Adv. Cryog. Eng. 23, 1978, pp-669-75.
4182. Miller, L. D., Phys. Rev., C, 14(2), Aug. 1976, pp-706-717.
4183. Miller, L. D., and Weber, H. J., Phys. Lett., B, 64(3), 27 Sept. 1976, pp-279-282.
4184. Miller, L. D., and Weber, H. J., AIP Conf. Proc. 33, 1976, pp-268-269.
4185. Miller, L. D., and Weber, H. J., Phys. Rev., C, 17(1), Jan. 1978, pp-219-226.
4186. Millett, R. J., and Wilson, E. J., AERE-Bib-127, Sept. 1959, p-37.
4187. Minamisono, T., Hugg, J. W., et al., Phys. Lett., B, 61 (2), 15 March 1976, pp-155-157.
4188. Minamisono, T., and Ramsay, D., INTDS Proc., Los Alamos, U. S. A., Oct. 19-21, 1976.
4189. Minamisono, T. Hugg, J. W., et al., Phys. Rev., C, 14(6), Dec. 1976, pp-2335-2337.
4190. Miracle, D. E., Hagen, E. C., et al., Phys. Rev., C, 16 (1), July 1977, pp-111-119.
4191. Mitchell, I. V., ACNTDS, CRNL, Oct. 1-3, 1974.
4192. Mitchell, R. J., Ragland, T. V., et al., Phys. Rev., C, 16(4), Oct. 1977, pp-1605-1608.
4193. Mito, A., Komura, K., et al., OULNS-69-1, 1968, p-22.
4194. Mitrofanova, A. V., Noga, V. I., et al., Yad. Fiz. 25(5), May 1977, pp-926-929.
4195. Miura, H., and Torizuka, Y., Phys. Rev., C, 16(4), Oct. 1977, pp-1688-1691.
4196. Mizumoto, M., and Takekoshi, H., Tokyo Univ. Nucl. Eng. Rsch. Lab., 1975, pp-67-75.
4197. Mizumoto, M., Macklin, R. L., et al., Phys. Rev., C, 17 (2), Feb. 1978, pp-522-528.
4198. Mlekodaj, R. L., Spejewski, E. H., et al., Nucl. Instr. and Meth. 139, 1 Dec. 1976, pp-299-303.
4199. Moak, C. D., et al., Rev. Sci. Instr. 30, 1959, p-694.
4200. Moake, G. L., Gutay, L. J., et al., Phys. Rev. Lett., ISSN 0031-9007, 43(13), 24 Sept. 1979, pp-910-913.
4201. Moake, G. L., Debevec, P. T., et al., LA-7892C, July 1979, pp-326-338.
4202. Moalem, A., Benenson, W., et al., Phys. Lett., B, 61(2), 15 March 1976, pp-167-170.
4203. Moalem, A., Benenson, W., et al., Israel Physical Soc., Jerusalem, 1976 Annual Meeting, p-11.
4204. Moalem, A., Gaillard, Y., et al., Phys. Rev., C, ISSN 0556-2813, 20(4), Oct. 1979, pp-1593-1596.
4205. Mobley, R., USAEC, TID-25473(1), 1969, pp-1-2.
4206. Modena, I., Montelatici, V., et al., LNF-66/1, Jan. 21, 1966, p-8, Nucl. Instr. and Meth. 44(1), 1966, p-175-8.
4207. Modirnia, M., Centre d'Etudes Nucleaires, Bordeaux-1 Univ., 33 (France), These (3e Cycle), 1976, p-46.

4208. Moffa, P. J., Dover, C. B., et al., Phys. Rev., C, 16(5), Nov. 1977, pp-1857-1864.
4209. Mohler, J. B., Electroplating and Related Processes, Chem. Pub. Co., Inc., NY, 1969, p-106.
4210. Molen, A. V., Janecke, J., et al., Univ. of Mich., Ann Arbor, ORO-4856-26, 1975, pp-413-414.
4211. Mollenauer, J. F., Rev. Sci. Instr. 36, July 1965, p-1044.
4212. Moller, W., and Williams, J. S., Nucl. Instr. and Meth. 157(2), 1978, pp-205-11.
4213. Mollet, W., Biel, J., et al., Phys. Rev. Lett. 39(26), 26 Dec. 1977, pp-1646-1649.
4214. Monahan, J. E., Elwyn, A. J., et al., Nucl. Phys., A, 269 (1), 28 Sept. 1976, pp-61-73.
4215. Moneti, G., and Montelatici, V., Nucl. Instr. and Meth. 15, March 1962, pp-207-8.
4216. Monies, J. A., Manes, K. R., et al., UCRL-80453, CONF-780324-1, Report 1978, p-18.
4217. Monigold, G., McDaniel, D. F., et al., Sci. and Ind. Appl. of Small Accel., 4th Conf., 1976.
4218. Moniz, E. J., Nixon, G. D., et al., High-Energy Phys. Nucl. Struct., Proc. Int. Conf., 3rd, 1970, pp-321-3.
4219. Monteiro, A. M. M., Aragao, F. R. F., et al., Rev. Bras. Fis. 7(2), Aug. 1975, pp-197-208.
4220. Montelatici, V., LNF-72/14, 16 Feb. 1972, p-56.
4221. Montelatici, V., Nucl. Instr. and Meth. 29, Sept. 1964, pp-121-4.
4222. Moore, G. E., and Kemper, K. W., Phys. Rev., C, 14(3), Sept. 1976, pp-977-991.
4223. Morales, J. R., Romero, J. L., et al., Nucl. Instr. and Meth. 119(1), 1 July 1974, pp-91-92.
4224. Moran, J. P., Wachman, H. Y., et al., MIT, 1967, p-20.
4225. Morand, C., Agard, M., et al., Phys. Rev., C, 13(6), June 1976, pp-2182-2188.
4226. Morand, C., Berthet, B., et al., Z. Phys., A, 278(2), Aug. 1976, pp-189-199.
4227. Morand, C., Berthet, B., et al., Grenoble-1 Univ., 38 (France), Inst. des Sciences Nucleaires, 1976, p-28.
4228. Morando, M., Stranieri, A., et al., Phys. Lett., B, 64(2), 13 Sept. 1976, pp-140-142.
4229. Moreh, R., and Samuel, D., Rev. Sci. Instr. 33, 1962, pp-1292-3.
4230. Moreh, R., and Shahal, O., Nucl. Phys., A, 262(2), 17 May 1976, pp-221-230.
4231. Moret, H., and Verheyen, F., Nucl. Instr. and Meth. 102, 1972, pp-575-580.
4232. Moret, H., and Verheyen, F., TISRMNM, ORNL, Oct. 5-8, 1971.
4233. Moret, H., and Verheyen, F., 3rd Int. Symp. on Rsch. Mat. for Nucl. Measure., Gatlinburg, TN, 5 Oct. 1971.
4234. Moret, H., and Louwerix, E., Microbalance for Ultrahigh-Vacuum Appl., Vacuum Microbalance Tech. 5, 1966, p-59.

4235. Moret, H., Louwerix, E., et al., Vacuum Microbalance Tech. 7, 1970, p-173.
4236. Moret, H., PSPSITF, AERE, Oct. 20 & 21, 1965.
4237. Moret, H., Nucl. Instr. and Meth. 15, May 1962, pp-357-8.
4238. Moret, H., AERE-R-5097, Paper 25, p-7
4239. Moretti, A., Yokosawa, A., et al., Rev. Sci. Instr. 38, Sept. 1967, pp-1335-6.
4240. Moretti, A., Suwa, S., et al., Proc. of 2nd Int. Symp. on Polar. Phenom. of Nucleons, Huber, P., and Schopper, H. (eds.), Basel and Stuttgart, Birkhaeuser Verlag, 1966, pp-128-31.
4241. Moretti, A., Suwa, S., et al., Experientia, Suppl. 12, 1966, pp-128-31.
4242. Moretto, L. G., and Sventek, J. S., LBL-5006, CONF-760424-8, March 1976, p-62.
4243. Moretto, L. G., Galin, J., et al., Nucl. Phys., A, 259(1), 8 March 1976, pp-173-188.
4244. Moretto, L. G., Cauvin, B., et al., Phys. Rev. Lett. 36 (18), 3 May 1976, pp-1069-1072.
4245. Morgan, G. L., ORNL/TM-5531, Aug. 1976, p-55.
4246. Morgan, G. L., and Perey, F. G., Nucl. Sci. and Eng. 61(3), Nov. 1976, pp-337-345.
4247. Morgan, G. L., and Perey, F. G., ORNL/TM-5829, June 1977, p-44.
4248. Morgan, G. R., and Fletcher, N. R., Phys. Rev., C, 16(1), July 1977, pp-167-176.
4249. Morgan, T. L., and Freeman, J., A. Rep. Schuster Lab., Univ. Manchester, VIC, 1975.
4250. Morgan, T. L., INTDS Proc., Berkeley, U. S. A., Oct. 19-20, 1977.
4251. Morgan, T. L., LBL-7950, Aug. 1978, pp-139-140.
4252. Moreh, R., and Samuel, D., Rev. Sci. Instr. 33,1962, pp-1292-3.
4253. Moorhead, R. D., and Poppa, H., Proc. 27th Annual EMSA Meeting, St. Paul, MN, Claitors Publ. Co., Baton Rouge, 1969.
4254. Morimoto, K., LBL-500, 1971, pp-33-41.
4255. Morochko, V. P., Yakushin, B. F., et al., Svar. Proizvod. ISSN 0491-6441(8), Aug. 1979, pp-3-6.
4256. Morokhovskii, V. L., Kasilov, V. I., et al., Physical-Technical Inst., Kharkov, KHFTI-73-26, 1973, pp-58-60.
4257. Morozov, A. M., and Shalnikov, A. I., Pribory i Tekhn. Eksperim. 6, Nov.-Dec. 1963, pp-169-70.
4258. Morris, C. L., Boudrie, R. L., et al.,Univ. Virginia, Charlottesville, VA.
4259. Morris, C. L., Thiessen, H. A., et al., Phys. Rev. Lett. 39(23), 5 Dec. 1977, pp-1455-1457.
4260. Morris, C. L., Piffaretti, J., et al., Phys. Lett., B, ISSN 0370-2693, 86(1), 10 Sept. 1979, pp-31-33.

4261. Morris, C. L., and Thornton, S. T., Nucl. Instr. and Meth. 96(2), 1971, pp-281-3.
4262. Morris, J. M., and Ophel, T. R., Aust. Nat. Univ., Canberra, Rsch. School of Physical Sci., ANU-P-414, p-4.
4263. Morrison, R. T., and Mincey, E. K., Proc. 6th Int. Cyc. Conf., Vancouver, Canada, July 18-21, 1972.
4264. Mors, P. M., Cienc. Cult. 27(7), July 1975, p-11.
4265. Mors, P. M., Herscovitz, V. E., et al., Rev. Bras. Fis. 7(2), Aug. 1975, pp-283-296.
4266. Morsch, H. P., Becker, T., et al., Nucl. Instr. and Meth. 68, Feb. 1, 1969, pp-39-41.
4267. Morsch, H. P., Peterson, J. F., et al., ANL/PHY-76-2(Vol. 2), May 1976, pp-699-705.
4268. Morsch, H. P., Decowski, P., et al., Phys. Rev. Lett. 37(5), 2 Aug. 1976, pp-263-265.
4269. Morsch, H. P., and Ellis, P. J., Phys. Lett., B, 64(4), 11 Oct. 1976, pp-386-388.
4270. Morsch, H. P., Rogge, M., et al., Phys. Rev., C, ISSN 0556-2813, 20(4), Oct. 1979, pp-1600-1602.
4271. Mortimer, A. R., and Stokoe, J. R., RHEL/R-237, Aug. 1971, p-84.
4272. Morton, T. H., BNWL-1051(Pt. 3), pp-28-30.
4273. Motz, H. T., Europ-Am. Nucl. Data Comm., Contract W-7405-eng-36, April 13-14, 1967, p-153.
4274. Moskvin, L. N., and Tsaritsyna, L. G., At. Energ. 24, April 1968, pp-383-4.
4275. Mostafa, Golam, A. B. M., and Solaija, T. J., Nucl. Instr. and Meth. 93(1), 1971, pp-153-6.
4276. Mostafa, A. B. M. G., Nucl. Instr. and Meth. 129(1), 1 Nov. 1975, pp-251-255.
4277. Mostafa, A. B. M. G., Nucl. Instr. and Meth. 125(4), 1975, pp-493-6.
4278. Motobayashi, T., Kohno, I., et al., AIP Conf. Proc. ISSN 0094-243X(47), 1978, pp-722-723.
4279. Mougey, J., Paris-11 Univ., 91-Orsay(France), These, 1976, p-251.
4280. Mougey, J., CEA-CR-7 (nd), pp-355-384.
4281. Mougey, J., Bernheim, M., et al., Nucl. Phys., A, 262(3), 24 May 1976, pp-461-492.
4282. Moulton, J., Wozniak, G., et al., LBL-5075, 1975, pp-126-7.
4283. Mouraille, H., CEA-N-1117, Feb. 1970, p-11.
4284. Moyer, A. R., Phys. Rev. C5, 1972, p-1678.
4285. Moyle, R. A., Georgetown Univ., Washington, D.C. (U. S. A.), Thesis (Ph. D), 1975, p-191.
4286. Mrozowski, S., Bull. Acad. Poland I, 1930, p-464.
4287. Mrozowski, S., Z. Physik 73, 1932, p-826.
4288. Muehlen, H. J., and Cleff, B., Nucl. Instr. and Meth. 155 (1), 1978, pp-193-4.

4289. Muellen, G., and Aumann, D. C., Nucl. Instr. and Meth. 128(3), 15 Oct. 1975, pp-425-428.
4290. Mueller, D., and Brunner, G., Isotopen Tech. 2, Sept. 1962, pp-257-61.
4291. Mueller, D., Kashy, E., et al., Phys. Lett., B, 59(3), 10 Nov. 1975, pp-233-235.
4292. Muenzenberg, G., Schoett, H. J., GSI-79-11, Oct. 1979, p-158.
4293. Mueschenborn, G., AERE-R-5097, Paper 7, p-4.
4294. Muggleton, A. H. F., INTDS Newsletter, Feb. 1980.
4295. Muggleton, A. H. F., Report 1979, ANU-P-719, p-86.
4296. Muggleton, A. H. F., J. Phys., E, 12(9), Sept. 1979, pp-780-807.
4297. Muggleton, A. H. F., PSPSITF, AERE, Oct. 21 & 21, 1965.
4298. Muggleton, A. H. F., INTDS Newsletter, Aug. 1980.
4299. Muggleton, A. H. F., and Howe, F. A., Nucl. Instr. and Meth. 28, 1964, p-242.
4300. Muggleton, A. H. F., and Parsons, C. T., Nucl. Instr. and Meth. 27, 1964, p-357.
4301. Muggleton, A. H. F., PSPSITF, AERE, Oct. 20 & 21, 1965.
4302. Muggleton, A. H. F., AERE-R-5097, Paper 18, p-5.
4303. Muggleton, A. H. F., and Howe, F. A., Nucl. Instr. and Meth. 13, 1961, p-211.
4304. Muggleton, A. H. F., and Howe, F. A., Nucl. Instr. and Meth. 12, June 1961, pp-192-4.
4305. Mughabghab, S. F., Natl. Bur. of Stands. (U. S.), Spec. Publ. 425, Oct. 1975, pp-795-798.
4306. Mulder, K., van Dantzig, R., et al., Nucl. Instr. and Meth. 92, 1971, pp-161-72.
4307. Müller, G., and Aumann, D. C., Nucl. Instr. and Meth. 128, 1975, pp-425-8.
4308. Multhauf, L. G., Tirsell, K. G., et al., Phys. Rev., C, 13(2), Feb. 1976, pp-771-789.
4309. Münch, G., Rev. Sci. Instr. 35, 1964, p-524.
4310. Münzel, H., and Lange, J., Proc. of a Conf. held at St. Catherine's Coll., Oxford, 22-23 Sept. 1969.
4311. Murphy, D. L., and Markowitz, S. S., LBL-4000, July 1975, pp-123-124.
4312. Murphy, J. J., Gehrhardt, H. J., et al., Nucl. Phys., A, 277(1), 7 Feb. 1977, pp-69-76.
4313. Murphy, M. J., Davids, C. N., et al., Phys. Rev., C, 17 (5), May 1978, pp-1574-1582.
4314. Murphy, P. D., Taki, T., et al., J. Am. Chem. Soc. ISSN 0002-7863, 101(15), 18 July 1979, pp-4055-4058.
4315. Müschenborn, G., PSPSITF, AERE, Oct. 20 & 21, 1965.
4316. Müschenborn, G., Vakuum Technik 20(7), 1971, p-197.
4317. Musgrove, A. R. de L., Allen, B. J., et al., Nucl. Phys., A, 259(3), 22 March 1976, pp-365-377.

4318. Musgrove, A. R. de L., Allen, B. J., et al., Nucl. Phys., A, 270(1), 12 Oct. 1976, pp-108-140.
4319. Musinski, D. L., Henderson, T. M., et al., Appl. Phys. Lett. 34(4), 15 Feb. 1979, pp-300-302.
4320. Musset, P., and Queru, P., Ind. At. 8(11), 1964, pp-67-77.
4321. Mustafa, M. G., Phys. Lett., B, 60(1), 22 Dec. 1975, pp-15-18.
4322. Myhrer, F., and Thomas, A. W., Nucl. Phys., A, ISSN 0375-9474, 326(2-3), 10 Sept. 1979, pp-497-507.
4323. Nablo, S. V., and King, W. J., IEEE Conf., Paper 64, 1964.
4324. Nadejdin, V. S., Petrov, N. I., et al., JINR-E-1-10820, 1977, p-10.
4325. Nadezhdin, V. S., Petrov, N. I., et al., Pis'ma Zh. Ehksp. Teor. Fiz. 26(2), 20 July 1977, pp-123-126.
4326. Nadkarni, M. N., Mayankutty, P. C., et al., Bhabha Atomic Rsch. Ctr., Bombay (India), BARC-912, 1976, p-6.
4327. Naegeli, H., Meyer, H. O., et al., Helv. Phys. Acta 51(1), 18 May 1978, p-110.
4328. Nagamine, K., Nishida, N., et al., 2nd Int. Conf. on Polar. Targets, Berkeley, CA, 30 Aug. 1971.
4329. Nagamine, K., Nucl. Instr. and Meth. 78, 1970, pp-285-94.
4330. Nagarajan, M. A., Strayer, M. R., et al., Phys. Lett., B, 68(5), 4 July 1977, pp-421-423.
4331. Nagatani, K., Towsley, C. W., et al., Phys. Rev., C, 14(6), Dec. 1976, pp-2133-2137.
4332. Nagel, P., and Koshel, R. D., Phys. Rev., C, 14(4), Oct. 1976, pp-1667-1670.
4333. Nagel, P. B., Ohio Univ., Athens (U. S. A.), Thesis (Ph. D), 1976, p-195.
4334. Nagel, W., and Stehle, H., BMFT-FBK-73-2, March 1973, pp-80-83.
4335. Nagl, A., Cannata, F., et al., Phys. Rev., C, 12(5), Nov. 1975, pp-1586-1588.
4336. Nagl, A., and Uberall, H., AIP Conf. Proc. 33, 1976, pp-612-613.
4337. Nagl, A., and Ueberall, H., Phys. Lett., B, 63(3), 2 Aug. 1976, pp-291-294.
4338. Nagl, A., Cannata, A. N. F., et al., Acta Phys. Austriaca 48(3), 1978, pp-267-281.
4339. Nagorcka, B. N., Symons, G. D., et al., Aust. J. Phys. 30 (2), April 1977, pp-149-165.
4340. Nagorcka, B. N., Symms, G. D., et al., Aust. Natl. Univ., Canberra, ANU-P-618, Oct. 1976, p-28.
4341. Nagorcka, B. N., and Newton, J. O., J. Phys., G, Nucl. Phys., ISSN 0305-4616, 3(11), Nov. 1977, pp-1565-75.
4342. Nagornyj, S. I., Inopin, E. V., Ukr. Fiz. Zh. ISSN 0503-1265, 24(5), May 1979, pp-591-598.
4343. Nagy, S., Flynn, K. F., et al., Phys. Rev., C, 17(1), Jan. 1978, pp-163-171.

4344. Nahabetian, R., Lyon-1 Univ., 69-Villeurbanne (France), Inst. de Physique Nucleaire, These (D. es S.), LYCEN-7840, 1978, p-65.
4345. Nair, K. G., Voit, H., et al., Phys. Rev., C, 12(5), Nov. 1975, pp-1575-1585.
4346. Nair, K. G., Hamm, M., et al., Texas A and M Univ., College Station, ORO-4856-26, 1975, pp-421-422.
4347. Nair, K. G., Hamm, M., et al., Texas A and M Univ., College Station, ORO-4856-26, 1975, pp-423-424.
4348. Nair, K. G., Dueck, P., et al., Erlangen-Nuernberg Univ., Erlangen, Physikalisches Inst., Annual Report 1975.
4349. Nair, K. G., and Kubo, K., Phys. Rev., C, 15(5), May 1977, pp-1774-1778.
4350. Nakada, A., Haik, N., et al., CEA-CONF-4090, 1977, p-1.
4351. Nakahara, H., Yanokura, M., et al., J. Inorg. and Nucl. Chem. 38(2), 1976, pp-203-204.
4352. Nakamura, H., and Noya, H., Nucl. Phys., A, 264(1), 21 June 1976, pp-54-62.
4353. Nakamura, H., Kawamura, T., et al., Proc. 7th Int. Vac. Cong. and the 3rd Int. Conf. on Solid Surf., Vienna, Austria(2), Sept. 1977, pp-1325-1327.
4354. Nakamura, K., Nucl. Phys., A, 259(2), 15 March 1976, pp-301-316.
4355. Nakamura, K., Hiramatsu, S., et al., Nucl. Phys., A, 268(3), 21 Sept. 1976, pp-381-407.
4356. Nakamura, K., Hiramatsu, S., et al., Nucl. Phys., A, 271(2), 2 Nov. 1976, pp-221-234.
4357. Nakamura, T., Yoshida, M., et al., Nucl. Instr. and Meth. ISSN 0029-554X, 151(3), 15 May 1978, pp-493-503.
4358. Nakamura, T., Hirayama, H., Proc. Int. Conf. Photonuclear React. Appl., CONF-730301, 1973, pp-257-8.
4359. Nakayama, K., Cybulska, E. W., et al., Cienc. Cult. 29(7), July 1977, pp-271-272.
4360. Namboodiri, M. N., Chulick, E. T., et al., Nucl. Phys., A, 263(3), 14 June 1976, pp-491-499.
4361. Nann, H., and Wildenthal, B. H., Phys. Rev. Lett. 37(17), 25 Oct. 1976, pp-1129-1131.
4362. Nann, H., Kashy, E., et al., Phys. Rev., C, 14(6), Dec. 1976, pp-2338-2339.
4363. Nann, H., Mueller, D., et al., Phys. Rev., C, 14(6), Dec. 1976, pp-2089-2094.
4364. Nann, H., Chien, W. S., et al., Phys. Rev., C, 15(6), June 1977, pp-1959-1966.
4365. Nann, H., and Wildenthal, B. H., Phys. Rev., C, 17(3), March 1978, pp-916-926.
4366. DiNapoli, V., Rosa, G., et al., J. Inorg. and Nucl. Chem. 38(1), 1976, pp-1-5.
4367. Naqib, I., and McDaniels, D. K., Rev. Sci. Instr. 31, Dec. 1960, pp-1358-60.

4368. Nardi, E., and Zinamon, Z., Phys. Rev., A, 18(3), 1978, pp-1246-9.
4369. DiNardo, R. P., and Goland, A. N., J. Opt. Soc. Am. 61, 1964, p-1321.
4370. Nashiyama, I., and Teranishi, E., Denski Gijutsu Sogo Kenkyujo Iho 37(4), 1973, pp-482-9.
4371. Nathan, A. M., Olness, J. W., et al., Phys. Rev., C, 17(3), March 1978, pp-1008-1025.
4372. Nathan, A. M., and Sandorfi, A. M., Phys. Rev. Lett. 40(19), 8 May 1978, pp-1252-1255.
4373. Natowitz, J. B., Pement, F., et al., Rev. Sci. Instr. 37, Jan. 1966, pp-121-2.
4374. Natowitz, J. B., Texas Agricultural and Mech. Univ., College Station, Contract AT(40-1)-3785, Jan. 1, 1969, p-17.
4375. Natowitz, J. B., and Namboodiri, M. N., Phys. Rev., C, 12 (5), Nov. 1975, pp-1678-1679.
4376. Natowitz, J. B., Namboodiri, M. N., et al., Phys. Rev. Lett. 40(12), 20 March 1978, pp-751-754.
4377. Naude, W. J., Peisach, M., et al., Proc. of 3rd Conf. on Accel. Targ. Designed for the Prod. of Neutrons, Liege (Belgium), Sept. 18-19, 1967, pp-261-82.
4378. Navinsek, B., Pozar, F., et al., J. Sci. Instr. 40, April 1963, pp-201-2.
4379. Nawrocki, D., Private comm. from Brookhaven National Lab.
4380. Naylor, H., and White, R. E., Nucl. Instr. and Meth. 144 (2), 15 July 1977, pp-331-335.
4381. Naylor, H., Nucl. Instr. and Meth. 76, 1969, p-357.
4382. Nazarova, T. S., and Rozen, A. A., Prib. Tekh. Eksp. 3, 1974, p-226.
4383. Nease, D. L., Cornell Univ., Ithaca, NY, 1976, p-96.
4384. Need, J. L., NASA-TN-D-3116, Nov. 1965, p-25.
4385. Need, J. L., New England Nuclear, U. S. A., Report 1977.
4386. Need, J. L., Nucl. Instr. and Meth. 42(2), 1967, pp-283-5.
4387. Need, J. L., TEICCTA, IU, U. S. A., Sept. 18-21, 1978.
4388. Neese, R. E., Guidry, M. W., et al., Phys. Lett., B, ISSN 0370-2693, 85(2-3), 13 Aug. 1979, pp-201-205.
4389. Nefkens, B. M. K., 2nd Int. Conf. on Polar. Targets, Berkeley, CA, 30 Aug. 1971.
4390. Neganov, B., Int. Conf. on Instr. for High Energy Phys., Dubna, USSR, 8 Sept. 1970, pp-575-84.
4391. Neilson, G. C., Panar, J. D., et al., Nucl. Instr. and Meth. 76, 1969, pp-75-6.
4392. Neirinckx, R. D., Int. J. Appl. Radiat. Isot. 27(1), Jan. 1976, pp-1-4.
4393. Neirinckx, R. D., Int. J. Appl. Radiat. Isot. 28(6), June 1977, pp-561-562.
4394. Neirinckx, R. D., Int. J. Appl. Radiat. Isot. 28(9), Sept. 1977, pp-808-9.

4395. Neirincky, R. D., Int. J. Appl. Radiat. Isot. 28(9), Sept. 1977, pp-802-804.
4396. Nellis, D. O., Hudspeth, E. L., et al., Pub. Health Serv., Washington, D. C., Nov. 1967, p-47.
4397. Nelson, J., Rev. Sci. Instr. 37(12), 1966, pp-1670-1.
4398. Nelson, L. R., and Duck, J. M., Proc. held at St. Catherine's Coll., Oxford, 22-23 Sept. 1969.
4399. Nelson, R. S., Proc. Conf. held at St. Catherine's Coll., Oxford, 22-23 Sept. 1969.
4400. Nelson, R. S., The Observation of Atomic Collisions in Crystalline Solids, (J. Wiley & Sons), 1968.
4401. Nelson, R. S., Proc. of 6th Int. Cyc. Conf., Vancouver, Canada, July 18-21, 1972.
4402. Nemashkalo, A. A., Likhachev, V. P., et al., Kharkov Fiziko-Tekhnicheskij Inst., Nuclear Sci. and Eng. Prob., KFTI-75-11, 2(14), 1975, pp-21-22.
4403. Nemashkalo, A. A., Likhachev, V. P., et al., KFTI-77-9, 2(19), 1977, pp-62-63.
4404. Nemashkalo, A. A., Afanas'ev, N. G., et al., KFTI-77-9, Voprosy atomnoj nauki i tekhniki 2(19), 1977, pp-58-61.
4405. Nemets, O. F., Ostapenko, A. A., et al., Ukr. Fiz. Zh. 22(2), Feb. 1977, pp-246-256.
4406. Nemets, O. F., Gofman, Yu. V., et al., Leningrad Nauka 1978, p-237.
4407. Nesci, P., and Amos, K., Nucl. Phys., A, 284(2), 4 July 1977, p-239-256.
4408. Nesterov, B. V., IAEA, Vienna (Austria), Int. Nucl. Data Comm., March 1978, p-22.
4409. Nethaway, D. R., Van Konynenburg, R. A., et al., UCRL-52024, 16 Feb. 1976, p-13.
4410. Nethaway, D. R., Van Konynenburg, R. A., et al., UCRL-79557, CONF-770523-14, May 2, 1977, p-20.
4411. Nethaway, D. R., Prindle, A. L., et al., Phys. Rev., C, 17(4), April 1978, pp-1409-1413.
4412. Neudachin, V. G., Romanovskij, E. A., et al., Yad. Fiz. ISSN 0011-0027, 29(2), Feb. 1979, pp-342-349.
4413. Neumann, B., Buschmann, J., et al., Inst. fuer Angewandte Kernphysik, F. R. (Germany), Annual Report, KFK-2868, Oct. 1979, pp-36-38.
4414. Neumann, B., Buschmann, J., et al., KFK-2868, Oct. 1979, pp-39-40.
4415. Neuhausen, R., Lightbody, J. W., Jr., et al., Nucl. Phys., A, 263(2), 7 June 1976, pp-249-260.
4416. Neuzil, E. F., J. Inorg. and Nucl. Chem. 38(12), 1976, pp-2153-2156.
4417. Newman, E., INTDS Proc., Boston, U. S. A., Oct. 1-3, 1979.
4418. Newman, E., INTDS Newsletter, Feb. 1980.
4419. Newman, E., INTDS Proc., Garching, FRG., Sept. 11-14, 1978.
4420. Newman, Prog. Nucl. Med. 4, 1978, p-47.

4421. Newman, E., Bell, W. A., Jr., et al., Nucl. Instr. and Meth. 139(1), 1976, pp-87-93.
4422. Newman, M., Cauvin, B., et al., LBL-5075, 1975, pp-120-23.
4423. Newton, J. P. Judish, et al., ORNL-3191, pp-89-91.
4424. Nezrick, F. A., and Ermelov, P., Natl. Accel. Lab., Batavia, IL, JINR-D-5805, Report 1971, pp-657-63.
4425. Ngo, C., Peter, J., et al., Nucl. Phys., A, 267(1), 16 Aug. 1976, pp-181-189.
4426. Nguyen, V. S., Gondrand, J. C., et al., Inst. des Sciences Nucleaires (France), ISN-75-46, 1975, p-11.
4427. Nguyen, V. S., Darves-Blanc, R., et al., Europ. Conf. on Nucl. Phys. with Heavy Ions, 6 Sept. 1976.
4428. Nicholas, D. J., Banks, P. H. T., et al., Nucl. Instr. and Meth. 88, 1970, pp-69-71.
4429. Nicholas, D. J., Williams, W. G., et al., Nucl. Instr. and Meth. 87, 1970, pp-301-2.
4430. Nicholas, D. J., Banks, P. H. T., et al., RHEL/R-201, Oct. 1970, p-24.
4431. Nicholas, D. J., Read, S. F. J., et al., 3rd Int. Symp. on Polar. Phenom. in Nucl. Reactions, Madison, WI, Aug.-Sept. 1970, pp-893-5.
4432. Nicholson, G. A., HW-55353, Contract W-31-109-eng-52, March 14, 1958, p-4.
4433. Nicholson, K. P., and Quealy, W. R., At. Energy Comm. Rsch. Estab., Lucas Heights, New South Wales, March 1961, p-21.
4434. Nickel, F., Hartmann, W., et al., Nucl. Instr. and Meth. 167(1), 1979, pp-175-7.
4435. Nickel, F., Hartmann, W., et al., INTDS Proc., Garching, FRG., Sept. 11-14, 1978.
4436. Nickel, F., Marx, D., et al., Nucl. Instr. and Meth. 134 (1), 1 April 1976, pp-11-14.
4437. Nickel, F., GSI-Bericht 73-7, 1973, p-1.
4438. Nickerson, R. A., UCRL-51737, Contract W-7405-eng-48, 15 Jan. 1975, p-15.
4439. Niedzwiedzuk, K., Vtyurin, V. A., et al., Acta Phys. Slovaca 25(2-3), 1975, pp-211-213.
4440. Nielsen, K. O., Nucl. Instr. and Meth. 1, 1957, p-289.
4441. Nielsen, K. O., and Skilbreid, O., Nucl. Instr. 2, 1958, pp-15-33.
4442. Niessner, W., and Ilgen, K., Z. Phys., A, 277(2), May 1976, pp-189-194.
4443. Nieukerke, K., Aust. Inst. of Nucl. Sci. and Eng., Lucas Heights, 7th AINSE Nucl. Phys. Conf., 1978, p-50.
4444. Nigh, H. E., J. Appl. Phys. 34, 1963, p-3323.
4445. Niinikoski, T. O., AIP Conf. Proc. High Energy Phys. Polariz. Beams Polariz. Targets, 1979, pp-62-9.
4446. Niinikoski, T. O., and Udo, F., Nucl. Instr. and Meth. 134(2), 15 April 1976, pp-219-233.

4447. Niinikoski, T. O., 2nd Int. Conf. on Polar. Targets, Berkeley, CA, 30 Aug. 1971, LBL-500, pp-107-12.
4448. Niinikoski, T. O., LBL-500, 30 Aug. 1971, pp-113-16.
4449. Nikitin, V. A., JINR-P-1476, 1963, p-14.
4450. Nikolaev, V. I., Instr. Exp. Tech. 17(4)(Part 1), July-Aug. 1974, pp-959-962.
4451. Nikolov, K., Popov, Kh., et al., Izv. Fiz. Inst. s Aneb, Bulgar, Akad. Nauk 13, 1965, pp-175-84.
4452. Nishi, M., Takiyama, M., et al., Jap. J. Appl. Phys. 5, March 1966, pp-215-17.
4453. Nishiyama, T., Phys. Lett., B, 63(2), 19 July 1976, pp-165-167.
4454. Nitschke, J. M., Nucl. Instr. and Meth. 138(3), 1 Nov. 1976, pp-393-406.
4455. Nitschke, J. M., LBL-4000, July 1975, p-376.
4456. Nitschke, J. M., Nucl. Instr. and Meth. 78, 1970, pp-45-69.
4457. Nobes, M., Nucl. Instr. and Meth. 72, 1969, pp-231-2.
4458. Nobes, M., J. Sci. Instr. 42, 1965, p-753.
4459. Nobes, M., J. Sci. Instr. 38, 1961, p-410.
4460. Nobes, M., Nucl. Instr. and Meth. 72(2), 1969, pp-231-2.
4461. Nobles, R., Rev. Sci. Instr. 28, Nov. 1957, pp-962-3.
4462. Noggle, T. S., TISRMNM, ORNL, Oct. 5-8, 1971.
4463. Noggle, T. S., and Dale, E. B., Bull. Am. Phys. Soc. II, 15, 1970, pp-13-40.
4464. Noggle, T. S., and Barrett, J. H., Phys. Statias. Solidi. 36, 1969, p-761.
4465. Noguchi, M., Radioisotopes 28(10), 1979, pp-658-67.
4466. Nolan, P. J., Lister, C. J., et al., INTDS Proc., Garching, FRG., Sept. 11-14, 1978.
4467. Nolan, P. J., Lister, C. J., et al., Nucl. Instr. and Meth. 167(1), 1979, pp-17-20.
4468. Nolan, P. J., et al., J. Phys., G: Nucl. Phys. 1, 1975, p-L33.
4469. Nolen, J. A., Jr., Curtin, M. S., et al., Nucl. Instr. and Meth. 150(3), 15 April 1978, pp-581-583.
4470. Nolen, J. A., Jr., and Austin, S. M., Phys. Rev., C, 13 (5), May 1976, pp-1773-1779.
4471. Nolen, J. A., Jr., and Miller, P. S., 7th Int. Conf. on Cyc. and Their Appl., Zurich (Switz.), Aug. 19-22, 1975.
4472. Nolen, R. L., Jr., Schneggenburger, R. G., et al., Intetial Confinement Fusion, Dig. Tech. Pap. Top. Meet. (1), TuE8, 1977, p-4.
4473. Nomura, T., Tosello, C., et al., Europ. Phys. Soc., Geneva (Switz.), Europ. Conf. on Nucl. Phys. with Heavy Ions, 6-10 Sept. 1976., p-107.
4474. Nonaka, I., and Sugai, I., Nucl. Instr. and Meth. 158 (2-3), 15 Jan. 1979, pp-453-8.

4475. Nonaka, I., and Sugai, I., Tokyo Univ. (Japan), Inst. for Nucl. Study, Tanashi-shi, Midori-Cho, Tokyo, INSJ-159, May 31, 1978.
4476. Nonaka, I., and Sugai, I., INS-TL-125, May 30, 1974.
4477. Nonaka, I., and Sugai, I., INSJ-132, 15 Nov. 1971, p-40.
4478. Norbury, J., Aust. Inst. of Nucl. Sci. and Eng., Lucas Heights, 7th AINSE Nucl. Phys. Conf., 1978, p-6.
4479. Norimatsu, T., Hashizume, K., et al., Technol. Rep. Osaka Univ. 27(1337-1363), 1977, pp-185-92.
4480. Northcliffe, L. C., and Schilling, R. F., "Range and Stopping Power for Heavy Ions", Nucl. Data Tables 7(3-4), Jan. 1970, Academic Press, NY.
4481. Northrop, J. A., and Stokes, R. H., Los Alamos Sci. Lab., LAMS-1522, Contract W-7405-eng-36, Dec. 10, 1956, p-25.
4482. Nosek, M. V., Atamanova, N. M., et al., Izv. Akad. Nauk Kaz. SSR, Ser. Khim. 25(3), May 1975, pp-61-65.
4483. Nosil, J., Spaventi, S., et al., Eur. J. Nucl. Med. 2(1), 1977, pp-1-8.
4484. Notarrigo, S., Porto, F., et al., Nuovo Cim., A, 35(1), 1 Sept. 1976, pp-33-44.
4485. Noyes, H. Pierre, SLAC-PUB-256, Contract AT(04-3)-515, Jan. 1967, p-51.
4486. Nozaki, T., Meeting on At. En. Rsch. with Heavy Ions, Tokai, Ibaraki (Japan), 25 Oct. 1973, pp-236-239.
4487. Nunn, A. D., and Waters, S. L., Int. J. Appl. Radiat. Isot. 26(12), 1975, pp-731-5.
4488. Nunn, A. D., Nucl. Instr. and Meth. 99(2), 1972, pp-251-4.
4489. Nurmia, M., Helsinki Univ., EUR-185.e(Rev.), pp-241-5.
4490. Nurushev, S. B., and Folomeshkin, V. N., Yadern. Fiz. 13, Feb. 1971, pp-424-5.
4491. Nurzynski, J., Gruebler, W., et al., 7th AINSE Nucl. Phys. Conf., 1978, p-65.
4492. Oberacker, V., Soff, G., et al., Phys. Rev. Lett. 36(17), 26 April 1976, pp-1024-1027.
4493. Oberacker, V., Greiner, W., et al., Phys. Rev., C, 20(4), ISSN 0556-2813, Oct. 1979, pp-1453-1466.
4494. Oblozinsky, P., and Ribansky, I., Fyz. Cas. 20(2), 1970, pp-105-9.
4495. Oblozinsky, P., Acta Phys. Slovaca 25(2-3), 1975, pp-214-17.
4496. O'Brien, H. A., Jr., Grant, P. M., et al., LA-UR-76-2127, CONF-760371-2, 1976, p-11.
4497. O'Brien, H. A., Jr., Ogard, A. E., et al., Recent Adv. Nucl. Med., Proc. World Congr. Nucl. Med., 1st, 1974, pp-201/1-20/5.
4498. O'Brien, H. A., Jr., and Schillaci, M. E., Meet. on Neut. Sources and Appl., Augusta, GA, 18 April 1971, CONF-710402(Vol. 2), pp-I.128-35.
4499 O'Brien, J. T., Crannell, H., et al., Phys. Rev., C, 9(4), 1974, pp-1418-29.

4500. O'Connell, J. S., Proc. of June Workshop in Intermed. Energy Electromagnetic Interactions with Nuclei, held at MIT, June 13-24, 1977, pp-11-45.
4501. Odehnal, M., Neel, P., et al., Rev. Phys. Appl. 6(1), March 1971, pp-59-63.
4502. O'Donnell, F. R., and Adair, H. L., Nucl. Instr. and Meth. 102, 1972, pp-501-2.
4503. O'Donnell, F. R., and Adair, H. L., 3rd Int. Symp. on Rsch. Mat. for Nucl. Measurements, Gatlinburg, TN, 5 Oct. 1971, pp-166-80.
4504. O'Donnell, F. R., and Adair, H. L., ORNL-4510, Contract W-7405-eng-26, March 1970, p-11.
4505. Oelert, W., Djaloeis, D., et al., AIP Conf. Proc. ISSN 0094-243X, (47), 1978, pp-708-709.
4506. Oelrich, I. C., Krein, K., et al., Phys. Rev., C, 14(2), Aug. 1976, pp-563-572.
4507. Oertzen, W. von, and Flynn, E., Hahn-Meitner-Inst., Berlin, ORO-4856-26, 1975, p-159.
4508. Oeschler, H., Hagemann, G. B., et al., Nucl. Phys., A, 266(1), 2 Aug. 1976, pp-262-268.
4509. Oeschler, H., and Sim, K. S., Nucl. Phys., A, ISSN 0375-9474, 325(2-3), 20-27 Aug. 1979, pp-463-480.
4510. Oganessian, Yu. Ts., and Pensionzhkevich, Yu. E., JINR-E-7-9187, 1976, p-28.
4511. Oganesyan, Yu. Ts., and Pleve, A. A., Leningrad Nauka, 1978, p-228.
4512. Ogawa, S., Takiguchi, K., et al., J. Vac. Sci. Technol. 8, 1971, p-192.
4513. Ohashi, H., Nakahara, K., et al., Kakuriken Kenkyu Hokoku 8(2), Dec. 1975, pp-256-265.
4514. Ohlsen, G. G., and Keaton, P. W., Jr., LBL-500, 1971, pp-367-9.
4515. Ohlsen, G. G., Hardekopf, R. A., et al., Phys. Rev., C, 14(5), Nov. 1976, pp-1688-1694.
4516. Ohmura, H., Ishimatsu, T., et al., Phys. Lett., B, 73(2), 13 Feb. 1978, pp-145-148.
4517. Ohnuki, Y., Oyo Butsuri 35, Feb. 1966, pp-128-31.
4518. Ohtomo, S., Okamoto, K., et al., JAERI-Memo-4023, May 1970.
4519. Oikawa, S., and Shoda, K., Nucl. Phys., A, 277(2), 14 Feb. 1977, pp-301-316.
4520. Okabe, S., and Abe, Y., Hokkaido Univ., Sapparo, ORO-4856-26, 1975, pp-81-82.
4521. Okabe, S., and Abe, Y., AIP Conf. Proc. ISSN 0094-243X, (47), 1978, pp-564-565.
4522. Okano, K., Kawase, Y., et al., Annu. Rep. Res. React. Inst., Kyoto Univ. 12, 1979, pp-146-51.
4523. Olin, A., Alexander, T. K., et al., Phys. Rev. D8, 1973, p-1633.

4524. Olivas, V., Stoner, O. J., Jr., INTDS Proc., Berkeley, U. S. A., Oct. 19-20, 1977.
4525. Olivo, M. A., and Bailey, G. M., Nucl. Instr. and Meth. 57, Dec. 1967, pp-353-4.
4526. Olmer, C., Buenerd, M., et al., LBL-5075, 1975, pp-111-15.
4527. Olmer, C., Buenerd, M., et al., LBL-5075, 1975, pp-102-07.
4528. Olmer, C., Yale Univ., New Haven, CT (U. S. A.), Thesis (Ph. D), 1975, p-227.
4529. Olmer, C., Zeidman, B., et al., Phys. Rev. Lett. ISSN 0031-9007, 43(9), 27 Aug. 1979, pp-612-615.
4530. Olsen, A. R., Sease, J. D., et al., Trans. Am. Nucl. Soc. 9, June 1966, p-66.
4531. Olsen, D. K., de Saussure, G., et al., Am. Nucl. Soc. Annual Meet., CONF-760622-19, 1976, p-6.
4532. Olsen, D. K., de Saussure, G., et al., Trans. Am. Nucl. Soc. 27, 1977, pp-870-871.
4533. Olsen, D. K., Morgan, G. L., et al., ORNL/TM-6832, Sept. 1979, p-120.
4534. Omelaenko, A. S., KFTI-77-9, Voprosy Atomnoj Nauki i Tekhniki 2(19), 1977, pp-20-22.
4535. O'Neil, J. A., Advan. Cryog. Eng. 14, 1969, pp-423-9.
4536. Ooi, T., Katori, K., et al., Rikagaku Kenkyusho Hokoku 52(4), July 1976, pp-145-154.
4537. Oona, H., and Rickel, D. G., Rev. Sci. Inst. 38, July 1967, p-980.
4538. Oosterkamp, W. J., Trans. Am. Nucl. Soc. 13, Nov. 1970, pp-736-7.
4539. Ophel, T. R., Martin, P. H., et al., Nucl. Phys., A, 173, 1971, pp-609-33.
4540. Ophel, T. R., ANU-P-502, 1970, p-39.
4541. Ophel, T. R., Frawley, A. D., et al., Aust. Natl. Univ., Canberra Rsch Sch. of Physical Sci., ANU-P-621, July 1976, p-27.
4542. Ophel, T. R., Frawley, A. D., et al., Nucl. Phys., A, 273 (2), 30 Nov. 1976, pp-397-409.
4543. Orihara, H., Takahashi, M., et al., Nucl. Phys., A, 267(2), 23 Aug. 1976, pp-276-284.
4544. Orlova, O. A., Brukhertzajfer, Kh., et al., Yad. Fiz. ISSN 0044-0027, 30(9), Sept. 1979, pp-618-625.
4545. Ormrod, J. H., Nucl. Instr. and Meth. 95(1), 1971, pp-49-51.
4546. Oselka, M., Gindler, J. E., et al., Int. J. Appl. Radiat. Isot. 28(9), 1977, pp-804-5.
4547. Oset, E., Phys. Lett., B, 65(1), 25 Oct. 1976, pp-46-50.
4548. Oset, E., AIP Conf. Proc. 33, 1976, pp-318-319.
4549. Oset, E., AIP Conf. Proc. ISSN 0094-243X, 54(1), 15 July 1979, pp-364-365.
4550. Oshima, Y., Yotsumoto, K., et al., JAERI-1190, July 1970, p-29.
4551. Osland, P., Rej, A. K., Phys. Rev., C, 13(6), June 1976, pp-2421-2432.

4552. Ost, R., Sanderson, N. E., et al., Europ. Phys. Soc., Geneva, (Switz.), Europ. Conf. on Nucl. Phys. with Heavy Ions, 6-10 Sept. 1976, p-45.
4553. Ost, R., Sanderson, N. E., et al., Europ. Phys. Soc., Geneva, (Switz.), Europ. Conf. on Nucl. Phys. with Heavy Ions, 6-10 Sept. 1976, p-158.
4554. Osterfeld, F., Hnizdo, V., et al., Phys. Lett., B, 68(4), 20 June 1977, pp-319-322.
4555. Osterlund, J. W., and Smythe, R., Colo. Univ., Boulder, Contract AT(11-1)-535, 1963, p-12.
4556. Osterman, P., Waldschmidt, M., et al., Nucl. Instr. and Meth. 72, 1969, pp-226-9.
4557. Ostrander, N. C., Calif. Rsch. and Develop. Co., Berkeley, CA, LRL-14, Feb. 26, 1957, LWS-2232, p-19.
4558. Otsubo, T., Asada, I., et al., Nucl. Instr. and Meth. 124(2), 1975, pp-325-7.
4559. Otteson, O. H., McAdams, R. E., et al., AIP Conf. Proc. ISSN 0094-243X, (47), 1978, pp-644-45.
4560. Otteson, O. H., McAdams, R. E., et al., AIP Conf. Proc. ISSN 0094-243X, 54(1), 15 July 1979, pp-282-83.
4561. Otto, G., Zentralinstitut fuer Isotopen- und Strahlenforschung, Zentralinstitut fuer Kernforschung, Rossendorf bei Dresden, Annual Report 1976, pp-7-9.
4562. Otto, R. J., Fowler, M. M., et al., LBL-5075, 1975, pp-165-8.
4563. Otto, R. J., Fowler, M. M., et al., LBL-5075, 1975, pp-168-173.
4564. Ouichaoui, S., Ngo, C., et al., Europ. Phys. Soc., Geneva, (Switz.), Europ. Conf. on Nucl. Phys. with Heavy Ions, 6-10 Sept. 1976, p-112.
4565. Outlaw, D. A., Mitchell, G. E., et al., Nucl. Phys., A, 269(1), 28 Sept. 1976, pp-99-111.
4566. Oyer, A. T., LA-6599-T, Dec. 1976, p-112.
4567. Paans, A. M. J., Vaalburg, W., et al., IEEE Trans. on Nucl. Sci., NS-26(2), April 1979.
4568. Paans, A. M. J., Vaalburg, W., et al., TEICCTA, IU, U. S. A., Sept. 18-21 1978.
4569. Pacheco, C., Stark, C., et al., LA-8109-MS, Dec. 1979, p-10.
4570. Paine, B., Kennett, S., et al., AINSE Nucl. Phys. Conf., 7th, 9-11 Feb. 1976, p-19.
4571. Paine, B. M., Kennett, S. R., et al., 7th AINSE Nucl. Phys. Conf., 6-8 Feb. 1978, p-9.
4572. Pais, A., Phys. Rev. Lett. 19, Aug. 28, 1967, pp-544-6.
4573. Pal, D., and Shibani, D., Proc. of Nucl. Phys. and Solid State Phys. Symp., Calcutta, Dec. 22-26, 1975, Vol. 18B, pp-59-62.
4574. Pal, S., Srivastava, D. K., et al., Proc. of Nucl. Phys. and Solid State Phys. Symp., Calcutta, Dec. 22-26, 1975. Vol. 18B, p-30.

4575. Pal, S., Srivastava, D. K., et al., AIP Conf. Proc. ISSN 0094-243X, (47), 1978, pp-578-79.
4576. Palcos, M. C., Radicella, R., et al., Com. Nac. de En. At., Buenos Aires, 1962, p-12.
4577. Palmberg, P. W., Rhodin, T. N., et al., Appl. Phys. Lett. 11, 1967, p-33.
4578. Palmer, D. C., Brown Univ., Providence, RI (U. S. A.), Thesis (Ph. D), 1975, p-99.
4579. Palmer, D. W., Skofronick, J. G., et al., Phys. Rev. 130, 1963, p-1153.
4580. Pan, Y. L., and Bailey, D. S., UCRL-78472, Report 1976, p-17.
4581. Panagiotou, A. D., Nuovo Cim., A, 37(4), 21 Feb. 1977, pp-370-84.
4582. Panarin, M. V., Krasnov, N. N., et al., Isotopenpraxix 10(11-12), 1974, pp-428-31.
4583. Pandey, M. S., State Univ. of NY, Albany (U. S. A.), Thesis (Ph. D), 1975, p-136.
4584. Pandian, S., and Preiss, I. L., Nucl. Phys., A, 282(1), 16 May 1977, pp-169-180.
4585. Papadopoulos, C. T., Hartas, A. G., et al., Nucl. Phys., A, 254(1), 1 Dec. 1975, pp-93-109.
4586. Papadopoulos, C. T., Hartas, A. G., et al., Phys. Rev., C, 15(6), June 1977, pp-1987-2005.
4587. Pape, A. J., and Markowitz, S. S., Phys. Rev., C, 13(6), June 1976, pp-2116-2121.
4588. Paradellis, T., Nucl. Instr. and Meth. 140(1), 1977, pp-205-9
4589. Paradellis, T., Nucl. Phys., A, 279(2), 28 March 1977, pp-293-316.
4590. Pardo, R. C., Davids, C. N., et al., Phys. Rev., C, 15(5), May 1977, pp-1811-1821.
4591. Park, J. Y., Scheid, W., et al., ANL/PHY-76-2(Vol. 2), May 1976, pp-715-722.
4592. Park, J. Y., Greiner, W., et al., Phys. REv., C, 16(6), Dec. 1977, pp-2276-2290.
4593. Parker, W., and Gullholmer, W., Chalmers Tekniska Hogskola, Goteborg, (Sweden), AERE-R-5097, Paper 16, p-3.
4594. Parker, W., and Gullholmer, W., AERE-R-5097, Paper 16, p-3.
4595. Parker, W., and Gullholmer, W., AERE-R-5097, Paper 10, p-2.
4596. Parker, W., and Gullholmer, W., PSPSITF, AERE, Oct. 20-21, 1965.
4597. Parker, W. C., and Slatin, H., 1965 Alpha, Beta and Gamma-Ray Spectroscopy (ed. K. Siegbahn, Amsterdam, North Holland), pp-379-407.
4598. Parker, W. C., DeCroes, M., et al., Nucl. Instr. and Meth. 7, 1960, p-22.
4599. Parker, W. C., and Falk, R., Nucl. Instr. and Meth. 16, 1962, pp-355-7.

4600. Parker, W. C., and Grunditz, Y., Nucl. Instr. and Meth. 22, 1963, p-369
4601. Parker, W. C., Bildstein, H., et al., Nucl. Instr. and Meth. 26, 1964a, b, c, pp-51-60, 61-65, 314-316.
4602. Parker, W., and Baumgartner, H., Nature 203(4946), 1964, pp-715-16.
4603. Parker, W., Methods in the Prep. of Radio. Mat., Dissertation thesis, Chalmers Univ. of Tech., Gothenburg, Sweden, 1965.
4604. Parker, R. H., and Frazer, R. F., Nucl. Instr. and Meth. 53, 1967, p-330.
4605. Parkins, M. F., IEEE Trans. Nucl. Sci., NS-14, June 1967, pp-950-4.
4606. Parkins, W. E., Proc. Inf. Meet. Accel.-Breed. 1977, pp-30-9.
4607. Parks, N. J., and Krohn, K. A., Int. J. Appl. Radiat. Isot. 29(12), 1978, pp-754-7.
4608. Parks, P. B., Beard, P. M., et al., Rev. Sci. Instr. 35, May 1964, pp-549-57.
4609. Parks, P. B., III, Duke Univ., Durham, NC, Thesis, 1963, p-83.
4610. Parnell, C. J., Page, B. C., et al., Phys. Med. Biol. 20 (1), Jan. 1975, pp-125-127.
4611. Parnell, C. J., Meet. on Neut. Srcs. and Appl., Augusta, GA, 18 April 1971, pp-II.7-15.
4612. Parthasarathy, R., CEA-CR-7 (nd), 1975, pp-195-202.
4613. Partridge, R. A., Brown, R. E., et al., Univ. of Minnesota, Minneapolis, ORO-4856-26, 1975, pp-225-226.
4614. Parvez, A., and Becker, M., Trans. Am. Nucl. Soc. 26, June 1977, pp-594-95.
4615. Parzhitskij, S. S., Popov, Yu. P., et al., JINR-R-15-9649, 1976, p-10.
4616. Pascal, P., Nouveau Traité de Chimie Minérale (Masson and Co., Paris, 1965) Tome IX, p-176.
4617. Pascal, P., Nouveau Traité de Chimie Minérale (Masson and Co., Paris, 1965), Tome VIII, p-331.
4618. Pascal, P., Nouveau Traité de Chimie Minérale 15 (Masson, Paris, 1958).
4619. Paschopoulos, I., Sinclair, D., et al., Nucl. Phys., A, 277(2), 14 Feb. 1977, pp-358-64.
4620. Pasechnik, M. V., Fedorov, M. V., et al., Proc. of 4th All-Union Conf. on Neut. Phys., Nejtronnaya fizika, 1977, pp-279-83.
4621. Pashley, D. W., Phil. Mag. 4, 1959, p-324.
4622. Pashley, D. W., Recent Progr. Surf. Sci. 3, 1970, p-23, Adv. Phys. 5, 1956, p-174.
4623. Passerieux, J. P., and Fouan, J. P., CEA-CONF-1155, Int. Symp. on Nucl. Elect., Versailles, France, p-3.
4624. Pastukhova, Z. V., Pribory i Tekhn. Eksperim. 8, Nov.-Dec. 1962, pp-126-7.

4625. Patnaik, B. K., and Dhere, N. G., Inst. Fis., Pontif. Univ. Catol., Rio de Janeiro, Brazil, PUC-TN-22/75, Report 1975, p-17.
4626. Patrick, B. H., Bowey, E. M., et al., Proc. of 3rd Conf. on Accel. Targ. Designed for the Prod. of Neut., Liege (Belgium), Sept. 18-19, 1967, pp-291-304.
4627. Patterson, H. W., and Thomas, R. H., Particle Accel. 2(2), April 1971, pp-77-104.
4628. Pauwels, J., and Van Audenhove, J., INTDS Proc., Boston, U. S. A., Oct. 1-3, 1979.
4629. Pauwels, J., and Tjoonk, J., Nucl. Instr. and Meth. 167(1), 1979, pp-77-9.
4630. Pauwels, J., Van Craen, J., et al., INTDS Proc., Garching, FRG., Sept. 11-14, 1978.
4631. Pauwels, J., and Tjoonk, J., INTDS Proc., Garching, FRG., Sept. 11-14, 1978.
4632. Pauwels, J., and Van Audenhove, J., INTDS Newsletter, Aug. 1980.
4633. Pauwels, J., and Van Audenhove, J., INTDS Proc., Boston, U. S. A., Oct. 1-3, 1979.
4634. Peck, A. A., INTDS Newsletter, July 1979.
4635. Peck, A. A., INTDS Proc., Argonne, U. S. A., Oct. 1975.
4636. Pedersen, T., England, T., et al., Nucl. Phys., A, 293 (1-2), 12-19 Dec. 1977, pp-10-28.
4637. Pedroni, E., Gabathuler, K., et al., AIP Conf. Proc. 33, 1976, pp-25-33.
4638. Peek, N. F., and Hegedues, F., Int. J. Appl. Radiat. Isot. ISSN 0020-708X, 30(10), Oct. 1979, pp-631-635.
4639. Peelle, R. W., and de Sassure, G., ORNL, CONF-770321-6, 1977, p-8.
4640. Peeters, J. M., and Guillaume, M. A., Liege Univ. (Belgium), May 1967, p-70.
4641. Peetermans, A., and Baarli, J., Environ. Surveill. Around Nucl. Install., Symp., Proc., Warsaw, Poland, VI PAP, IAEA-SM-180/10, 1973 Nov. 5-9, pp-433-448.
4642. Pelliccioni, M., ORNL-TR-1897, LNF-66/14, p-10.
4643. Peng, J. C., Pittsburgh Univ., PA (U. S. A.), Thesis (Ph. D), 1975, p-116.
4644. Peng, J. C., Maher, J. V., et al., Nucl. Phys., A, 264(2), 28 June 1976, pp-312-340.
4645. Peng, J. C., Stein, N., et al., Phys. Rev. Lett ISSN 0031-9007, 43(10), 3 Sept. 1979, pp-675-678.
4646. Peng, Y. K., INTDS Newsletter, Jan. 1979.
4647. Peng, Y. K., INTDS Proc., Boston, U. S. A., Oct. 1-3, 1979.
4648. Penner, S., Fivozinsky, S. P., et al., NBS, Wash., D. C. (U. S. A.), June 1975, p-82.
4649. Pennington, E. M., Poenitz, W. P., et al., Trans. Am. Nucl. Soc. 26, June 1977, pp-591-92.
4650. Perez, R. B., and de Saussure, G., NBS (U. S.), Spec. Publ. 425, Oct. 1975, pp-623-626.

4651. Perris, A. G., Lane, R. O., et al., Int. J. Appl. Radiat. and Isot. 25(1), Jan. 1974, pp-19-23.
4652. Perry, D. R., and Hargreaves, D. M., Proc. of Conf. on Radiat. Prot. in Accel. Environ., Chilton, England, March 1969, pp-3-7.
4653. Perry, F. C., Mix, L. P., et al., Appl. Phys. Lett. 34(4), 15 Feb. 1979, pp-251-253.
4654. Perry, F. C., and Widner, M. M., Appl. Phys. Lett. 29(5), 1976, pp-282-4.
4655. Perry, W. L., and Gallant, J. L., INTDS Proc., Chalk River, Canada, Oct. 1-3, 1974.
4656. Persiani, P. J., Pennington, E. M., et al., NBS (U. S.), Spec. Publ. 425, Oct. 1975, pp-708-711.
4657. Persiani, P. J., Becker, W., et al., ANL, CONF-770523-1, Report 1977, p-28.
4658. Pesso, A., Farrelly, E., et al., Rev. Bras. Fis. 8(3), 1978, pp-724-33.
4659. Peter, J., Ngo, C., et al., Nucl. Phys., A, 279(1), 21 March 1977, pp-110-124.
4660. Peter, J., Berlanger, M., et al., Z. Phys., A, 283(4), Oct. 1977, pp-413-14.
4661. Peters, J. M., and Del Fiore, G., Radiochem. Radioanal. Lett. ISSN 0079-9483, 40(4), 21 Sept. 1979, pp-235-49.
4662. Peters, J. M., Radiochem. Radioanal. Lett. 43(2-3), 1980, pp-159-72.
4663. Peters, J. M., Guillaume, M., et al., Nucl. Instr. and Meth. 80, 1970, pp-351-3.
4664. Peters, J. M., and Guillaume, M., Liege Univ., Belgium, Dec. 11, 1968, p-40.
4665. Peters, J. M., Proc. of 3rd Conf. on Accel. Targ. Des. for the Prod. of Neut., Liege (Belgium), Sept. 18-19, 1967, pp-41-59.
4666. Petersen, J. F., Dehnhard, D., et al., Phys. Rev., C, 15 (5), May 1977, pp-1719-1725.
4667. Peterson, C., and Laubert, R., IEEE Trans. Nucl. Sci., NS-24(3), June 1977, pp-1542-1544.
4668. Peterson, G. A., Int. Conf. on Nucl. Phys. with Electromag. Inter., Mainz, FRG., 5-9 June 1979, p-6.
4669. Peterson, J. R., Univ. of Tenn., Knoxville, TN, Private Communication
4670. Peterson, R. E., and Hassberger, J. A., Trans. Am. Nucl. Soc. 34, 1980, pp-34, 35-6.
4671. Peterson, R. J., and Ristinen, R. A., Nucl. Phys., A, 276(1), 17 Jan. 1977, pp-61-71.
4672. Peterson, R. J., LA-6926-C, Aug. 1977, pp-71-95.
4673. Peterson, R. J., Emigh, R. A., et al., Nucl. Phys., A, 290(1), 14 Oct. 1977, pp-155-172.
4674. Peterson, R. J., Anderson, R. E., et al., Colo. Univ., Boulder (U. S. A.), Tech. Prog. Report, Nov. 1, 1977, pp-41-45.

4675. Petricek, V., Nucl. Instr. and Meth. 58, Jan. 1968, pp-111-16.
4676. Petrov, V. I., Instr. Exp. Tech. 2, 1961, pp-383-4.
4677. Petrov, V. I., and Oparin, E. M., Pribory i Tekhn. Eksperim. 5, Sept.-Oct. 1962, pp-38-40.
4678. Petrykhin, V. I., Prokoshkin, Yu. D., et al., Pribory i Tekhn. Eksperim. 2, March-April 1964, pp-22-3.
4679. Pettit, A. W., IEEE Trans. Nucl. Sci., NS-26(1), Feb. 1979.
4680. Petty, D. T., Ikossi, P. G., et al., Phys. Rev., C, 14(3), Sept. 1976, pp-908-11.
4681. Petukhov, A. K., Petrov, G. A., et al., Pis'ma Zh. Ehksp. Teor. Fiz. ISSN 0370-274X, 30(7), 5 Oct. 1979, pp-470-74.
4682. Pfeiffer, J., and Riepe, G., INTDS Newsletter, Feb. 1980.
4683. Pfeiffer, L., and Keaton, P. W., Jr., Nucl. Instr. and Meth. 40, March 1966, pp-357-8.
4684. Pham, D. L., Grenoble-1 Univ., 38 (France), Inst. des Sciences Nucleaires, 1976, p-1.
4685. Pham, D. L., AIP Conf. Proc. ISSN 0094-243X, 47, 1978, pp-734-735.
4686. Philipp, G., and Hofmann, A., Nucl. Instr. and Meth. 37, Nov. 1965, pp-313-17.
4687. Philis, C., and Young, P. G., CEA-R-4712, Dec. 1975, p-51.
4688. Philis, C., and Bersillon, O., CEA-N-1875, April 1976, pp-123-25.
4689. Phillips, D., and Pringle, J. P. S., Nucl. Instr. and Meth. 135(2), 1976, pp-389-90.
4690. Phillips, D., Nucl. Instr. and Meth. 116, 1974, p-195.
4691. Phillips, G. C., Contract AT(40-1)-1316, Conf. on Polar. Targ. and Srcs., Saclay, France, Dec. 2, 1966, p-18.
4692. Phillips, T. W., Nucl. Sci. and Eng. ISSN 0029-5639, 69(3), March 1979, pp-375-77.
4693. Phillips, T. W., and Johnson, R. G., Phys. Rev., C, ISSN 0556-2813, 20(5), Nov. 1979, pp-1689-1699.
4694. Phillips, W. R., Phys. Rev. 110, 1950, pp-1408-13.
4695. Picker, H. S., and Haftel, M. I., Phys. Rev., C, 14(4), Oct. 1976, pp-1293-1297.
4696. Picot, A., Rev. Sci. Instr. 50(8), 1979, pp-1021-2.
4697. Pieper, M., Technische Univ., Muenchen, FRG., Fachbereich Physik Diss. (D. Sc.), 22 Feb. 1978, p-66.
4698. Pierce, W. B., 3rd Int. Cryo. Eng. Conf., Berlin, Germany, 25 May 1970, pp-287-302.
4699. Piffaretti, J., Corfu, R., et al., Phys. Lett., B, 67(3), 11 April 1977, pp-289-91.
4700. Piffaretti, J., Corfu, R., et al., Phys. Lett., B, 71(2), 21 Nov. 1977, pp-324-326.
4701. Pilt, A. A., Cornell, J. C., et al., Oxford Univ., ORO-4856-26, 1975, pp-415-16.
4702. Pilt, A. A., AIP Conf. Proc. ISSN 0094-243X, 47, 1978, pp-173-184.

4703. Pimentel, C. A., Amaral, L. Q., et al., Rev. Bras. Fis. 5 (3), Dec. 1975, pp-285-303.
4704. Pinston, J. A., Roussille, R., et al., Nucl. Phys., A, 264 (1), 21 June 1976, pp-1-12.
4705. Pinston, J. A., Roussille, R., et al., Nucl. Phys., A, 270 (1), 12 Oct. 1976, pp-61-73.
4706. Pinston, J. A., Mampe, W., et al., Inst. Max von Laue-Paul Langevin, 38-Grenoble (France), KFK-2868, Oct. 1979
4707. Pirart, C., Bosman, M., et al., Phys. Rev., C, 17(2), Feb. 1978, pp-810-12.
4708. Pirner, H. J., Phys. Rev., C, 14(4), Oct. 1976, pp-1665-6.
4709. Pirner, H. J., Phys. Lett., B, 69(2), 1 Aug. 1977, pp-170-172.
4710. Pisano, D. J., and Parker, P. D., Phys. Rev., C, 14(2), Aug. 1976, pp-475-90.
4711. Pittaway, L. G., Brit. J. Appl. Phys. 15, Aug. 1964, pp-967-82.
4712. Pitthan, R., and Buskirk, F. R., Phys. Rev., C, 16(3), Sept. 1977, pp-983-87.
4713. Pivarc, J., Atomki Kozl. 18(2), 1976, pp-463-78.
4714. Pivarc, J., Wiss. Z. Tech. Univ., Dresden, 21(4), 1972, pp-694-6.
4715. Pivovar, L. I., and Levchenko, Yu. Z., Instr. Exp. Tech. 4, July-Aug. 1970, pp-986-7.
4716. Plagnol, E., Doubre, H., et al., Inst. de Physique Nucleaire, Orsay (France), Nov. 1976, p-12.
4717. Plasil, F., ORNL, CONF-770115-1, 1977, p-12.
4718. Plasil, F., Ferguson, R. L., et al., Phys. Rev. Lett. 40 (18), 1 May 1978, pp-1164-66.
4719. Platchkov, S. K., Bellicard, J. B., et al., Phys. Lett., B, ISSN 0370-2693, 86(1), 10 Sept. 1979, pp-1-4.
4720. Plattner, G. R., Bornand, M., et al., Phys. Rev. Lett. 39(3), 18 Aug. 1977, pp-127-30.
4721. Plattner, R., Anderson, J., et al., Nucl. Instr. and Meth. 64, Sept. 15, 1968, pp-192-6.
4722. Plch, J., and Zderadicka, J., Czech. J. Phys. 27(8), 1977, pp-865-72.
4723. Plendl, H. S., Maghami, B., et al., AIP Conf. Proc. ISSN 0094-243X, 47, 1978, pp-648-649.
4724. Pleticha-Lansky, R., Biol. Listy. 41(1), Feb. 1976, pp-39-55.
4725. Plicht, J. van der, Harakeh, M. N., et al., Nederlandse Natuurkundige Vereniging, Amsterdam, Spring Meet. of the Dutch Phys. Soc., March 1978, p-25.
4726. Pil'ts, V., Rumf, I., Gosudarstvennyj Komitet po Ispol'zovaniyu Atomnoj Ehnergii SSSR, Obninsk, Fiziko-Ehnergeticheskij Inst., 1976, p-22.
4727. Pluchery, M., Centre D'etudes Nucleaires, Aug. 1965, p-15.

4728. Pobereskin, M., Langendorfer, W., et al., Battelle Columbus Labs., Ohio (U. S. A.), 31 Jan. 1975, p-210.

4729. Podgorsak, E. B., Rawlinson, J. A., et al., Am. J. Roent., Radium Ther. Nucl. Med. 121(4), Aug. 1974, pp-873-82.

4730. Poenitz, W. P., NBS (U. S.), Spec. Publ. 425, Oct. 1975, pp-901-04.

4731. Poenitz, W., Pennington, E., et al., ANL/NDM-32, Oct. 1977, p-100.

4732. Poenitz, W. P., Nucl. Sci. and Eng. 64(4), Dec. 1977, pp-894-97.

4733. Poggenburg, J. K., Kobisk, E. H., et al., Trans. Am. Nucl. Soc. 11, Nov. 1968, pp-453-4.

4734. Pogodin, V. I., Prokof'ev, D. D., et al., Instr. Exp. Tech. 19(2)Pt. 1, March-April 1976, pp-332-33.

4735. Poiani, G., Nuovo Cimento (1), 2, Suppl., 1964, pp-355-82.

4736. Polak, P., Inst. voor Kernphysisch Onderzoek, Amsterdam (Netherlands), 27 Oct. 1976, p-84.

4737. Polandov, A. G., Marin, B. V., et al., Izv. Akad. Nauk Arm. SSR, Fiz. 14(3), 1979, pp-223-6.

4738. Polane, J. H., Nederlandse Natuurkundige Vereniging, Utrecht, Sectie Kernfysica, Autumn Meet., Oct. 1977, p-12.

4739. Poletti, A. R., Sjoreen, T. P., et al., Phys. Rev., C, ISSN 0556-2813, 20(5), Nov. 1979, pp-1768-1774.

4740. Polikanov, S. M., and Sletten, G., Nucl. Phys. A151, 1970, p-656.

4741. Poling, J. E., Iowa Univ., Iowa City (U. S. A.), Thesis (Ph. D), 1975, p-83.

4742. Poling, J. E., Norbeck, E., et al., Phys. Rev., C, 13(2), Feb. 1976, pp-648-660.

4743. Pollock, R. E., IEEE Trans. Nucl. Sci., NS-24(3), 1977, pp-1505-8.

4744. Polucci, G. M., Koehler, A. M., et al., Nucl. Instr. and Meth. 71, 1969, pp-218-20.

4745. Polyukhov, V. G., Syuzev, V. N., et al., At. Energ. (USSR), 33(3), Sept. 1972, pp-773-74.

4746. Pomorski, L., Volkov, V. V., et al., JINR-P13-2965, 1966, p-17.

4747. Ponomarev, L. I., Somov, L. N., et al., Yad. Fiz. ISSN 0044-0027, 29(1), Jan. 1979, pp-133-37.

4748. Ponpon, J. P., Siffert, P., et al., "Thin dE/dX Detectors of Uniform Thickness Made on Epitaxial Silicon", NIM 112, 1973, pp-465-67.

4749. Pontius, P. E., and Cameron, J. M., NBS Monograph 103, Realistic Uncertainties and the Mass Measurement Process, An Illustrated Review, Aug. 15, 1967.

4750. Pontius, P. E., NBS Tech. Note 228, Measure. Philos. of the Pilot Prog. for Mass Calib., May 6, 1966.

4751. Poon, S. J., and Durand, J., Calif. Inst. of Tech., Pasadena (U. S. A.), W. M. Keck Lab. of Eng. Mat., July 1976, p-48.
4752. Popli, R., Grau, J. A., et al., Phys. Rev., C, ISSN 0556-2813, 20(4), Oct. 1979, pp-1350-1371.
4753. Popov, Yu. P., Sukhovoj, A. M., et al., Issledovanij po Atomnoj Nauke i Tekhnike, Moscow (USSR), 1976, No. 4, pp-50-54.
4754. Poppa, H., Moorhead, R. D., et al., Nucl. Instr. and Meth. 102, 1972, pp-521-3.
4755. Poppa, H., Moorhead, R. D., et al., TISRMNM, ORNL, Oct. 5-8, 1971.
4756. Poppa, H., Heinemann, Kl., et al., J. Vac. Sci. Tech. 8, 1971, p-471, H. Poppa, and A. G. Elliot, Surf. Sci. 24, 1971, p-149.
4757. Poppa, H., J. Appl. Phys. 38, 1967, p-3883.
4758. Poppa, H., J. Vac. Sci. Tech. 2, 1965, p-42.
4759. Poppe, C. H., Anderson, J. D., et al., Phys. Rev., C, 14(2), Aug. 1976, pp-438-445.
4760. Poppema, O. J., Nijgh, G. J., et al., Nederlandse Natuurkundige Vereniging, Amsterdam, Spring Meet. of the Dutch Phys. Soc., March 1978, p-30.
4761. Porile, N. T., Purdue Univ., Lafayette, IN, Dept. of Chem., 28 April 1969, p-24.
4762. Porile, N. T., Dropesky, B. J., et al., Phys. Lett., B, 67(1), 14 March 1977, pp-43-45.
4763. Porile, N. T., Purdue Univ., Lafayette, IN (U. S. A.), Dept. of Chem., Aug. 1979, p-24.
4764. Porile, N. T., Fortney, D. R., et al., Phys. Rev. Lett. ISSN 0031-9007, 43(13), 24 Sept. 1979, pp-918-921.
4765. Porter, L. E., and Shepard, C. L., Nucl. Instr. and Meth. 117(1), 1974, pp-1-4.
4766. Possoz, A., Deschepper, Ph., et al., Phys. Lett., B, 70(2), 26 Sept. 1977, pp-265-268.
4767. Postma, H., Leiden, Universiteit, KR-64(Sect. XXX), p-10.
4768. Potemans, M., Nucl. Eng. Int. ISSN 0029-5507, 23(267), Feb. 1978, pp-52-57.
4769. Potter, K. G., Sc. Rsch. Council, Chilton (UK), Rutherford Lab., June 1976, p-20.
4770. Potts, C. W., and Marcks, D. F., IEEE Trans. Nucl. Sci., NS-16, June 1969, pp-642-7.
4771. Pougheon, F., Inst. de Physique Nucleaire, Orsay (France), 1977, p-30.
4772. Pougheon, F., Roussel, P., et al., Nucl. Phys., A, ISSN 0375-9474, 325(2-3), 20-27 Aug. 1979, pp-481-509.
4773. Poulis, J. A., Poldervaart, L. J., et al., Appl. Sci. Res. B6, 1956, p-124.
4774. Povelites, J. G., LASL, AERE-R-5097, Paper 5, p-7.
4775. Povelites, J. G., PSPSITF, AERE, Oct. 20-21, 1965.

4776. Powell, C. F., Oxley, J. H., et al., Vapor Deposition, 1966 (New York: Wiley).
4777. Prade, H., Hagemann, U., et al., Zentralinstitut fuer Isotopen- und Strahlenforschung, Zentralinstitut fuer Kernforschung, Rossendorf bei Dresden (FRG.), Annual Report 1976, pp-36-37.
4778. Prade, H., Hagemann, U., et al., Int. Symp. on High-Spin States and Nucl. Structure, Dresden, 19-24 Sept. 1977, pp-23-24.
4779. Prakash, S., Manohar, S. B., et al., Int. J. Appl. Radiat. Isotop. 22, Feb. 1971, pp-128-9.
4780. Preedom, B. M., Darden, C. W., et al., Phys. Rev., C, 17(4), April 1978, pp-1402-1408.
4781. Preedom, B. M., Corfu, R., et al., Nucl. Phys., A, ISSN 0375-9474, 326(2-3), 10 Sept. 1979, pp-385-400.
4782. Prescott, C. Y., Stanford Linear Accel. Ctr., CA (U. S. A.), SLAC-PUB-2401, 1979, p-12.
4783. Preskitt, C. A., Neill, J. M., et al., Gulf Gen. At., Inc., San Diego, CA, GA-9027, Contract AT(04-3)-167, 7 Oct. 1968, p-26.
4784. Preston, M. K., Sease, J. D., et al., Contract W-7405-eng-26, 14th Conf. on Remote Syst. Tech., Pitts., PA, 1966, p-29.
4785. Pretorius, R., and Coetzee, P., J. Radioanal. Chem. 12(1), 1972, pp-301-11.
4786. Price, D. L., and Meister, H., Nucl. Instr. and Meth. 84, 1970, pp-61-6.
4787. Price, P. B., Stevenson, J., et al., Phys. Rev. Lett. 39 (4), 25 July 1977, pp-177-180.
4788. Prindle, A. L., Sisson, D. H., et al., Phys. Rev., C, ISSN 0556-2813, 20(5), Nov. 1979, pp-1824-1830.
4789. Prior, R. M., Moss, A. J., Jr., et al., Radiology 119(2), May 1976, pp-463-465.
4790. Prokhorova, L. I., Platonov, V. P., et al., IAEA, Vienna (Austria), Int. Nucl. Data Comm., June 1976, p-14.
4791. Prokopenko, V. S., Sklyarenko, V. D., et al., Izv. Akad. Nauk SSSR, Ser. Fiz. 40(6), June 1976, pp-1289-1293.
4792. Prokopets, G. A., and Murzin, A. V., Izv. Akad. Nauk SSSR, Ser. Fiz. 40(10), Oct. 1976, pp-2256-2258.
4793. Pronko, J. G., Lindgren, R. A., et al., Nucl. Instr. and Meth. 91, 1971, pp-659-61.
4794. Pronko, J. G., Bardin, T. T., et al., Phys. Rev., C, 13(2), Feb. 1976, pp-608-13.
4795. Pronko, P. P., Hardy, W. R., et al., Radiat. Eff. 10(1-2), 1971, pp-79-85.
4796. Prosdocimi, A., and Deruytter, A. J., J. Nucl. Energy A/B 17, 1963, p-83.
4797. Provo, J. L., 25th Nat. Vac. Symp., San Francisco, CA (U. S. A.), 27 Nov. 1978.

4798. Provo, J. L., 9th Ann. New Mexico Am. Vac. Soc. Symp., Albuquerque, NM, 1974.
4799. Prudnikov, I. A., and Tronov, B. N., 3rd All-union Conf. on Using Charged Particles Accel. in Nat. Economy, summaries of reports, 1979.
4800. Pruess, K., Delic, G., et al., ANL/PHY-76-2(2), May 1976, pp-723-731.
4801. Pruess, K., and Lichtner, P., Nucl. Phys., A, 291(2), 14 Nov. 1977, pp-475-509.
4802. Prümmer, R. A., Thin Solid Films 45, 1977, p-205.
4803. Prümmer, R. A., Int. Report for GSI, Feb. 1976.
4804. Prussin, S. G., Lanier, R. G., et al., Phys. Rev., C, 16(3), Sept. 1977, pp-1001-1009.
4805. Puehlhofer, F., Nucl. Phys., A, 280(1), 11 April 1977, pp-267-84.
4806. Puehlhofer, F., Schneider, W. F. W., et al., Phys. Rev., C, 16(3), Sept. 1977, pp-1010-1019.
4807. Puigh, R. J., Dyer, P., et al., Phys. Lett., B, ISSN 0370-2693, 86(1), 10 Sept. 1979, pp-24-28.
4808. Pulker, H. K., Z. Angew. Phys. 20, 1966, p-537.
4809. Pulker, H. K., and Jung, E., Thin Solid Films 4, 1969, p-219.
4810. Pulsen, A., NBS (U. S.), Spec. Publ. 493, Oct. 1977, pp-165-169.
4811. Purcell, J. E., and Meder, M. R., Phys. Rev., C, 16(1), July 1977, pp-76-79.
4812. Purser, F. O., Gould, C. R., et al., BNL-NCS-22500, March 1977, pp-276-286.
4813. Put, L. W., and Paans, A. M. J., Nucl. Phys., A, 291(1), 7 Nov. 1977, pp-93-125.
4814. Puumalainen, P., Aeystoe, J., et al., Jyvaeskylae Univ. (Finland), Dept. of Phys., March 1973, p-11.
4815. Puymbroeck, J. Van, and Gijbels, R., J. Inorg. and Nucl. Chem. 38(5), 1976, pp-957-960.
4816. Pywell, R. E., AINSE Nucl. Phys. Conf., 9-11 Feb. 1976, p-13.
4817. Qaim, S. M., Weinreich, R., et al., Inst. J. Appl. Radiat. Isot., ISSN 0020-708X, 30(2), 1979 Feb., pp-85-95.
4818. Rabenstein, D., Proc. of 2nd Int. Symp. on Neut. Capture Gamma Ray Spectros. and Related Topics, Petten, the Netherlands, 2-6 Sept. 1974, pp-584-88.
4819. Rad, F. N., Conzett, H. E., et al., Phys. Rev. Lett. 35 (17), 27 Oct. 1975, pp-1134-36.
4820. Rad, F. N., Sasanuma, T., et al., Phys. Rev. Lett. 40(6), 6 Feb. 1978, pp-368-371.
4821. Radomski, M., Phys. Rev., C, 14(5), Nov. 1976, pp-1704-08.
4822. Raduta, A. A., Ceausescu, V., et al., Nucl. Phys., A, 272 (1), 9 Nov. 1976, pp-11-20.
4823. Radutskij, G. M., and Sidorov, A. A., Leningrad Nauka 1978, p-250.

4824. Rae, J., AERE-M-3033, Report 1979, p-15.
4825. Raedt, J. A. G. de, Holthuizen, A., et al., Nederlandse Natuurkundige Vereniging, Amsterdam, Spring Meet. of the Dutch Phys. Soc., March 1978, p-28.
4826. Rafelski, J., Phys. Rev., C, 13(5), May 1976, pp-2086-88.
4827. Ragan, Hemmendinger, et al., BNL-NCS-22500, March 1977, p-114.
4828. Raghunathan, K., Ph. D. Thesis, Northwestern Univ.
4829. Rahman, M. M., New Mexico State Univ., Las Cruces (U. S. A.), Thesis (Ph. D), 1976, p-85.
4830. Rajagopalan, M., and Ramaniah, M., Proc. Nucl. Radiat. Chem. Symp., 2nd, 1966, pp-65-71.
4831. Raich, D. G., Yale Univ., New Haven, CT (U. S. A.), Thesis (Ph. D), 1976, p-119.
4832. Raisbeck, G. M., Lestringuez, J., et al., Centre de Spectrometrie Nucleaire et de Spectrometrie de Masse, Orsay (France), 1975, p-3.
4833. Ramachandran, G., Keshavamurthy, R. S., et al., Proc. Nucl. Phys. Solid State Phys. Symp. 21B, 1978, pp-135-7.
4834. Ramachandran, G., Murthy, R. S., et al., Phys. Lett., B, 87B(3), 1979, pp-252-6.
4835. Raman, S., Kim, H. J., et al., Nucl. Instr. and Meth. 81, 1970, pp-331-3.
4836. Raman, S., Slaughter, G. G., et al., Proc. of 2nd Int. Symp. on Neut. Capt. Gamma Ray Spectros. and Related Topics, Petten, the Netherlands, 2-6 Sept. 1974.
4837. Raman, S., Mizumoto, M., et al., Phys. Rev. Lett. 40(20), 15 May 1978, pp-1306-1309.
4838. Raman, S., Mizumoto, M., et al., ORNL-5306, Sept. 1977, pp-113-114.
4839. Ramavataram, K., Rangacharyulu, C., et al., Phys. Rev., C, 17(5), May 1978, pp-1583-1587.
4840. Ramsay, D., Stanford Univ., CA (U. S. A.), AECL-5503, April 1976, pp-151-58.
4841. Ramsay, D., INTDS Proc., Berkeley, U. S. A., Oct. 19-20, 1977.
4842. Ramsay, D., INTDS Proc., Los Alamos, U. S. A., Oct. 19-21, 1976.
4843. Ramsay, D., INTDS Newsletter, Jan. 1979.
4844. Ramsay, D., Nucl. Instr. and Meth. 167(1), 1979, pp-41-4.
4845. Ramsay, D., INTDS Proc., Garching, FRG., Sept. 11-14, 1978.
4846. Ramsay, D., INTDS Proc., Chalk River, Canada, Oct. 1-3, 1974.
4847. Ramsay, W. D., Birchall, J., et al., Nucl. Instr. and Meth. 169(3), 1980, pp-369-72.
4848. Ramstein, B., Jeanperrin, C., et al., J. Phys. 37(6), June 1976, pp-651-657.

4849. Ramstein, B., Rosier, L. H., et al., Nantes Univ. (France), Lab. de Spectroscopie Nucleaire, LSNN-78-05, 1978, p-22.
4850. Ramstroem, E., and Wiedling, T., Nucl. Phys., A, 272(1), 9 Nov. 1976, pp-259-268.
4851. Ramstroem, E., and Wiedling, T., Int. Conf. on the Inter. of Neut. with Nuclei, Vol. II, CONF-760715-P2, 1976, pp-1461-65.
4852. Randall, D. H., and Smith, M. L., Nature 175, 1955, p-1041.
4853. Ranft, J., Leipzig Univ., 1965, p-84.
4854. Ranft, J., and Ranft, G., Phys. Rev. Lett. 36(16), 19 April 1976, pp-988-90.
4855. Rao, G. N., and Guenther, C., Phys. Rev., C, 17(3), March 1978, pp-1266-1267.
4856. Rao, K. S., and Susila, S., AIP Conf. Proc. ISSN 0094-243X, 54(1), 15 July 1979, pp-430-31.
4857. Raquet, E., and Timm, U., Nucl. Instr. and Meth. 67, Jan. 15, 1969, pp-309-17.
4858. Ratner, G. L., Hornstra, F., et al., IEEE Trans. on Nucl. Sci., NS-16(3), 1969, pp-170-1.
4859. Rattazzi, G. U., Schmitt, R. P., et al., LBL-8151, 1978, pp-106-108.
4860. Ratynski, W., and Kopecky, J., Proc. of the 2nd Int. Symp. on Neut. Capt. Gamma Ray Spectros. and Related Topics, Petten, the Netherlands, 2-6 Sept. 1974.
4861. Rautenbach, L. W., Steyn, J., et al., Proc. of a Conf. held at St. Catherine's Coll., Oxford, 22-23 Sept. 1969.
4862. Ravn, H. L., Nucl. Instr. and Meth. 139, 1 Dec. 1976, pp-281-290.
4863. Ravn, H. L., Sundell, S., et al., Nucl. Instr. and Meth. 123(1), 1975, pp-131-44.
4864. Ravn, H. L., Sundell, S., et al., Proc. Int. EMIS Conf. Low Energy Ion Accel. Mass Sep., 8th, 1973, pp-432-44.
4865. Ray, L., Blanpied, G. S., et al., Phys. Rev. Lett. 40(24), 12 June 1978, pp-1547-1549.
4866. Ray, L., Blanpied, G. S., et al., Phys. Rev., C, ISSN 0556-2813, 20(4), Oct. 1979, pp-1236-43.
4867. Ray, L., Phys. Rev., C, ISSN 0556-2813, 20(5), Nov. 1979, pp-1857-1872.
4868. Razbudej, V. F., Vertebnyj, V. P., et al., Proc. of 4th All-union Conf. on Neut. Phys., 1977, pp-276-280.
4869. Reay, N. W., ANL/HEP-7208(Vol. 3), 1971, pp-1140-5.
4870. Rebak, M., Sellschop, J. P. F., et al., Nucl. Instr. and Meth. 167(1), 1979, pp-115-24.
4871. Rebak, M., Sellschop, J. P. F., et al., INTDS Proc., Garching, FRG., Sept. 11-14, 1978.
4872. Redish, E. F., Maryland Univ., College Park (U. S. A.), Dept. of Phys. and Astron., July 1976, p-32.

4873. Redstone, R., and Rowland, M. C., Nature 201, March 14, 1964, pp-1115-26.
4874. Redwine, R. P., AIP Conf. Proc. ISSN 0094-243X, 54(1), 15 July 1979, pp-501-14.
4875. Reed, R. E., and Grisham, W. B., Nucl. Instr. and Meth. 102, 1972, pp-513-519.
4876. Reed, R. E., and Grisham, W. B., TISRMNM, ORNL, Oct. 5-8, 1971.
4877. Reedy, R. C., Herzog, G. F., et al., Earth Planet. Sci. Lett. ISSN-0012-821X, 44(2), Aug. 1979, pp-341-348.
4878. Regimbart, R., Wozniak, G. J., et al., LBL-8151, 1978, pp-90-91.
4879. Regnier, S., Phys. Rev., C, ISSN 0556-2813, 20(4), Oct. 1979, pp-1517-1527.
4880. Rehm, K. E., Essel, H., et al., Phys. Lett., B, ISSN 0370-2693, 86(3-4), 8 Oct. 1979, pp-256-59.
4881. Rehm, K. E., Henning, W., et al., ANL-78-66, 1978, pp-46-8.
4882. Rehm, K. E., Henning, W., et al., Phys. Rev. Lett. 40(23), 5 June 1978, pp-1479-1482.
4883. Rehm, K. E., Henning, W., et al., ANL, CONF-771059-5, 1977, p-5.
4884. Rehm, K. E., Korner, H. J., et al., Phys. Rev., C, 12(6), Dec. 1975, pp-1945-1961.
4885. Rehm, K. E., Essel, H., et al., Z. Phys., A, 293(2), Nov. 1979, pp-119-121.
4886. Reich, K. H., Kerntechnik 4, March 1962, pp-94-9.
4887. Reid, A. F., Rev. Sci. Instr. 18, July 1947, pp-501-3.
4888. Reid, P. G. E., Sott, M., et al., Proc. Conf. Hyperfine Interactions (Asilomar, 1967).
4889. Reimer, G. M., Radiochem. Radioanal. Lett. 21(6), 1975, pp-339-342.
4890. Reimer, L., Elektronenmikroskopische Untersuchungs-und Präparations-methoden, Springer-Verlag, Berlin, Heidelberg, 1967.
4891. Reiter, W. L., Breunlich, W. H., et al., Nucl. Phys., A, 249(1), 8 Sept. 1975, pp-166-172.
4892. Rekalo, M. P., Gakh, G. I., et al., AN Ukrainskoj SSR, Kharkov, Fiziko-Tekhnicheskij Inst., KFTI-79-13, 1979, p-22.
4893. Rekawek, H., Mielczarek, M., et al., Nukleonika 21(3), 1976, pp-303-308.
4894. Rème, H., Toulouse, France, Université, 1962, p-50.
4895. Renner, C., Harvey, J. A., et al., ORNL-5306, Sept. 1977, pp-115-16.
4896. Rethmeier, J., and Meulen, D. R. Van der, Nucl. Instr. and Meth. 24, Oct. 1963, pp-349-52.
4897. Reus, U., Habbestad-Waetzig, A. M., et al., Europ. Phys. Soc., Geneva (Switz.), Europ. Conf. on Nucl. Phys. with Heavy Ions, 6-10 Sept. 1976, p-171.

4898. Reus, U., Habbestad Waetzig, A. M., et al., Phys. Rev. Lett. 39(4), 25 July 1977, pp-171-74.
4899. Revell, R. S. M., and Agar, A. W., Br. J. Appl. Phys. 6, 1955, pp-23-5.
4900. Reynolds, B. J., INTDS Newsletter, Jan. 1979.
4901. Reynolds, J., and Morgan, T., INTDS Proc., Argonne, U. S. A., Oct. 19-20, 1975.
4902. Reynolds, J. B., AERE-R-5097, Paper 21, p-3.
4903. Reynolds, J. B., and Boreham, D., AERE-R-5097, Paper 3, p-6.
4904. Reynolds, J. B., and Boreham, D., PSPSITF, AERE, Oct. 20 & 21, 1965.
4905. Rhoads, M., Stoner, J. O., et al., INTDS Proc., Berkeley, U. S. A., Oct. 19-21, 1977.
4906. Rhodin, T. N., Palmberg, P. W., et al., Molecular Processes on Solid Surfaces, McGraw-Hill, NY, 1968.
4907. Rhyne, J. J., Ames Lab., Iowa State Univ., Ames, U. S. A., Thesis (Ph. D), Report IS-T-21, July 1965.
4908. Ricci, E., Nucl. Instr. and Meth. 114(3), 1974, pp-477-81.
4909. Rich, B., Lembares, N., et al., Int. Symp. on Radiopharm., Atlanta, GA (U. S. A.), 12 Feb. 1974, pp-174-179.
4910. Richard, P., Sci. and Ind. Appl. of Small Accel., 4th Conf., 1976.
4911. Richards, H. T., Nucl. Spectr., ed. F. Ajzenberg-Selove (Academic Press, London, 1960), Part A, p-110.
4912. Richardson, J. R., 7th Int. Conf. on Cyc. and Their Appl., Zurich (Switz.), August 19-22, 1975.
4913. Richaud, J. P., Nucl. Instr. and Meth. 167(1), 1979, pp-97-100.
4914. Richaud, J. P., INTDS Proc., Garching, FRG., Sept. 11-14, 1978.
4915. Richter, A., Proc. of Int. Symp. on Interaction Studies in Nuclei held in Mainz, FRG., 17-20 Feb. 1975, pp-191-232.
4916. Richter, U., Maschinenmarkt 79, 1973.
4917. Rickards, J., Nucl. Instr. and Meth. ISSN 0029-554X, 152 (2-3), 15 June 1978, pp-585-587.
4918. Rickel, D. G., Roberson, N. R., et al., Nucl. Phys., A, 256(1), 5 Jan. 1976, pp-152-162.
4919. Rickertsen, L. D., Schneider, M. J., et al., Phys. Lett., B, 60(1), 22 Dec. 1975, pp-19-24.
4920. Rickertsen, L. D., Satchler, G. R., et al., ANL/PHY-76-2 (2), May 1976, pp-733-740.
4921. Ridley, B. W., and Chang, H. H., Colo. Univ., Boulder (U. S. A.), Dept. of Phys. and Astrophys., 1 Nov. 1975, pp-55-60.
4922. Riedel, J. T., Przegl. Elektron. 6, Oct. 1965, pp-492-7.
4923. Rieger, H., and Kim, K., Report 1979, UCRL-15113, p-63.
4924. Riehs, P., and Thomas, B. W., Proc. of 2nd Int. Symp. on Gamma Ray Spectro. and Related Topics, Petten, the Netherlands, 2-6 Sept. 1974, pp-300-305.

4925. Riel, W. D., INTDS Newsletter, Jan. 1979.
4926. Riel, W. D., Nucl. Instr. and Meth. 167(1), 1979, pp-179-81.
4927. Riel, W. D., INTDS Proc., Garching, FRG., Sept. 11-14, 1978.
4928. Riel, W. D., INTDS Proc., Chalk River, Canada, Oct. 1-3, 1974.
4929. Riel, W. D., INTDS Proc., Berkeley, U. S. A., Oct. 19-21, 1977.
4930. Riel, W. D., INTDS Newsletter, August 1974.
4931. Rieppo, R., Keinaenen, J. K., et al., J. Inorg. and Nucl. Chem. 38(11), 1976, pp-1927-1928.
4932. Rigaud, F., Petit, G. Y., et al., Nucl. Phys., A, ISSN 0375-9474, 326(1), 3 Sept. 1979, pp-26-36.
4933. Rigoleur, C. Le, Arnaud, A., et al., NBS (U. S.), Spec. Publ. 425, Oct. 1975, pp-953-956.
4934. Rihet, Y., Genoux-Lubain, A., et al., Centre de Recherches Nucleaires, CRN-PN-76-29, 1976, p-11.
4935. Rihet, Y., Costa, G., et al., Phys. Rev., C, ISSN 0556-2813, 20(4), Oct. 1979, pp-1583-1584.
4936. Riley, P. J., Bjork, C. W., et al., Phys. Rev., C, 17(5), May 1978, pp-1881-1884.
4937. Rinat, A. S., and Thomas, A. W., AIP Conf. Proc. 33, 1976, pp-450-451.
4938. Rinat, A. S., and Thomas, A. W., Nucl. Phys., A, 282(3), 30 May 1977, pp-365-388.
4939. Ripin, B. H., Decoste, R., et al., Phys. Fluids 23(5), 1980, pp-1012-30.
4940. Riska, D. O., and Radomski, M., Phys. Rev., C, 16(6), Dec. 1977, pp-2105-2116.
4941. Riska, D. O., Phys. Lett., B, ISSN 0370-2693, 86(2), 24 Sept. 1979, pp-151-153.
4942. Risler, R., Grueebler, P. A., et al., Helv. Phys. Acta 50(5), 30 Nov. 1977, p-579.
4943. Ritchie, A. I. M., J. Phys., D, 9(1), 11 Jan. 1976, pp-15-26.
4944. Ritson, D. M., AIP Conf. Proc. 30, 1976, pp-75-82.
4945. Rivet, M. F., Bimbot, R., et al., Inst. de Physique Nucleaire (France), IPNO-RC-79-08, 1979, p-51.
4946. Roach, K. E., Mullett, W. L., et al., Southern Nucl. Eng., Inc., Dunedin, FL, Contract AT(29-2)-2741, Feb. 1972, p-89.
4947. Robb, A. D., and Schier, W. A., Nucl. Instr. and Meth. 91, 1971, pp-13-14.
4948. Robeau, P., and Vermeulen, J., Cryogenics 11(6), Dec. 1971, pp-478-82.
4949. Roberson, P. L., Long, D. D., et al., Phys. Lett., B, 70 (1), 12 Sept. 1977, pp-35-38.
4950. Robertson, B. C., Alberta Univ., Edmonton, Canada, Nucl. Rsch. Ctr., March 1970, p-10.

4951. Robertson, R. G. H., Khoo, T. L., et al., Phys. Rev., C, 17(5), May 1978, pp-1535-1539.
4952. Robertson, W. D., J. Vac. Sci. Tech. 8, 1971, p-403.
4953. Robinson, D. C., J. Phys., E, 3, May 1970, pp-408-9.
4954. Robinson, M. T., Phys. Rev. 179, 1969, p-327.
4955. Robinson, P. S., Nucl. Instr. and Meth. 40, 1966, p-136.
4956. Robinson, R. G., Bradshaw, D., et al., Int. J. Appl. Radiat. Isot. ISSN 0020-708X, 28(12), Dec. 1977, pp-919-923.
4957. Robinson, R. L., Bair, J. K., et al., Phys. Rev., C, 14(6), Dec. 1976, pp-2126-2132.
4958. Robinson, R. L., Kim, H. J., et al., Phys. Rev., C, 16(6), Dec. 1977, pp-2268-2275.
4959. Robinson, T. B., and Edwards, V. R. W., Nucl. Phys., A, ISSN 0029-5582, 301(1), 22 May 1978, pp-36-52.
4960. Robrish, P., Symp. on Polar. at High Energy, Argonne, IL, 15 April 1970, pp-73-96.
4961. Roche, C. T., Clark, R. G., et al., Phys. Rev., C, 14(2), Aug. 1976, pp-410-418.
4962. Roche, G., Landaud, G., et al., AIP Conf. Proc. ISSN 0094-243X, 47, 1978, pp-650-51.
4963. Rockwood, S. D., and Rabideau, W. W., 8th Int. Conf. on Quantum Elect., San Francisco, 1974.
4964. Roeckl, E., Kirchner, R., et al., BNL-50847, July 1978, pp-331-352.
4965. Roemer, J., and Ross, R., Isotopenpraxis 3, 1967, pp-341-2.
4966. Rogers, D. W. O., Dixon, W. R., et al., Nucl. Phys., A, 281(2), 2 May 1977, pp-345-353.
4967. Rogers, J. W., and McElroy, W. N., Hanford Eng. Dev. Lab., Richland, WA (U. S. A.), Aug. 1976, pp-ANC.11-ANC.13.
4968. Röhm, H. F., Münzel, H., et al., Nucl. Instr. and Meth. 113, 1973, pp-101-107.
4969. Rohwer, R. K., and Urizar, M. J., INTDS Proc., Los Alamos (U. S. A.), Oct. 19-21, 1976.
4970. Rolfs, C., Gorres, J., et al., Nucl. Instr. and Meth. 157 (1), 15 Nov. 1978, pp-19-27.
4971. Rolfs, C., AIP Conf. Proc. ISSN 0094-243X, 47, 1978, pp-197-220.
4972. Rolfs, C., Proc. of Int. Workshop on Gross Properties of Nuclei and Nucl. Excitations, Hirschegg, Austria, 15-27 Jan. 1979, pp-183-197.
4973. Roll, C., Gorres, J., et al., Nucl. Instr. and Meth. 157 (1), 15 Nov. 1978, pp-19-27.
4974. Rollin, D. F., and Robinson, J. E., Nucl. Instr. and Meth. 123(3), 1 Feb. 1975, pp-465-470.
4975. Romain, R., Rev. Sci. Instr. 50(12), 1979, pp-1561-3.
4976. Roman, R. D., Guigler, R. M., et al., IEEE Trans. Nucl. Sci., NS-16, June 1969, pp-633-6.
4977. Ronningen, R. M., Vanderbilt Univ., Nashville, TN (U. S. A.), Thesis (Ph. D), 1975, p-220.

4978. Ronningen, R. M., Hamilton, J. H., et al., Phys. Rev., C, 16(6), Dec. 1977, pp-2208-2217.
4979. Ronningen, R. M., Piercey, R. B., et al., Phys. Rev., C, 16(6), Dec. 1977, pp-2218-2222.
4980. Ronningen, R. M., Baker, F. T., et al., Phys. Rev. Lett. 40(6), 6 Feb. 1978, pp-364-367.
4981. Roos, M., Rose, H. P., et al., Rev. Sci. Instr. 36(4), April 1965, pp-544-6.
4982. Roos, P. G., Smith, S. M., et al., Nucl. Phys., A, 255(1), 15 Dec. 1975, pp-187-203.
4983. Roos, P. G., Goldberg, D. A., et al., Nucl. Phys., A, 257(2), 2 Feb. 1976, pp-317-332.
4984. Roos, P. G., Maryland Univ., College Park (U. S. A.), Dept. of Phys. and Astron., Nov. 1976, p-44.
4985. Roos, P. G., Chant, N. S., et al., Phys. Rev., C, 15(1), Jan. 1977, pp-69-83.
4986. Roos, P. G., Chant, N. S., et al., Phys. Rev. Lett. 40(22), 29 May 1978, pp-1439-1443.
4987. Roos, P. G., AIP Conf. Proc. 36, 1977, pp-32-50.
4988. Ropero, M., Junta de Energia Nuclear, Madrid (Spain), JEN-399, 1978, p-55.
4989. Rosario-Garcia, E., and Benenson, R. E., Nucl. Phys., A, 275(2), 10 Jan. 1977, pp-453-463.
4990. Rose, A., Nucl. Instr. and Meth. 35, 1965, pp-165-6.
4991. Rosen, J. L., AIP Conf. Proc. 26, 1975, pp-265-88.
4992. Rosenblatt, B. G., Jr., and Kibler, M. G., Calif. Rsch. and Dev. Co., Livermore Rsch. Lab., Livermore, Contract AT(11-1)-74, Jan. 9, 1953, p-89.
4993. Rossi, B. B., and Staub, H. H., Ionization Chambers and Counters (McGraw-Hill, NY, 1949), p-210.
4994. Rossi, J., Truong, C. M., et al., Radiat. Eff. 44(1-4), 1979, pp-213-17.
4995. Rossi, J., Truong, C. M., et al., Nucl. Instr. and Meth. 153(1), 1978, pp-285-8.
4996. Rost, E., CEA-CR-2 (nd), 8-10 May 1974, pp-159-170.
4997. Rost, E., and Shepard, J. R., Phys. Lett., B, 59(5), 8 Dec. 1975, pp-413-415.
4998. Rost, H., Eyrich, W., et al., Erlangen-Nuernberg Univ., Erlangen (Germany, F. R.), Physikalisches Inst. Annual Report 1976, pp-9-10.
4999. Rost, M., Reisse, G., et al., Wiss. Z. - Tech. Hochsch. Karl-Marx-Stadt 20(7), 1978, pp-827-33.
5000. Rotbard, G., Vergnes, M., et al., Phys. Rev., C, 16(5), Nov. 1977, pp-1825-1834.
5001. Rotbard, G., LaRana, G., et al., Phys. Rev., C, 18(1), July 1978, pp-86-95.
5002. Rotter, H., Heiser, C., et al., Zentralinstitut fuer Kernforschung, Rossendorf bei Dresden (FRG.), Annual Report 1975, pp-30-31.

5003. Roubeau, P. M., Univ. Grenoble, Faculte des Sciences, France, Thesis, 1967, p-136.
5004. Roubeau, P., CEN, Comm. Energ. At., Saclay, France, Report 1971, LBL-500, pp-47-55.
5005. Roubeau, P., 2nd Int. Conf. on Polar. Targ., Berkeley, CA, 30 Aug. 1971, pp-47-55.
5006. Roubeau, P., and Vermeulen, J., Cryogenics 11(6), 1971, pp-478-82.
5007. Roubeau, P., Ezratty, J., et al., Nucl. Instr. and Meth. 82, 1970, pp-323-4.
5008. Roubeau, P. M., and Der Nigohossian, G., Int. Inst. of Refrig. Comm. I Meet., Boulder, CO, 1967, pp-383-8.
5009. Roubeau, P. M., and Der Nigohossian, G., Annexe Bull. Inst. Int. Froid. 66(5), pp-383-8.
5010. Routledge, K. J., and Stowell, M. J., Thin Solid Films 6, 1970, p-40.
5011. Rovner, L. H., and Chen, K. Y., J. Nucl. Mater. 63, Dec. 1976, pp-307-12.
5012. Rowland, F. S., and Wolfgang, R. L., Rev. Sci. Instr. 29, 1958, pp-210-14.
5013. Rowley, N., Inst. de Physique Nucleaire, Paris (France), March 1977, p-13.
5014. Rowley, N., Doubre, H., et al., Phys. Lett., B, 69(2), 1 Aug. 1977, pp-147-150.
5015. Roy, A., Iyengar, K. V. K., et al., Proc. of Nucl. Phys. and Solid State Phys. Symp., Bombay, 17B, Dec. 27-31, 1974, pp-98-100.
5016. Roy, G., and Riebeek, N., Nucl. Instr. and Meth. 71, 1969, pp-234-6.
5017. Roy, R., Seiler, F., et al., LBL-8151, 1978, pp-59-61.
5018. LeRoy, N. R., McKisson, R. L., et al., Calif. Rsch. and Dev. Co., LRL, Livermore, CA, Contract AT(11-1)-74, March 7, 1957, p-21.
5019. Roynette, J. C., Doubre, H., et al., Phys. Lett., B, 67 (4), 25 April 1977, pp-395-398.
5020. Roynette, J. C., Doubre, H., et al., Inst. de Physique Nucleaire, Paris, 1977, p-10.
5021. Ruchti, R. C., AIP Conf. Proc. 30, 1976, pp-88-93.
5022. Rudak, Eh. A., Soroka, A. V., et al., Leningrad Nauka, 1978, p-51.
5023. Rudolf, G., Centre Nat. de la Recherche Scientifique, Cent. de Rech. Nucleaires, Strasbourg (France), These (D. es S.), 1975, p-122.
5024. Rudstam, G., Nucl. Instr. and Meth. 139(1), 1976, pp-239-49.
5025. Rudstam, G., Bruninx, E., et al., Phys. Rev. 126(5), June 1, 1962, pp-1852-7.
5026. Ruff, O., Z. Anorg. Chem. 129, 1923, p-267.

5027. Rupp, A. F., Radiopharm. Label. Comp., Proc. Symp. 1, 1973, pp-223-37.
5028. Rurarz, E., and Sulik, A., Nukleonika 14, 1969, pp-1165-9.
5029. Russell, F. M., RHEL, LBL-500, Report 1971, pp-89-98.
5030. Russell, F. M., RHEL/R 225, 1971, p-25.
5031. Russell, F. M., 2nd Int. Conf. on Polar. Targ., Berkeley, CA, 30 Aug. 1971, pp-89-98.
5032. Russell, F. M., RHEL/R-199, Sept. 1970, p-11.
5033. Russell, G. J., Trans. Am. Nucl. Soc. 27, 1977, pp-861-2.
5034. Russo, P., Schmitt, R. P., et al., Nucl. Phys., A, 281(3), 9 May 1977, pp-509-532.
5035. Rust, N. J. A., Clover, M. R., et al., AIP Conf. Proc. ISSN 0094-243X, 47, 1978, pp-718-719.
5036. Ruth, T. J., Wolf, A. P., IEEE Trans. Nucl. Sci., NS-26 (1), Pt. 2, Feb. 1979, pp-1710-12.
5037. Rutledge, L. L., Jr., and Hiebert, J. C., Phys. Rev., C, 13(3), March 1976, pp-1072-1082.
5038. Rutledge, L. L., Jr., Macias, E. S., et al., Phys. Rev., C, 13(6), June 1976, pp-2166-2181.
5039. Ryan, J. F., NASA Access. UCRL-14120, 1965, pp-182-7.
5040. Ryan, N. E., Henderson, F., et al., Nature 180, 1957, p-1406.
5041. Ryan, N., J. Electrochem. Soc. 106, 1959, p-388.
5042. Ryan, N., J. Electrochem. Soc. 107, 1960, p-397.
5043. Ryan, P. J. P., 7th AINSE Nucl. Phys. Conf., Lucas Heights, Australia, 1978, p-3.
5044. Rykova, R. N., and Sokolov, L. I., ITEF-67, 1977, p-11.
5045a. Sadkovskij, V. S., Feofilov, G. A., et al., Leningrad Nauka, 1978, p-186.
5045b. Saettel, M. A., INTDS Proc., Los Alamos, U. S. A., Oct. 19-21, 1977.
5046. Saettel, M. A., INTDS Proc., Berkeley, U. S. A., Oct. 19-20, 1977.
5047. Saettel, M. A., INTDS Proc., ANL, U. S. A., Sept. 30-Oct. 2, 1975.
5048. Saettel, M. A., INTDS Newsletter, August 1980.
5049. Sagara, K., Motobayashi, T., et al., Nucl. Phys., A, 273 (2), 30 Nov. 1976, pp-493-504.
5050. Sagara, K., Hara, M., et al., J. Phys. Soc. Jpn. 42(3), March 1977, pp-732-737.
5051. Sagara, K., Motobayashi, T., et al., Nucl. Phys., A, 299(1), 17 April 1978, pp-77-91.
5052. Saghai, B., Villeurbanne (France), Inst. de Physique Nucleaire, 4th Biennial Session of Nucl. Phys., La Toussuire, 28 Feb.-4 March 1977, pp-S15.1-S15.2.
5053. Saha, A., Scholten, O., et al., Phys. Lett., B, ISSN 0370-2693, 85(2-3), 13 Aug. 1979, pp-215-218.
5054. Saha, S. K., Helppi, H., et al., Phys. Rev., C, 16(6), Dec. 1977, pp-2159-2164.

5055. Sahlin, H. L., Ann., NY, Acad. Sci. 25, May 8, 1975 for Meet., NY, NY, March 4-7, 1974, pp-238-272.
5056. Sahlin, H. L., Law. Liv. Lab., UCID-17328, Report 1976, p-30.
5057. Sailor, V. L., BNL-13624, May 1, 1969, p-89.
5058. Saint-Laurent, F., Conjeaud, M., et al., Nucl. Phys., A, ISSN 0375-9474, 327(2), 24 Sept. 1979, pp-517-532.
5059. Saint-Simon de, M., Lessard, L., et al., Phys. Rev., C, 14(6), Dec. 1976, pp-2185-2188.
5060. Sakai, M., and Rester, A. C., At. Data Nucl. Data Tables 20(5), Nov. 1977, pp-441-474.
5061. Sakamoto, N., Matsuki, S., et al., Phys. Lett., B, ISSN 0370-2693, 83(1), 23 April 1979, pp-39-42.
5062. Sakharov, E. S., and Chuchalin, I. P., At. Energ. (USSR) 29, Aug. 1970, pp-125-7.
5063. Saladin, J. X., Lee, I. Y., et al., Phys. Rev., C, 14(3), Sept. 1976, pp-992-995.
5064. Salaita, G. N., and Eapen, P. K., NBS (U. S.), Spec. Publ. 425, Oct. 1975, pp-712-715.
5065. Salmi, U., Wagschal, J. J., et al., Proc. of a Seminar-Workshop, Oak Ridge, TN, ORNL/RSIC-42, Feb., 1979, pp-49-59.
5066. Salomon, M., Baer, H., et al., LA-7892C, July 1979, pp-16-22.
5067. Saloner, D. A., and Toepffer, C., Nucl. Phys., A, 283(1), 6 June 1977, pp-108-130.
5068. Saloner, D. A., and Toepffer, C., Nucl. Phys., A, 283(1), 6 June 1977, pp-131-148.
5069. Saltmarsh, M. J., 3rd Conf. on the Use of Small Accel., Denton, TX (U. S. A.), 21 Oct. 1974, p-18.
5070. Saltmarsh, M. J., Ludemann, C. A., et al., Symp. on Neut. Stds. and Appl., Gaithersburg, MD (U. S. A.), 28-31 March 1977, p-22.
5071. Saltykov, L. S., Izv. Akad. Nauk SSSR, Ser. Fiz. 31, Feb. 1967, pp-260-3.
5072. Salvini, G. A., Nuovo Cimento, Supp. 24(1), 1963, pp-1-388.
5073. Sampson, T. E., and Carpenter, J. M., Nucl. Instr. and Meth. 50, 1967, pp-179-80.
5074. Samsonov, G. V. (ed.), Handbook of the Physiochem. Prop. of the Elements, Plenum, NY, 1968.
5075. Samuel, R. L., Ind. At. 14(7-8, 21-31), July-Aug. 1970.
5076. Samuelson, L. E., Rickey, F. A., et al., Nucl. Phys., A, ISSN 0029-5582, 301(1), 22 May 1978, pp-159-178.
5077. Sanada, J., Nucl. Instr. and Meth. 57, Dec. 1967, pp-58-60.
5078. Sanborn, C. B., Irvine, J. W., et al., J. Chem. Phys. 12, 1944, pp-132-4.
5079. Sander, V., Bukow, H. H., et al., INTDS Newsletter, Jan. 1979.

5080. Sander, U., Bukow, H. H., et al., Nucl. Instr. and Meth. 167(1), 1979, pp-35-9.
5081. Sander, U., Bukow, H. H., et al., J. Phys. 40, 1979, pp-C1-301.
5082. Sander, U., Bukow, H. H., INTDS Newsletter, July 1979.
5083. Sander, U., and Bukow, H. H., INTDS Proc., Garching, FRG., Sept. 11-14, 1978.
5084. Sanders, J. H., and Wapstra, A. H. (eds.), Atomic Masses and Fund. Constants, Vol. 5, Plenum Publ. Corp., NY, 1976, pp-81-87.
5085. Sanders, S. J., Martz, L. M., et al., Phys. Rev., C, ISSN 0556-2813, 20(5), Nov. 1979, pp-1743-1753.
5086. Sandifer, W. C., and Taherzader, IEEE Trans. Nucl. Sci., NS-15(6), Dec. 1968, pp-336-45.
5087. Sandorfi, A. M., Kilius, L. R., et al., Nucl. Instr. and Meth. 136(2), 15 July 1976, pp-395-396.
5088. Sandorfi, A. M., Kilius, L. R., et al., Phys. Rev. Lett. 38(25). 20 June 1977, pp-1463-1466.
5089. Sandorfi, A. M., Kilius, L. R., et al., Phys. Rev. Lett. 40(19), 8 May 1978, pp-1248-1252.
5090. Sandorfi, A. M., BNL-26598, CONF-790743-6, 1979, p-42.
5091. Sannes, F., Trischuk, J., et al., Nucl. Instr. and Meth. 70, April 15, 1969, pp-235-6.
5092. Santana, A. M., Coelho, H. T., et al., Phys. Rev., C, 16 (5), Nov. 1977, pp-1785-1791.
5093. Santoro, R. T., Wachter, J. W., et al., Nucl. Instr. and Meth. 93(2), 1971, pp-371-3.
5094. Santoro, R. T., and Weaver, H., Rev. Sci. Instr. 36, Jan. 1965, p-98.
5095. Santoro, R. T., Peelle, R. W., et al., ORNL-3360, pp-272-80.
5096. Santry, D. C., and Werner, R. D., IEEE Trans. Nucl. Sci., NS-26(1), Pt. 2, Feb. 1979, pp-1335-1337.
5097. Santry, D. C., INTDS Proc., Chalk River, Canada, Oct. 1-3, 1974.
5098. Santry, D. C., and Sitter, C. W., Proc. Int. Conf. on Electromag. Isot. Separators (Marburg), Bundesministerium fur Bildung und Wissenchaft, Report K70-28, 1970, p-505.
5099. Santry, D. C., and Werner, R. D., J. Nucl. Eng. 27, 1973, p-409,
Selin, E., Arnell, S. E., et al., Nucl. Instr. and Meth. 56, 1967, p-218.
5100. Santry, D. C., AECL-5503, April 1976, pp-1-13.
5101. Saphier, D., Ilberg, D., et al., Trans. Am. Nucl. Soc. 22, Nov. 1975, pp-671-72.
5102. Saphier, D., Ilberg, D., et al., Nucl. Sci. and Eng. 62 (4), April 1977, pp-660-694.
5103. Sapir, L., INTDS Newsletter, July 1979.

5104. Sapir, L., Nucl. Instr. and Meth. 167(1), 1979, pp-161-3.
5105. Sapir, L., INTDS Proc., Garching, FRG., Sept. 11-14, 1978.
5106. Sapper, F., Stuttgart Univ., Inst. fuer Kernenergetik, Stuttgart Univ. (TH) (Germany, F. R.), May 1976, p-134.
5107. Sarantites, D. G., Urbon, J., et al., Phys. Rev., C, 14(4), Oct. 1976, pp-1412-1428.
5108. Sarantites, D. G., Barker, J. H., et al., ORNL-5306, Sept. 1977, pp-46-49.
5109. Sarantites, D. G., Westerberg, L., et al., Phys. Rev., C, 17(2), Feb. 1978, pp-601-621.
5110. Sarma, N., Nucl. Instr. 2, 1958 May, pp-361-2.
5111. Sasao, M., and Torizuka, Y., Phys. Rev., C, 15(1), Jan. 1977, pp-217-232.
5112. Satchler, G. R., Ford, J. L. C., Jr., et al., Phys. Lett., B, 60(1), 22 Dec. 1975, pp-43-46.
5113. Satchler, G. R., ORNL-5137, May 1976, pp-133-152.
5114. Satchler, G. R., Halbert, M. L., et al., Nucl. Phys., A, 298(2), 3 April 1978, pp-313-332.
5115. Sauer, J. M., Rev. Sci. Instr. 36, Sept. 1965, p-1374.
5116. Saurbrey, C., Z. f. Physik 155, 1959, p-206.
5117. Saussure, G. de, and Macklin, R. L., ORNL/TM-6161, Dec. 1977, p-52.
5118. Saussure, G. de, Olsen, D. K., et al., ORNL/TM-6152, Jan. 1978, p-99.
5119. Savage, C. D., Naval Postgrad. Sch., Monterey, CA, June 1971, p-44.
5120. Savin, I. A., and Zacek, J., JINR-E-1-12502, 1979, p-11.
5121. Sawicki, M., Phys. Lett., B, 68(1), 9 May 1977, pp-43-46.
5122. Sayer, R. O., Eichler, E., et al., Phys. Rev., C, 17(3), March 1978, pp-1026-1033.
5123. Scaife, W. A., Hanley, P. R., et al., INTDS Proc., Argonne, U. S. A., Sept. 30-Oct. 2, 1975.
5124. Schädel, M., Kratz, J. V., et al., Phys. Rev. Lett. 41, 1978, p-469.
5125. Schaeffer, M., Strasbourg-1 Univ., 67 (France), Centre de Recherches Nucleaires, These (D. es S.), 1975, p-134.
5126. Schaeffer, M., Degre, A., et al., Nucl. Phys., A, 275(1), 4 Jan. 1977, pp-1-12.
5127. Schaeffer, R., and Pearson, R. K., J. Am. Chem. Soc. 91, 1969, p-2153.
5128. Schaeffer, R., Kernforschungszentrum Karlsruhe G.m.b.H. (Germany, F. R.), Inst. fuer Angewandte Kernphysik, KFK-2830, 2-4 May 1979, pp-98-122.
5129. Schaeffer, W., and Rauch, F., Nucl. Phys., A, 291(1), 7 Nov. 1977, pp-165-182.
5130. Schaer, H., Trautmann, D., et al., Helv. Phys. Acta 50(1), 4 Feb. 1977, pp-29-47.
5131. Schaich, R. W., ORNL-TM-3225, Contract W-7405-eng-26, Dec. 1970, p-23.

5132. Schapira, J. P., Maison, J. M., et al., Phys. Rev., C, 17 (5), May 1978, pp-1588-1601.
5133. Schatz, G., and Schulz, F., Kernforschungszentrum, Karlsruhe (West Germany), Zyklotron-Laboratorium, June 1970, p-9.
5134. Scheffler, K., 2nd Int. Conf. on Polar. Targ., Berkeley, CA, 30 Aug. 1971, pp-271-9.
5135. Scheffler, K., Nucl. Instr. and Meth. 82, 1970, pp-205-7.
5136. Scheidemann, O., and Porile, N. T., Phys. Rev., C, 14(4), Oct. 1976, pp-1534-44.
5137. Schermer, R. I., Conf. on Polar. Targ. and Srcs., Saclay, France, Dec. 5-9, 1966, pp-357-66.
5138. Scherzer, B. M. U., Behrisch, R., et al., J. Nucl. Mater. 63, Dec. 1976, pp-100-105.
5139. Scheu, K. W., and Gee, T., INTDS Proc., Berkeley, U. S. A., Oct. 19-20, 1977.
5139a. Schue, K. W., and Gee, T., INTDS Proc., Boston, U. S. A., Oct. 1-3, 1980.
5140. Schier, W. A., Barnes, B. K., et al., Nucl. Phys., A, 254(1), 1 Dec. 1975, pp-80-92.
5141. Schiffer, J. P., Int. Symp. on Collectivity of Medium and Heavy Nuclei, Tokyo (Japan), 20-25 Sept. 1976, p-18.
5142. Schilling, K. D., Andrejtscheff, W., et al., Zentralinstitut fuer Kernforschung, Rossendorf bei Dresden (FRG.), Annual Report, 1975, pp-47-48.
5143. Schiller, S. A., and Eck, J. S., Z. Phys., A, 275(3), Dec. 1975, pp-255-259.
5144. Schioett, H. E., Kgl. Dan. Vidensk. Selsk., Mat.-Fys. Medd. 35(9), 1966, p-20.
5145. Schlenk, B., and Valek, A., ATOMKI (At. Kut. Intex.) Kozlem. 10, April 1968, pp-71-6.
5146. Schlenk, B., Papp, I., et al., ATOMKI Kozlem. 8, Sept. 1966, pp-232-6.
5147. Schlenker, M., and Baruchel, J., Proc. Conf. on Neut. Scattering, CONF-760601-P2, 1 Sept. 1976, pp-1136-42.
5148. Schmelzbach, P. A., Grueebler, W., et al., Helv. Phys. Acta 49(5), 29 Oct. 1976, p-790.
5149. Schmelzbach, P. A., Hardekopf, R. A., et al., LA-6791-MS, May 1977, p-11.
5150. Schmelzbach, P. A., Hardekopf, R. A., et al., BNL-NCS-22500, March 1977, p-125, 133-134.
5151. Schmelzbach, P. A., Hardekopf, R. A., et al., Phys. Rev., C, 17(1), Jan. 1978, pp-16-23.
5152. Schmidt, K. H., Faust, W., et al., Verh. Dtsch. Phys. Ges. 6, 1977, pp-1012-13.
5153. Schmidt, P. H., Castellano, R. N., et al., "Deposition and Evaluation of Thin Films by DC Ion Beam Sputtering", Solid State Tech., July 1972.

5154. Schmidt-Ott, W. D., Mlekodaj, R. L., et al., Nucl. Instr. and Meth. 108(1), 1973, pp-13-21.
5155. Schmitt, R. P., Russo, P., et al., LBL-5075, 1975, pp-136-7.
5156. Schmitt, R. P., LBL-7168, May 1978, p-189.
5157. Schmitt, R. P., Rattazzi, G. U., et al., LBL-8151, 1978, pp-93-94.
5158. Schmitt, R. P., Wozniak, G. J., et al., LBL-8151, 1978, pp-96-97.
5159. Schmoranzer, H., Grabe, H., et al., Appl. Phys. Lett. 26 (8), 15 April 1975, pp-483-85.
5160. Schneider, D., and Stolterpoh, N., 10th Int. Conf. on the Phys. of Electronic and Atomic Collisions, Vol. 2, 1977, pp-1314-15.
5161. Schneider, M. J., Burch, J. D., et al., Colo. Univ., Dept. Astrophys., Boulder (U. S. A.), Tech. Prog. Report, Nov. 1, 1975, pp-25-28.
5162. Schneider, M. J., Rudolph, H., et al., Colo. Univ., Tech. Prog. Report, Nov. 1, 1975, pp-46-48.
5163. Schneider, R. J., and Goldberg, C. J., Int. J. Appl. Radiat. Isot. 27(3), 1976, pp-189-91.
5164. Schneider, R. L., Nat. Conf. on Stands. Labs., Annual Meet., Gaithersburg, MD, April 1966, p-10.
5165. Schneider, S., Erlangen-Nuernberg Univ., Erlangen, (FRG.), Diss. (D. Sc.), INIS-mf-5457, 23 Nov. 1977, p-144.
5166. Schneider, S., Eyrich, W., et al., INIS-mf-4348, Nov. 1977, Annual Report 1976, pp-3-4.
5167. Schneider, S., Eyrich, W., et al., INIS-mf-4348, Nov. 1977, Annual Report 1976, pp-4-5.
5168. Schneider, W., Marburg Univ., Fachbereich Physik (FRG.), Diss. (D. Sc.), 28 June 1973, p-73.
5169. Schneider, W. F. W., Kohlmeyer, B., et al., Z. Phys., A, 275(3), Dec. 1975, pp-249-253.
5170. Schneider, W., Kerntechnik 5, Aug. 1963, pp-334-8.
5171. Schneggenburger, R. G., Updegrove, W. S., et al., Rev. Sci. Instr. 49(11), 1978, pp-1543-4.
5172. Scholz, D., Gemmeke, H., et al., Nucl. Phys., A, 288(2), 26 Sept. 1977, pp-351-364.
5173. Schormann, R., Fischer, H., et al., Phys. Rev., C, 16(6), Dec. 1977, pp-2165-2166.
5174. Schrader, H., K. E. Pferdekämper, et al., Nucl. Instr. and Meth. 134, 1976, p-157.
5175. Schramel, P., Kerntechnik 12, Sept. 1970, pp-373-6.
5176. Schraube, H., Johlige, H., et al., 1st Symp. on Neut. Dosimetry in Biol. and Medicine, Neuherberg/Munich, Germany, 15 May 1972, pp-757-781.
5177. Schreiber, R. E. (ed.), LA-4759, Contract W-7405-eng-36, Sept. 1971, p-18.
5178. Schrieder, G., Mueller-Arnke, A., et al., Nucl. Phys., A, 279(3), 4 April 1977, pp-463-473.

5179. Schrils, R., Flynn, D. S., et al., Phys. Rev., C, ISSN 0556-2813, 20(5), Nov. 1979, pp-1706-1710.
5180. Schroeder, H., Kern, K. K., et al., Nucl. Phys., A, 269(1), 28 Sept. 1976, pp-74-86.
5181. Schroeder, L. S., Nucl. Instr. and Meth. 162(1-3)(Pt. II), 1-15 June 1979, pp-395-404.
5182. Schryver, S. De, Fiore, L., et al., LNF-68/70, Nov. 26, 1968, p-8.
5183. Schultheis, H., and Schultheis, R., AIP Conf. Proc. ISSN 0094-243X, 47, 1978, pp-512-513.
5184. Schulz, D., and Searcy, A., J. Chem. Phys. 36, 1962, p-3099.
5185. Schultz, F., and Bellemann, H., Proc. of 3rd Conf. on Accel. Targ. Designed for the Prod. of Neutrons, Liege (Belgium), Sept. 18-19, 1967, pp-103-12.
5186. Schutz, Y., Strasbourg-1 Univ., 67 (France), Centre de Recherches Nucleaires, These (3e Cycle), 1978, p-52.
5187. Schwarz, H., Rev. Sci. Instr. 32, 1961, p-194.
5188. Schwarzbach, E., and Michel, F., Nucl. Instr. and Meth. 62, June 1, 1968, p-112.
5189. Schwarzschild, A. Z., Auerbach, E. H., et al., ANL/PHY-76-2 (Vol. 2), May 1976, pp-753-760.
5190. Schwarzschild, A. Z., Auerbach, E. H., et al., BNL-21216, CONF-760424-17, 1976, p-8.
5191. Schweitzer, T., Seeliger, D., et al., Zentralinstitut fuer Kernforschung, Rossendorf bei Dresden (FRG.), Annual Report 1975, pp-11-12.
5192. Schweitzer, T., Seeliger, D., et al., Zenttalinstitut fuer Isotopen- und Strahlenforschung, Zentralinstitut fuer Kernforschung, Rossendorf bei Dresden (FRG.), Annual Report 1976, pp-11-12.
5193. Schwierczinski, A., Frey, R., et al., Phys. Rev. Lett. 35 (18), 3 Nov. 1975, pp-1244-1247.
5194. Scott, A., and Mathur, N. P., et al., Nucl. Phys., A, 285 (2), 25 July 1977, pp-222-234.
5195. Scott, A., Owais, M., et al., Nucl. Phys., A, 289(1), 10 Oct. 1977, pp-123-140.
5196. Scott, D. K., Buenerd, M., et al., LBL-5075, 1975, pp-89-91.
5197. Scott, V. D., and Owen, L. W., Brit. J. Appl. Phys. 10, Feb. 1959, pp-91-3.
5198. Scott, M. J., and Lindgren, R., Rev. Sci. Instr. 28, Dec. 1957, pp-1090-1.
5199. Seagren, H. E., ORNL-TM-793, Contract W-7405-eng-26, March 1964, p-40.
5200. Seale, W. A., Mariscotti, M., et al., Cienc. Cult. Suppl. 29(7), July 1977, p-271.
5201. Sealock, R. M., Oregon Univ., Eugene (U. S. A.), Thesis (Ph. D), 1975, p-78.
5202. Sealock, R. M., and Overley, J. C., Phys. Rev., C, 13(6), June 1976, pp-2149-2158.

5203. Sease, J. D., ORNL-TM-1095, Contract W-7405-eng-26, June 1965, p-50.
5204. Sedlacek, W. A., Nucl. Instr. and Meth. 99(3), 1972, pp-429-31
5205. Sedlacek, W. A., and Ryan, V. A., Anal. Chem. 40, 1968, p-678.
5206. Seelmann-Eggebert, W., ORNL-tr-2223, Atomwirt., Atomtech 6, 273-7, 1961, p-16.
5207. Seeman, J. M., Vacuum 17(3), 1967, pp-129-37.
5208. Segev, M., Caner, M., et al., Trans. Am. Nucl. Soc. 22, Nov. 1975, pp-679-680.
5209. Segre, E., Matscience Symposia on Theoretical Phys., Vol. 1, Ramakrishnan, Alladi (ed.), NY, Plenum Press, 1966, pp-121-7.
5210. Seibt, E., and Weddigen, C. H., Nucl. Instr. and Meth. 100 (2), 1972, pp-253-66.
5211. Seiler, F., Phys. Lett., B, 61(2), 15 March 1976, pp-144-46.
5212. Seiler, R. F., Cleland, M. R., et al., Rev. Sci. Instr. 38, July 1967, pp-972-4.
5213. Seiler, R. F., Cleland, M. R., et al., IEEE Trans. Nucl. Sci., NS-14, June 1967, pp-943-4.
5214. Seiler-Graef, H., Husar, D., et al., Verh. Dtsch. Phys. Ges. 4, 1978, p-843.
5215. Seitz, T. P., Woods, R., et al., Nucl. Instr. and Meth. 93(1), 1971, pp-125-9.
5216. Sekharan, K. K., Kentucky Univ., Lexington (U. S. A.), Thesis (Ph. D), 1976, p-141.
5217. Seki, M., Ogawa, M., et al., J. Nucl. Sci. Technol. 16(11), 1979, pp-838-46.
5218. Selin, E., Arnell, S. E., et al., Nucl. Instr. and Meth. 56, Nov. 1967, pp-218-28.
5219. Sellin, A. I., Sci. and Ind. Appl. of Small Accel., 4th Conf., 1976.
5220. Seltz, R., and Sharma, H. L., Minn. Univ., Minneapolis (U. S. A.), John H. Williams Lab. of Nucl. Phys., 1975 Sept., pp-115-117.
5221. Sens, J. C., Refaei, S. M., et al., Phys. Rev., C, 16(6), Dec. 1977, pp-2129-2134.
5222. Sercely, R. R., Peterson, R. J., et al., Phys. Rev., C, 17(6), June 1978, pp-1919-1923.
5223. Serov, V. I., Abramovich, S. N., et al., At. Energ. 42(1), Jan. 1977, pp-59-61.
5224. Seth, K. K., Saha, A., et al., Phys. Lett., B, 59(4), 24 Nov. 1975, pp-333-335.
5225. Seth, K. K., Baer, H., LA-7892C, July 1979, pp-201-221.
5226. Settel, M. A., INTDS Proc., Argonne, U. S. A., Sept. 30-Oct. 2, 1975.
5227. Sevast'yanov, Yu. G., Isotopenpraxis 12(12), 1976, pp-472-4.
5228. Sevast'yanov, Yu. G., J. Radioanal. Chem. 21(1), 1974, pp-247-57.

5229. Sevier, K., and Parker, W., Nucl. Instr. and Meth. 6, 1960, pp-218-9.
5230. Sexton, E. H., State Univ. of New York, Albany (U. S. A.), Thesis (Ph. D), 1975, p-138.
5231. Shaeffer, M. C., Barreto, F., et al., ORNL/MIT-258, 19 Oct. 1977, p-17.
5232. Shakin, C. M., AIP Conf. Proc. ISSN 0094-243X, 54(1), 15 July 1979, pp-550-560.
5233. Shamai, Y., Alster, J., et al., Phys. Rev. Lett., 36(2), 12 Jan. 1976, pp-82-85.
5234. Shamai, Y., Alster, J., et al., Israel Phys. Soc., Jerusalem, 1977 Annual Meeting, pp-15-16.
5235. Shamu, R., Haouat, G., et al., CEA-CONF-3429, 1975, p-20.
5236. Shani, G., Ann. Nucl. Energy 3(9/10), 1976, pp-431-35.
5237. Shanta, R., Proc. of Nucl. Phys. and Solid State Phys. Symp., Calcutta, Dec. 22-26, 1975, Vol. 18B, pp-72-4.
5238. Shapira, D., Stokstad, R. G., et al., Phys. Rev., C, 12(6), Dec. 1975, pp-1907-1917.
5239. Shapira, D., Vries, R. M. de, et al., Phys. Lett., B, 71 (2), 21 Nov. 1977, pp-293-296.
5240. Shapira, D., DeVries, R. M., et al., Phys. Rev. Lett. 40 (6), 6 Feb. 1978, pp-371-74.
5241. Shapira, D., Dayras, R., et al., Nucl. Instr. and Meth. 163(2-3), 1979, pp-325-30.
5242. Shapiro, G., UCRL-20655, Workshop on Part. Phys. at Intermed. Energ., Pasadena, CA, 29 March 1971, pp-117-28.
5243. Shapiro, G., Sci. Amer. 215, July 1966, pp-68-78.
5244. Shapiro, G., "Prog. in Nucl. Tech. and Instrum., Vol. I", Amsterdam, North-Holland Pub. Co., 1965, pp-173-220.
5245. Shapiro, G. (ed.), LBL-500, CONF-718015, 2 Sept. 1971, p-421.
5246. Shapiro, G., LBL-500, from 2nd Int. Conf. on Polar. Targ., Berkeley, CA, 30 Aug. 1971, pp-391-5.
5247. Shapiro, M. W., Cormack, A. M., et al., Phys. Rev. 138, May 24, 1965, pp-B823-30.
5248. Sharma, D. N., Iyer, M. R., et al., Phys. Rev., C, 14(1), July 1976, pp-181-194.
5249. Sharma, H. L., and Hintz, N. M., Phys. Rev., C, 13(6), June 1976, pp-2288-2317.
5250. Sharma, J. S., and Mitra, A. N., Nucl. Phys., A, 292(3), 5 Dec. 1977, pp-437-444.
5251. Sharpey-Schafer, J. F., et al., Nucl. Instr. and Meth. 135, 1976, p-583.
5252. Shaw, K. B., and Stevenson, G. R., IEEE Trans. Nucl. Sci., NS-16, June 1969, pp-570-4.
5253. Shcherbakov, Yu. A., AIP Conf. Proc. 33, 1976, pp-365-374.
5254. Shea, J. H., Durell, J. L., et al., Nucl. Phys., A, ISSN 0375-9474, 327(1), 17 Sept. 1979, pp-207-220.
5255. Shebeko, A. V., and Ganenko, V. B., Ukr. Fiz. Zh. 23(3), March 1978, pp-390-395.

5256. Shelton, A. V., Calif. Rsch. and Dev. Co., Livermore, CA, CRD-R-35, Feb. 5, 1953, Decl. Feb. 27, 1957, p-12.
5257. Shen, Y. H., and Hua, C. Y., Nucl. Instr. and Meth. 167(1), 1979, pp-139-43.
5258. Shen, Y. H., and Hua, C. Y., INTDS Proc., Garching, FRG., Sept. 11-14, 1978.
5259. Shepard, J. R., Zimmerman, W. R., et al., Colo. Univ., Dept. Phys. and Astrophys., Boulder (U. S. A.), Tech. Prog. Report and Proposal for Continuation of Contract, 1 Nov. 1975, pp-13-20.
5260. Shepard, J. R., Kunz, P. D., et al., Colo. Univ., Tech. Prog. Rep. and Prop. for Continuation of Contract, 1 Nov. 1975, pp-49-53.
5261. Sherman, J. D., Nucl. Instr. and Meth. 135(2), 1976, pp-391-2.
5262. Sherman, J. D., Hendrie, D. L., et al., Univ. of Calif., Berkeley, ORO-4856-26, 1975, pp-319-320.
5263. Sherman, J. D., Flynn, E. R., et al., Phys. Rev., C, 13(6), June 1976, pp-2122-2126.
5264. Sherman, J. D., Hendrie, D. L., et al., Phys. Rev., C, 13(1), Jan. 1976, pp-20-34.
5265. Sherman, N. K., Nucl. Instr. and Meth. 116(2), 1 April 1974, pp-301-15.
5266. Sherman, N. K., Ferdinande, H. M., et al., Phys. Rev. Lett. 35(18), 3 Nov. 1975, pp-1215-1219.
5267. Shera, E. B., Ritter, E. T., et al., Phys. Rev., C, 14(2), Aug. 1976, pp-731-747.
5268. Shevchenko, N. G., Buki, A. Yu., et al., Leningrad Nauka 1978, p-257.
5269. Shibata, T., Ejiri, H., et al., LBL-5075, 1975, pp-196-200.
5270. Shih, L. L., Hill, J. C., et al., Phys. Rev., C, 17(3), March 1978, pp-1163-1167.
5271. Shikata, E., J. Nucl. Sci. Technol. 7(9), 1970, pp-481-3.
5272. Shimizu, K., and Faessler, A., Phys. Rev., C, 17(5), May 1978, pp-1891-1894.
5273. Shimizu, R., Ikuta, T., et al., Jap. J. Appl. Phys. 9, Nov. 1970, pp-1429-30.
5274. Shimizu, R., and Miura, M., Technol. Rep. Osaka Univ. 16, Oct. 1966, pp-415-22.
5275. Shinozuka, T., Tanaka, Y., et al., Nucl. Phys., A, ISSN 0375-9474, 326(1), 3 Sept. 1979, pp-47-54.
5276. Shirk, D. G., Kansas Univ., Lawrence (U. S. A.), Thesis (Ph. D), 1975, p-124.
5277. Shirley, D. A., Annu. Rev. Nucl. Sci. 16, UCRL-16764, 1966, pp-89-118.
5278. Shkolnik, V., Dehnhard, D., et al., Proc. of Symp. on Macroscopic Features of Heavy-Ion Collisions, Vol. II, ANL/PHY-76-2(Vol. 2), May 1976, pp-761-766.

5279. Shkolnik, V., Dehnhard, D., et al., Phys. Lett., B, 74(3), 10 April 1978, pp-195-198.
5280. Shoda, K., Ohashi, H., et al., AIP Conf. Proc. 33, 1976, pp-604-605.
5281. Shope, T. B., Steuer, M. F., et al., Nucl. Phys., A, 260 (1), 29 March 1976, pp-95-108.
5282. Shortall, J. W., Calif. Rsch. and Dev. Co., Livermore, CA, Decl. Feb. 27, 1957, p-25.
5283. Shotter, A. C., Reid, J. M., et al., J. Phys., A, 6(6), June 1973, pp-874-877.
5284. Shotter, A. C., Gelbke, C. K., et al., Phys. Rev. Lett. ISSN 0031-9007, 43(8), 20 Aug. 1979, pp-569-72.
5285. Shreter, U., and Kalish, R., Nucl. Instr. and Meth. 166, 1979, pp-117-120.
5286. Shuck, A. B., Lotts, A. L., et al., Gen. Elect. Co., HW-SA-3755, Cont. AT(45-1)-1350, Aug. 27, 1964, p-140.
5287. Shuck, A. B., Hins, A. G., et al., ANL-66, Breeding-Gain Specimens for EBR-I Core IV, Sept. 1963.
5288. Shuck, A. B., and Mayfield, R. M., ANL-5499, The Process Equip. and Protective Encl. Designed for the Fuel Fabrication Fac. #350, Jan. 1956.
5289. Shyam, R., and Mukherjee, S., Phys. Rev., C, 13(6), June 1976, pp-2099-2104.
5290. Siciliano, E. R., LA-7892C, July 1979, pp-92-99.
5291. Sick, I., Bellicard, J. B., et al., Helv. Phys. Acta 48(4), 25 Nov. 1975, p-555.
5292. Sick, I., Bellicard, J. B., et al., Phys. Rev. Lett. 38 (22), 30 May 1977, pp-1259-1262.
5293. Sidenius, G., and Skilbreid, O., Proc. Int. Symp. on Separation of Radio. Isot., Springer-Verlag, 1961, p-243.
5294. Sidhu, G. S., and Czirr, J. B., NBS (U. S.), Spec. Publ. 425, Oct. 1975, pp-615-619.
5295. Sie, S. H., and Gebbie, D. W., Aust. Nat. Univ., Canberra, Dept. Nucl. Phys., June 1977, p-36.
5296. Siebel, R., Klauss, E. U., et al., Verh. Dtsch. Phys. Ges. 4, 1978, p-867.
5297. Siegert, G., Wollnik, H., et al., Phys. Rev., C, 14(5), Nov. 1976, pp-1864-1873.
5298. Siekmann, H., Gebauer, B., et al., Verh. Dtsch. Phys. Ges. 4, 1978, p-807.
5299. Sigel, R., Krause, H., et al., J. Phys., E, 2, Feb. 1969, pp-187-90.
5300. Sigg, R. A., Ark. Univ., Fayetteville (U. S. A.), Thesis (Ph. D), 1976, p-117.
5301. Sikkema, C. P., and Steendam, S. P., Nucl. Instr. and Meth. 124(1), 1975, pp-161-4.
5302. Silbar, R. R., AIP Conf. Proc. 33, 1976, pp-297-304.
5303. Silbert, M. G., LASL, N. Mex., LA-6239-MS, Feb. 1976, p-33.

5304. Silva, J. R., Hahn, L. R., et al., Proc. of Conf. held at St. Catherine's Coll., Oxford, 22-23 Sept. 1969.
5305. Silveira, E. F. Da, Proc. of Int. School of Phys. 'Enrico Fermi' Course 62 held at Varenna, Italy, 22 July-3 Aug. 1974, pp-588-591.
5306. Silvester, J. D., 7th Int. Conf. on Cyc. and Their Appl., Zurich (Switz.), Aug. 19-22, 1975.
5307. Silvester, J. D., Sugden, J., et al., Radiochem. Radioanal. Lett. 2, 1969, p-17.
5308. Simon, G. G., Schmitt, C., et al., Nucl. Instr. and Meth. 158(1), 1979, pp-185-91.
5309. Simon, G., and Frehaut, J., CEA-CONF-3308, 1975, p-12.
5310. Simon, G. G., Borkowski, F., et al., Phys. Rev. Lett. 37 (12), 20 Sept. 1976, pp-739-742.
5311. Simon, G. G., Borkowski, F., et al., Verh. Dtsch. Phys. Ges. 7, 1977, p-1063.
5312. Simon, G. G., Borkowski, F., et al., Verh. Dtsch. Phys. Ges. 4, 1978, p-789.
5313. Simon, J., Grimson, J., et al., 23 Conf. on Remote Syst. Technol., Proc. ANS Winter Meet., San Francisco, CA, Nov. 1975, pp-32-40.
5314. Simon, R. S., Banaschik, M. V., et al., Phys. Rev. Lett. 36(7), 16 Feb. 1976, pp-359-61.
5315. Simon, W. G., Univ. Alberta, Edmonton, Alberta, Canada, Report TRI-71-7, 1971, p-19.
5316. Simonsic, G. A., Report LA-6207, 1975, p-5.
5317. Simpson, J. J., Wilson, S. J., et al., Phys. Rev. Lett. 40 (3), 16 Jan. 1978, pp-154-157.
5318. Sinclair, D., Chait, B. T., et al., ANL/PHY-76-2(Vol. 2), May 1976, pp-767-774.
5319. Singer, K. D., Headley, S. C., et al., Phys. Rev., C, 15 (5), May 1977, pp-1662-1664.
5320. Singh, J. J., NASA, Langley Rsch. Ctr., Langly Station, VA, 2 Sept. 1967, p-25.
5321. Singh, J. J., NASA-TM-X-72014, Oct. 1974, p-18.
5322. Singh, P. P., Sadler, M., et al., Phys. Rev., C, 14(4), Oct. 1976, pp-1655-1658.
5323. Singh, R., and Knitter, H. H., Proc. of Nucl. Phys. and Solid State Phys. Symp., Bombay, Dec. 27-31, 1974, Vol. 17B, p-110.
5324. Singh, S. K., and Ahmad, I., Phys. Lett., B, 69(4), 29 Aug. 1977, pp-422-424.
5325. Singh, U. N., Rainwater, J., et al., Phys. Rev., C, 13(1), Jan. 1976, pp-124-127.
5326. Singh, U. N., Nakogome, Y., et al., Trans. Am. Nucl. Soc. 27, 1977, pp-866-868.
5327. Singh, U. N., Block, R. C., et al., Nucl. Sci. and Eng. 67(1), July 1978, pp-54-60.

5328. Singhal, R. P., Knight, E. A., et al., Phys. Lett., B, 68 (2), 23 May 1977, pp-133-135.
5329. Singhal, R. P., Purdie, H., et al., Nucl. Instr. and Meth. 73, 1969, pp-237-9.
5330. Singleton, R. M., Hendricks, C. D., et al., UCRL-80148, CONF-780202-11, Report 1977, p-6.
5331. Sit'ko, S. P., Izv. Akad. Nauk SSSR, Ser. Fiz. ISSN 0367-6765, 42(9), Sept. 1978, pp-1809-1822.
5332. Sivaramakrishnan, C. K., Proc. 2nd Chem. Symp., 1971, pp-165-70.
5333. Skaggs, L. S., Kuchnir, F. T., et al., INTDS Proc., Garching, FRG., Sept. 11-14, 1978.
5334. Skakun, N. A., and Strashinskii, A. G., Pribory i Tekh. Ekspt. 6(1), Jan.-Feb. 1961, p-180.
5335. Skakun, N. A., and Strashinskii, A. G., Pribory i Tekh. Ekspt. 6(1), Jan.-Feb. 1961, p-179.
5336. Skarnemark, G., Aronsson, P. O., et al., J. Inorg. and Nucl. Chem. ISSN 0022-1902, 39(9), 1977, pp-1487-93.
5337. Skidmore, M. R., Iodine-123 Planning Conf., Rockville, MD, U. S. A., 19 May 1975, p-8.
5338. Skold, K., Karlen, L., et al., Nucl. Instr. and Meth. 56, Nov. 1967, pp-305-8.
5339. Skorka, S., Naturwissenschaften, 40(23), 1953, p-605.
5340. Skrable, K. W., Rutgers State Univ., New Brunswick, NJ, 1970, p-306.
5341. Skrivankova, M., and Andrejsek, M., Jad. Energ. 22(4), April 1976, pp-145-146.
5342. Slaback, L. A., Jr., IEEE Trans. Nucl. Sci., NS-26(1), Pt. 2, Feb. 1979, pp-1584-1586.
5343. Slafford, G. T., Rev. Sci. Instr. 27, 1956, p-972.
Flinta, J., Vacuum 2, 1952, p-257.
Hill, H. A., Rev. Sci. Instr. 27, 1956, p-1086.
O'Bryan, H. M., Rev. Sci. Instr. 125, 1934.
5344. Slater, D. N., and Booth, W., Nucl. Phys., A, 267(1), 16 Aug. 1976, pp-1-12.
5345. Slater, D. N., and Booth, W., Nucl. Phys., A, 274(1-2), 7 Dec. 1976, pp-93-107.
5346. Slaus, I., Inst. Rudjer Boskovic, Zagreb, Int. Conf. on Interactions of Neutrons with Nuclei, Vol. I, 1976, pp-272-344.
5347. Slaus, I., Allas, R. G., et al., Nucl. Phys., A, 286(1), 8 Aug. 1977, pp-67-88.
5348. Slemmer, J., Albrecht, R., et al., Europ. Phys. Soc., Geneva (Switz.), Europ. Conf. on Nucl. Phys. with Heavy Ions, 6-10 Sept. 1976, p-156.
5349. Selptsov, G. N., and Solodovnikov, A. P., Prib. Tekh. Eksp. 2, March-April 1969, pp-208-10.
5350. Sletten, G., INTDS Proc., Boston, U. S. A., Oct. 1-3, 1979.

5351. Sletten, G., INTDS Proc., Chalk River, Canada, Oct. 1-3, 1974.
5352. Sletten, G., Niels Bohr Inst., Copenhagen (Denmark), AECL-5503, April 1976, pp-47-58.
5353. Sletten, G., and Knudsen, P., Nucl. Instr. and Meth. 102 (3), 1972, pp-459-63.
5354. Sletten, G., 3rd Int. Symp. on Rsch. Mat. for Nucl. Measurements, Gatlinburg, TN, 5 Oct. 1971, pp-116-21.
5355. Slobodian, T. J., INTDS Newsletter, July 1979.
5356. Slobodrian, R. J., Irshad, M., et al., Proc. of VII Int. Conf. on Few Body Prob. in Nucl and Particle Phys., Delhi, India, 29 Dec. 1975-3 Jan. 1976, pp-263-265.
5357. Slocombe, M. G., Newton, J. O., et al., Nucl. Phys., A, 275(1), 4 Jan. 1977, pp-166-188.
5358. Smith, A. B., Guenther, P., et al., ANL/NDM-29, June 1977, Contract W-31-109-eng-48, p-29.
5359. Smith, A., Guenther, P., et al., Nucl. Phys., A, ISSN 0375-9474, 332(3/4), Dec. 1979, pp-297-316.
5360. Smith, A. R., and Thomas, R. H., Nucl. Instr. and Meth. 137 (3), 15 Sept. 1976, pp-459-461.
5361. Smith, A. B., Guenther, P., ANL/NDM-29, Contract W-31-109-eng-48, June 1977, p-29.
5362. Smith, B. J., and Jain, A. K., UKAEA Rsch. Group, Harwell, AERE-R-8326, ISBN 0705803961, March 1976, p-7.
5363. Smith, D. L., and Meadows, J. W., Nucl. Sci. and Eng. 60 (3), July 1976, pp-319-322.
5364. Smith, D. L., and Meadows, J. W., Nucl. Sci. and Eng. 60 (2), June 1976, pp-187-192.
5365. Smith, D. L., and Meadows, J. W., ANL/NDM-37, Dec. 1977, p-71.
5366. Smith, D. L. E., Proc. of 3rd Conf. on Accel. Targ. Designed for the Prod. of Neutrons, Liege (Belgium), Sept. 18-19, 1967, pp-5-19.
5367. Smith. D. L. E., Atomic Weapons Rsch. Estab., Aldermaston (England), AWRE-O-52/67, Sept. 1967, p-17.
5368. Smith, G. J., Casten, R. F., et al., Verh. Dtsch. Phys. Ges. 4, 1978, p-848.
5369. Smith, G. R., Boudrie, R. L., et al., Phys. Lett., B, 72(2), 19 Dec. 1977, pp-176-178.
5370. Smith, J. L., Calif. Rsch. and Develop. Co., Livermore, CA, Feb. 27, 1957, p-23.
5371. Smith. L. R., Cameron, J., et al., INTDS Proc., Boston, U. S. A., Oct. 1-3, 1979.
5372. Smith, M. L. (ed.), AERE-R-5097, Dec. 1965, p-175.
5373. Smith, M. L., Electromagnetically Enriched Isot. and Mass Spectrometry, Butterworth, 1956, p-97.
5374. Smith, P. B., Phys. Rev., C, 13(5), May 1976, pp-2071-74.
5375. Smith, P. B., Groningen Rijksuniversiteit, AERE-R-5097, Paper 9, p-6.

5376. Smith, P. K., Keski, J. R., et al., E. I. du Pont de Nemours and Co., Aiken, SC (DP-11114), 1967, p-40.
5377. Smith, R. A., and Moniz, E. J., Nucl. Phys., B, B101(2), 1975, p-547.
5378. Smith, S. R., Frisch, D. H., et al., Nucl. Instr. and Meth. 86, 1970, pp-291-9.
5379. Smithells, C. J., Tungsten, A. Treatise on its Metallurgy, Properties and Appl., 3rd ed. (Chapman and Hall Ltd., London, 1952).
5380. Smits, J. W., Girisch, R., et al., Spring Meet., 26-27 April 1976, Rijksuniversiteit Groningen (Netherlands), Kernfysisch Versneller Inst., p-52.
5381. Smits, J. W., Kernfysisch Versneller Inst., 27 May 1977, p-180.
5382. Smits, J. W., de Meijer, R. J., et al., AIP Conf. Proc. ISSN 0094-243X, 47, 1978, pp-696-97.
5383. Smits, J. W., Siemssen, R. H., et al., AIP Conf. Proc. ISSN 0094-243X, 47, 1978, pp-698-99.
5384. Smolec, W., Burzynski, S., et al., Acta Phys. Slovaca 25 (2-3), 1975, pp-190-194.
5385. Smolec, W., Burzynaski, S., et al., Maryland Univ., College Park (U. S. A.), Dept. of Phys. and Astron., ORO-4856-26, 1975, pp-433-434.
5386. Smolec, W., Burzynski, S., et al., Nucl. Phys., A, 257(3), 9 Feb. 1976, pp-397-402.
5387. Smolyankina, T. G., Radkevich, I. A., et al., Cryogenics 7, Oct. 1967, p-298.
5388. Smolyankina, T. G., Radkevich, I. A., et al., Prib. Tekh. Eksp. 3, May-June 1966, pp-208-9.
5389. Smulders, P. J. M., and Boerma, D. O., Nederlandse Natuurkundige Vereniging, Amsterdam, Spring Meet., 26-27 April 1976, p-34.
5390. Sober, D. I., Ballagh, H. C., AIP Conf. Proc. 33, 1976, pp-65-67.
5391. Sobotka, L. G., Mathews, G. J., et al., Z. Phys., A, 292(2), Sept. 1979, pp-191-195.
5392. Sodd, J. V., Blue, W. J., et al., U. S. Dept. of HEW Publication BRH/DMRE 70-4, 1970.
5393. Sodd, J. V., Blue, W. J., et al., Proc. of a Conf. held at St. Catherine's Coll., Oxford, 22-23 Sept. 1969.
5394. Soeffge, F., Nucl. Instr. and Meth. 142(3), 1 May 1977, pp-399-402.
5395. Soga, F., Tanaka, M., et al., Tokyo Univ., Tanashi (Japan), Inst. for Nucl. Study, INS-TL-132, Nov. 1976, p-24.
5396. Soga, F., Hashimoto, Y., et al., Nucl. Phys., A, 288(3), 3 Oct. 1977, pp-504-528.
5397. Soini, H. E., and Hoenig, M. O., Advan. Cryog. Eng. 14, 1969, pp-473-5.
5398. Sokol, J. P., Notre Dame Univ., Ind. (U. S. A.), Thesis (Ph. D), 1975, p-124.

5399. Sokol, J. P., and Browne, C. P., Phys. Rev., C, 17(1), Jan. 1978, pp-388-391.

5400. Sokol, J. P., Nucl. Instr. and Meth. 99(2), 1972, p-379.

5401. Solin, L. M., Nemilov, Yu. A, et al., Yad. Fiz. ISSN 0044-0027, 29(2), Feb. 1979, pp-289-292.

5402. Solin, L. M., Nemilov, Yu. A., et al., Leningrad Nauka, 1978, p-206.

5403. Solomon, D., and Nolen, R., Proc. of 7th Int. Vacuum Cong. and 3rd Int. Conf. on Solid Surfaces, Vol. 3, Vienna, Austria, Sept. 1977, p-2717.

5404. Sommerkamp, P., Z. Angew. Phys. 28, 1970, p-220.

5405. Sommerkamp, P., Dissertation, Univ. München, 1968.

5406. Somerville, L. P., Nurmia, M. J., et al., LBL-5075, 1975, pp-39-40.

5407. Somerville, L. P., Nurmia, M. J., et al., LBL-8151, 1978, pp-39-40.

5408. Song, H. S., Kolata, J. J., et al., Phys. Rev., C, 16(4), Oct. 1977, pp-1363-1376.

5409. Soole, B. W., Phys. Med. Biol. 21(3), May 1976, pp-369-89.

5410. Soroka, J. M., City Univ., NY, NY (U. S. A.), 1979, p-297.

5411. Sorokin, A. A., and Ponomarev, V. N., Leningrad Nauka, 1978, p-258.

5412. Sorokin, A. F., Prib. Tekh. Eksp. 5, Sept.-Oct. 1969, pp-29-31.

5413. Sorokin, A. F., and Tsel'nik, F. A., Prib. Tekh. Eksp. 6, Nov.-Dec. 1969, pp-26-7.

5414. Soroko, L. M., JINR-P13-6388, 1972, p-20.

5415. Soroko, L. M., JINR-P13-5699, 1971, p-7.

5416. Sorriaux, A., de la Fourniere, B., et al., Nucl. Instr. and Meth. 62, June 1, 1968, pp-61-7.

5417. Soukup, J., Cameron, J. M., et al., Nucl. Instr. and Meth. 141(3), 1977, pp-409-11.

5418. Sourkes, A. M., Houdayer, A., et al., Phys. Rev., C, 13(2), Feb. 1976, pp-451-460.

5419. Southam, D. N., and King, J. D., Am. Phys. Soc. Summer Meet. in the West, Edmonton, Canada, Aug. 1963, p-13.

5420. Southon, J. R., Poletti, A. R., et al., Nucl. Phys., A, 267(2), 23 Aug. 1976, pp-263-275.

5421. Sowerby, B. D., and Sheppard, D. M., Nucl. Instr. and Meth. 60, April 1968, p-358.

5422. Spaa, J. H., J. Sci. Instr. 35, 1958 May, pp-175-8.

5423. Spaepen, J., Nucl. Data for Reactors, IAEA, Vol. I, 1967, p-241.

5424. Sparks, R. J., Lancman, H., et al., Nucl. Phys., A, 259(1), 8 March 1976, pp-13-19.

5425. Sparks, R. J., Rijksuniversiteit Utrecht (Netherlands), Afdeling Experimentele Natuurkunde, 12 April 1976, p-55.

5426. Sparks, R. J., Nucl. Phys., A, 265(3), 26 July 1976, pp-416-428.

5427. Sparks, R. J., Nucl. Phys., A, 265(3), 26 July 1976, pp-429-442.
5428. Sparrow, D. A., LA-7892C, Jan. 22-24, 1979, pp-53-65.
5429. Spear, R. H., Esat, M. T., et al., Aust. J. Phys. 30(2), April 1977, pp-133-148.
5430. Spedding, F. H., and Daane, A. H., Metals 6, 1954, p-504.
5431. Spehl, H., Z. Physik 168, 1962, pp-266-72.
5432. Speidel, R., and Holl, P., Optik 24, 1966, p-296.
5433. Spejewski, E. H., Naumann, R. A., et al., Nucl. Instr. and Meth. 84, 1970, pp-237-43.
5434. Spencer, R. R., and Beer, H., Nucl. Sci. and Eng. 60(4), Aug. 1976, pp-390-398.
5435. Spencer, R. R., NBS (U. S.), Spec. Publ. 425, Oct. 1975, pp-620-622.
5436. Spencer, R. R., and Macklin, R. L., Nucl. Sci. and Eng. 61(3), Nov. 1976, pp-346-355.
5437. Sperber, D., ANL/PHY-76-2(Vol. 2), May 1976, pp-775-781.
5438. Sperr, P., Henning, W., et al., Phys. Rev., C, 13(1), Jan. 1976, pp-447-449.
5439. Sperr, P., Braid, T. H., et al., ANL/PHY-76-2(Vol. 2), May 1976, pp-783-793.
5440. Sperr, P., Braid, T. H., et al., Europ. Phys. Soc., Geneva (Switz.), Europ. Conf. on Nucl. Phys. with Heavy Ions, 6-10 Sept. 1976, p-117.
5441. Spisak, M. J., Pitts. Univ., PA (U. S. A.), Thesis (Ph. D), 1977, p-189.
5442. Spitaleri, C., Lattuada, M., et al., Lett. Nuovo Cim. 21 (10), 11 March 1978, pp-345-350.
5443. Spivack, M. A., Rev. Sci. Instr. 41, Nov. 1970, pp-1614-16.
5444. Sprengel, D., Hoppe, W., et al., Verh. Dtsch. Phys. Ges. 4, 1978, p-948.
5445. Sprengel, D., Hoppe, W., et al., AIP Conf. Proc. ISSN 0094-243X, 47, 1978, pp-588-89.
5446. Springer, K., Proc. of Topical Seminar on Interactions of Elementary Particles with Nuclei, Trieste (Italy), Istituto Nazionale di Fisica Nucleare, 1970, pp-297-300.
5447. Springer, R. W., INTDS Proc., Los Alamos, U. S. A., Oct. 19-21, 1976.
5448. Srinivasa, R. K., AIP Conf. Proc. 33, 1976, pp-601-603.
5449. Stach, W., Graw, G., et al., Erlangen-Nuernberg Univ., Erlangen (FRG.), Physikalisches Inst., Annual Report 1975, pp-3-20.
5450. Stach, W., Bittner, G., et al., Verh. Dtsch. Phys. Ges. 4, 1978, p-878.
5451. Stadnyk, B. I., Kuritnyk, I. P., et al., Teplofiz. Vys. Temp. 13(2), March 1975, pp-403-406.
5452. Staebler, A., Buck, W., et al., Nucl. Phys., A, 275(2), 10 Jan. 1977, pp-269-279.

5453. Stahel, D. P., Wozniak, G. J., et al., LBL-5075, 1975, pp-85-88.
5454. Stahel, D. P., Wozniak, G. J., et al., Phys. Rev., C, 16(4), Oct. 1977, pp-1456-1466.
5455. Stahel, D. P., Wozniak, G. J., et al., LBL-5075, 1975, pp-79-81.
5456. Stahel, D. P., de Meijer, R. J., et al., LBL-8151, 1978, pp-10-11.
5457. Stang, L. G., Jr., BNL-841, pp-52-62.
5458. Starostov, B. I., Semenov, A. F., et al., Yad. Fiz. NIIAR-P-22(356), 1978, p-40.
5459. Staub, P., Baumgartner, E., et al., Helv. Phys. Acta 50(1), 4 Feb. 1977, pp-9-28.
5460. Steckelmacher, W., Thin Film Microelectronics (London: Chapman and Hall), 1965, pp-113-25.
5461. Steckelmacher, W., and English, J., Trans. 8th Nat. Vacuum Symp. (Oxford: Pergamon), 1962, p-852.
5462. Steckelmacher, W., Parisot, J. M., et al., Vacuum 9, 1959, p-171.
5463. Steers, G., LBL-7950, INTDS Proc., Berkeley, U. S. A., Oct. 19-20, 1977.
5464. Stefanini, A. M., Daly, P. J., et al., Nucl. Phys., A, 258 (1), 16 Feb. 1976, pp-34-42.
5465. Steffen, R. M., Purdue Univ., Lafayette, IN, Contract AT (11-1)-1746, June 15, 1968, p-99.
5466. Steffens, E., Dreves, W., et al., Nucl. Instr. and Meth. 143(3), June 15, 1977, pp-409-421.
5467. Steffens, E., Dreves, W., et al., AIP Conf. Proc. ISSN 0094-243X, 47, 1978, pp-748-49.
5468. Stein, N., Sunier, J. W., et al., Europ. Phys. Soc., Geneva (Switz.), Europ. Conf. on Nucl. Phys. with Heavy Ions, 6-10 Sept. 1976, p-54.
5469. Steinberg, R., and Alger, D. L., Div. Electron. Prod., Bur. Radiol. Health, Rockville, MD, Report 1972, p-44.
5470. Steinkilberg, W., Mack, G., et al., Verh. Dtsch. Phys. Ges. 4, 1978, p-766.
5471. Stelson, P. H., ORNL-TM-2364, Contract W-7405-eng-26, Sept. 1968, p-24.
5472. Stelts, M. L., Calif. Univ., Davis (U. S. A.), Thesis (Ph. D), 1975, p-134.
5473. Stelts, M. L., and Browne, J. C., Calif. Univ., Livermore (U. S. A.), Law. Livermore Lab., Status Report to ERDA Nuclear Data Comm., 9 April 1976, pp-5-6.
5474. Stelzer, F., Ber. Kernforschungsanlage Juelich, Jul-Conf-34, 1980, pp-185-195.
5475. Stelzer, K., Rauch, F., et al., Europ. Phys. Soc., Geneva (Switz.), Europ. Conf. on Nucl. Phys. with Heavy Ions, 6-10 Sept. 1976, p-97.

5476. Stelzer, K., Rauch, F., et al., Phys. Lett., B, 70(3), 10 Oct. 1977, pp-297-300.
5477. Stengl, G., and Vonach, H., Nucl. Instr. and Meth. 140(1), 1977, p-197.
5478. Stengl, G., Uhl, M., et al., Nucl. Phys., A, 290(1), 14 Oct. 1977, pp-109-127.
5479. Stephany, W. P., and Knoll, G. F., NBS (U. S.), Spec. Publ. 1, Oct. 1975, pp-236-239.
5480. Stephen, J., and Osborne, D. N., United Kingdom At. Energy Estab. Report AERE-R5430, 1967.
5481. Stephenson, E. J., Conzett, H. E., et al., LBL-5075, 1975, pp-59-61.
5482. Sterrenburg, W. A., Middelkoop, G. van, et al., Nucl. Phys., A, 275(1), 4 Jan. 1977, pp-48-60.
5483. Stetz, A. W., Cameron, J. M., et al., Nucl. Phys., A, 290 (2), 31 Oct. 1977, pp-285-293.
5484. Stevens, H. H., Jr., Proc. of Soc. for Exp. Stress Analysis, Vol. ii, Addison Wesley, 1944, p-139.
5485. Steward, S. A., Nickerson, R., et al., Radiat. Eff. Tritium Technol. Fusion React., Proc. Int. Conf. 4, 1976, pp-236-53.
5486. Steward, S. A., Nickerson, R., et al., UCRL-76761, Report 1975, p-19.
5487. Stewart, L., and Hale, G. M., BNL-NCS-50446, April 1975, pp-1-6.
5488. Steyn, J., Proc. of Reg. Conf. on Radiat. Prot., Vol. I, 1973, pp-299-310.
5489. Stig, B., Arkiv Fysik 2, 1950, pp-461-9.
5490. Stinson, J. D., INTDS Newsletter, Jan. 1979.
5491. Stinson, J. D., INTDS Proc., Berkeley, U. S. A., Oct. 19-20, 1977.
5492. Stinson, J. D., INTDS Proc., Los Alamos, U. S. A., Oct. 19-21, 1976.
5493. Stinson, J. D., INTDS Proc., Los Alamos, U. S. A., Oct. 19-21, 1976.
5494. Stinson, J. D., INTDS Newsletter, August 1979.
5495. Stinson, J. D., INTDS Proc., Chalk River, Canada, Oct. 1-3, 1974.
5496. Stirland, D. J., Appl. Phys. Lett. 8, 1966, p-326.
5497. Stock, R., Schneider, W. F. W., LBL-5075, 1975, pp-93-96.
5498. Stock, R., Jahnke, U., Phys. Rev., C, 14(5), Nov. 1976, pp-1824-1831.
5499. Stoeckmann, H. J., Ackermann, H., et al., Verh. Dtsch. Phys. Ges. 4, 1978, p-941.
5500. Stoeffl, W., Egidy, T. von, et al., Kernforschungsanlage Juelich G.m.b.H. (FRG.), Inst. fuer Kernphysik, Annual Report 1978, Pt. 1, p-52.
5501. Stokstad, R. G., Gomez del Campo, J., et al., ORNL, CONF-760424-3, 1976, p-8.

5502. Stokstad, R. G., del Campo, J. G., et al., ANL/PHY-76-2 (Vol. 2), May 1976, pp-795-801.
5503. Stokstad, R. G., Namboodiri, M. N., et al., Texas Agricultural and Mech. Univ., College Station (U. S. A.), Cyc. Inst. Angular Momentum Effects in Nuclear Reactions, Prog. Rep., Aug. 1, 1976-July 31, 1977, pp-12-15.
5504. Stokstad, R. G., del Campo, J. G., et al., Phys. Rev. Lett. 36(26), 28 June 1976, pp-1529-1532.
5505. Stokstad, R. G., Switkowski, Z. E., et al., Phys. Rev. Lett. 37(14), 4 Oct. 1976, pp-888-891.
5506. Stokstad, R. G., Wieland, R. M., et al., ORNL/TM-5935, June 1977, p-73.
5507. Stokstad, R. G., Dayras, R. A., et al., Phys. Lett., B, 70(3), 10 Oct. 1977, pp-289-291.
5508. Stoler, P., Nucl. Instr. and Meth. 93(2), 1971, pp-377-80.
5509. Stolk, C., and Tjon, J. A., Nucl. Phys., A, 295(3), 6 Feb. 1978, pp-384-404.
5510. Stolz, H. M., Univ. Mainz, Mainz, Ger., Report 1972, AED-CONF-71-400, pp-135-45.
5511. Stone, F. K., IEEE Trans. on Nucl. Sci., June 1967.
5512. Stone, F. K., and Force, J. R., Rev. Sci. Instr. 30(9), Sept. 1959, pp-787-93.
5513. Stone, R. R., Gregg, D. W., et al., J. Appl. Phys. 46(6), 1975, pp-2693-2700.
5514. Stoner, J. O., Jr., INTDS Newsletter, Feb. 1980.
5515. Stoner, J. O., Jr., INTDS Proc., Boston, U. S. A., Oct. 1-3, 1979.
5516. Stoner, J. O., Jr., and Bashkin, S., INTDS Proc., Boston, U. S. A., Oct. 1-3, 1979.
5517. Stoner, J. O., Jr., and Bashkin, S., Appl. Opt. 17, 1978, pp-321-4.
5518. Stoner, J. O., Jr., and Bashkin, S., INTDS Proc., Berkeley, U. S. A., Oct. 19-20, 1977.
5519. Stoner, J. O., Jr., and Bashkin, S., INTDS Proc., Los Alamos, U. S. A., Oct. 19-21, 1976.
5520. Stoner, J. O., Jr., J. Appl. Phys. 40, 1969, p-707.
5521. Stoner, J. O., Jr., Rathman, P. W., et al., INTDS Newsletter, August 1980.
5522. Stoner, J. O., Jr., and Bashkin, S., INTDS Newsletter, August 1980.
5523. Stotlar, S. C., Maggiore, C. J., et al., INTDS Proc., Los Alamos, U. S. A., Oct. 19-21, 1976.
5524. Stowell, M. J., Thin Solid Films 0, 1969, p-1.
5525. Strain, J. E., and Haywood, F. F., ORNL.
5526. Strang, R. M., and Ritter, R. C., Nucl. Instr. and Meth. 93(2), 1971, pp-221-6.
5527. Strassner, G., Karlsruhe Univ. (TH) (FRG.), Fakultaet fuer Physik, Diss. (D. Sc.), June 20, 1975, p-90.

5528. Straume, O., Loevhoeiden, G., et al., Nucl. Phys., A, 266 (2), 9 Aug. 1976, pp-390-412.
5529. Strizhak, V. I., and Primenko, G. I., Ukr. Fiz. Zh. 15, Jan. 1970, pp-169-71.
5530. Strizhak, V. I., Primenko, G. I., et al., At. Energ. 28, March 1970, pp-251-2.
5531. Strizhak, V. I., Primenko, G. I., et al., At. Energ. 28, March 1970, pp-249-51.
5532. Stroffolini, R., Phys. Rev. 104, 15 Nov. 1956, pp-1146-9.
5533. Strohal, P., Tehnika 18, Feb. 1963, pp-234-5.
5534. Strohbusch, U., Bennett, C. C., et al., Verh. Dtsch. Phys. Ges. 4, 1978, p-921.
5535. Strohe, H., and Ernst, J., Verh. Dtsch. Phys. Ges. 4, 1978, p-859.
5536. Strohmaier, B., Uhl, M., et al., Nucl. Sci. and Eng. 65(2), Feb., pp-368-384.
5537. Stromswold, D. C., Elliott, D. O., Jr., et al., Johns Hopkins Univ., Baltimore, MD (U. S. A.), Nuclear Moments and Nucl. Structure, Annual Progress Report, May 1, 1975-April 30, 1976, pp-24-32.
5538. Stromswold, D. C., Elliott, D. O., Jr., et al., Phys. Rev., C, 17(1), Jan. 1978, pp-143-154.
5539. Stronach, C. E., College of William and Mary, Williamsburg, VA (U. S. A.), Thesis (Ph. D), 1976, p-216.
5540. Stronach, C. E., Stith, J. H., et al., AIP Conf. Proc. 33, 1976, pp-330-331.
5541. Struble, G. L., Lanier, R. G., et al., Phys. Rev. Lett. 40(10), 6 March 1978, pp-615-618.
5542. Stubbins, W. F., Cincinnati Univ., Ohio (U. S. A.), AINSE Nucl. Phys. Conf., 9-11 Feb. 1976, p-49.
5543. Stubbins, F. W., IRE Trans. Nucl. Sci., NS-2(1), June 1955, pp-3-8.
5544. Studer, R. R., United Nucl. Ind., Inc., Richland, WA (U. S. A.), Contract E(45-1)-1350, 10 April 1967, p-8.
5545. Sturm, W. J., and Jones, R. J., Rev. Sci. Instr. 25, 1954, pp-392-3.
5546. Styczen, J., Chevallier, A., et al., Phys. Rev., C, 15(5), May 1977.
5547. Styczen, J., Chevallier, A., et al., Centre de Recherches Nucleaires (France), 1977, p-12.
5548. Suenkel, W., and Wildermuth, K., Univ., Tuebingen, FRG., ORO-4856-26, 1975, p-156.
5549. Sugai, I., Inst. for Nucl. Study, Univ. Tokyo, Tanashi, Tokyo, Jpn., INS-TL-121, 1972, pp-9-14.
5550. Sugai, I., INS-TL-131, 18 May 1976, p-31.
5551. Sugai, I., INS-J-149, 13 Jan. 1975, p-12.
5552. Sugai, I., INS-J-150, 1975, p-15.
5553. Sugai, I., INS-J-152, 1975, p-32.
5554. Sugai, I., Jpn. J. Appl. Phys. 15(4), 1976, pp-729-30.

5555. Sugai, I., Nucl. Instr. and Meth. 145(3), 15 Sept. 1977, pp-409-415.
5556. Sugai, I., INS-TL-134, 5 Dec. 1977, p-60.
5557. Sugai, I., Takaku, S., et al., Radioisotopes 27(6), 1978, pp-295-9.
5558. Sugai, I., Takaku, S., et al., INTDS Proc., Garching, FRG., Sept. 11-14, 1978.
5559. Sugai, I., Takaku, S., et al., Nucl. Instr. and Meth. 167 (1), 1979, pp-135-6.
5560. Sugai, I., INTDS Proc., Garching, FRG., Sept. 11-14, 1978.
5561. Sugai, I., INTDS Newsletter, July 1979.
5562. Sugimoto, K., Mizobuchi, A., et al., Jpn. J. Appl. Phys. 3, May 1964, pp-303-4.
5563. Sugiyama, Y., and Kikuchi, S., Nucl. Phys., A, 264(2), June 28, 1976, pp-179-187.
5564. Sujkowski, Z., Chmielewska, D., et al., Phys. Rev. Lett. ISSN 0031-9007, 43(14), 1 Oct. 1979, pp-998-1001.
5565. Sukhovitskij, E. Sh., Benderskij, A. R., et al., IAEA, Vienna (Austria), Int. Nucl. Data Comm., Jan. 1976, pp-43-47.
5566. Sukhovitskij, E. Sh., and Kon'shin, V. A., IAEA, Int. Nucl. Data Comm., Translation of Selected Papers Presented at 4th All-Union Conf. on Neut. Phys., Kiev, 18-22 April, 1977, pp-25-33.
5567. Sullivan, A. H., Health Phys. 23(2), Aug. 1972, pp-253-5.
5568. Sully, A. H., Chromium, Butterworth, London, 1954.
5569. Sully, E. A., and Brandes, E. A., Chromium, metallurgy of the Rarer Metals, 2nd ed., 1967, Butterworth & Co. Ltd., London
5570. Sun, C. R., Dhar, S., et al., Phys. Rev., D, 14(5), 1 Sept. 1976, pp-1188-1189.
5571. Sun, Shu-Hua, Su, Shih-Chun, et al., INTDS Proc., Boston, U. S. A., Oct. 1-3, 1979.
5572. Sundqvist, B., et al., IEEE Trans. Nucl. Sci., NS-24, 1977, p-652.
5573. Sunyar, A. W., Der Mateosian, E., et al., Phys. Lett., B, 62(3), 7 June 1976, pp-283-286.
5574. Surenyants, V. V., Sidorov, I. T., et al., Instr. Exp. Tech. 16(1), 1973, pp-14-17.
5575. Suter, M., Woelfli, W., et al., Helv. Phys. Acta 49(6), 20 Dec. 1976, pp-863-887.
5576. Suwa, S., Yokosawa, A., et al., V Int. Conf. on High Energy Accel., Rome, Comitato Nazionale per l'Energia Nucleare, 1966, pp-564-7.
5577. Suzuki, A., Shoda, K., et al., Nucl. Phys., A, 257(3), 9 Feb. 1976, pp-477-489.
5578. Suzuki, A., and Shoda, K., Nucl. Phys., A, 260(1), 29 March 1976, pp-172-188.

5579. Suzuki, H., Shinku 6, Nov. 1963, pp-450-5, Tokyo Univ.
5580. Suzuki, K. Wata, R., et al., Radioisotopes 26(2), 1977, pp-67-73.
5581. Suzuki, T., Baba, S., et al., Nucl. Instr. and Meth. 87(2), 1970, pp-311-12.
5582. Suzuki, Y., and Ando, T., Nucl. Phys., A, 295(3), 6 Feb. 1978, pp-365-383.
5583. Suzuki, Y., and Imanishi, B., AIP Conf. Proc. ISSN 0094-243X, 47, 1978, pp-500-501.
5584. Svalbe, I. D., 7th AINSE Nucl. Phys. Conf., 1978, p-29.
5585. Svensson, L. G., Sarantites, D. G., et al., Nucl. Phys., A, 267(1), 16 Aug. 1976, pp-190-204.
5586. Svoboda, K., J. Inorg. and Nucl. Chem. ISSN 0022-1902, 39 (12), 1977, pp-2121-2123.
5587. Svoboda, K., Proc. of a Conf. held at St. Catherine's Coll., Oxford, 22-23 Sept. 1969.
5588. Swann, C. P., Phys. Rev., C, 15(6), June 1977, pp-1967-1971.
5589. Swanson, H. F., Jr., Wash. Univ., Seattle (U. S. A.), Thesis (Ph. D), 1975, p-135.
5590. Sweeney, M. A., Appl. Phys. Lett. 29(4), 1976, pp-231-3.
5591. Sweeney, M. A., and Clauser, M. J., SAND-75-5666, Report 1975, p-11.
5592. Sweeney, M. A., SAND-75-5691, Report 1975, p-17.
5593. Swiniarski, R. de, Bagieu, G., et al., Inst. de Physique Nucleaire (France), March 1976, p-12.
5594. Swiniarski, R. de, Bagieu, G., et al., J. Phys. 37(10), Oct. 1976, pp-1125-1128.
5595. Swiniarski, R. de, Pham, D. L., et al., Can. J. Phys. 54 (1), 1 Jan. 1977, pp-43-54.
5596. Swiniarski, R. de, Resmini, F. G., et al., Helv. Phys. Acta 49(2), 18 May 1976, pp-227-240.
5597. Swiniarski, R. de, Sherman, J., et al., Helv. Phys. Acta 49(2), 18 May 1976, pp-241-257.
5598. Switkowski, Z. E., and Dayras, R. A., Nucl. Instr. and Meth. 128(1), 1975, pp-9-11.
5599. Switkowski, Z. E., Stokstad, R. G., et al., Nucl. Phys., A, 274(1-2), 7 Dec. 1976, pp-202-222.
5600. Switkowski, Z. E., Winkler, H., et al., Phys. Rev., C, 15(1), Jan. 1977, pp-449-451.
5601. Switkowski, Z. E., Stokstad, R. G., et al., Nucl. Phys., A, 279(3), 4 April 1977, pp-502-516.
5602. Switkowski, Z. E., Kennett, S. R., et al., Phys. Rev., C, 16(3), Sept. 1977, pp-1264-1267.
5603. Switkowski, Z. E., Wu, S. C., et al., Nucl. Phys., A, 289 (1), 10 Oct. 1977, pp-236-252.
5604. Switkowski, Z. E., Overley, J. C., et al., 7th AINSE Nucl. Phys. Conf., 1978, p-69.
5605. Switkowski, Z. E., Ryan, C., et al., 7th AINSE Nucl. Phys. Conf., 1978, p-61.

5606. Symons, T. J. M., Doll, P., et al., LBL-8151, 1978, pp-74-75.
5607. Szabo, J., Boedy, Z. T., et al., Nucl. Phys., A, 289(2), 17 Oct. 1977, pp-526-532.
5608. Szalay, S., and Smorjai, E., Nucl. Instr. and Meth. 49(2), 1967, pp-355-6.
5609. Szalok, M., and Peter, I., Fiz. Kut. Int. Kozlem 13, 1965, pp-355-61.
5610. Tabor, S. L., Eisen, Y., et al., ANL/PHY-76-2 (Vol. 2), May 1976, pp-803-810.
5611. Tabor, S. L., Young, K. C., Jr., et al., Phys. Rev., C, 13 (6), June 1976, pp-2262-2268.
5612. Tabor, S. L., Eisen, Y., et al., Phys. Rev., C, 16(2), Aug. 1977, pp-673-678.
5613. Taft, E. A., and Philipp, H. R., Phys. Rev. 138, 1965, p-A197.
5614. Taibert, W. L., Jr., Wohn, F. K., et al., Int. Conf. on Electromag. Isot. Separators and Related Ion Aceel., Kibbutz-Kiryat-Anavim, Israel, 10-13 May 1976, p-35.
5615. Tait, N. R. S., Tolfree, D. W. L., et al., Nucl. Instr. and Meth. 166(3), 1979, pp-333-7.
5616. Tait, N. R. S., and Tolfree, D. W. L., UKAEA, Harwell, Oxon OX11 ORA, England
5617. Tait, N. R. S., Armitage, B. H., et al., INTDS Proc., Garching, FRG., Sept. 11-14, 1978.
5618. Takahashi, N., Matsuo, T., et al., Conf. Ser. - Inst. Phys. 38, Low-Energy Ion Beams, 1978, pp-44-9.
5619. Takahashi, N., Itahashi, T., et al., Jpn. J. Appl. Phys. 15(5), May 1976, pp-861-863.
5620. Takahashi, T., Phys. Rev., C, 16(2), Aug. 1977, pp-529-36.
5621. Takai, M., Kambara, T., et al., J. Phys. Soc. Jpn. 43(1), July 1977, pp-17-24.
5622. Takagui, E. M., and Dietzch, O., Cienc. Cult. 27(7), July 1975, p-23.
5623. Takayanagi, S., Katsuta, M., et al., Nucl. Instr. and Meth. 45(2), 1966, pp-345-6.
5624 Takeda, G., Nippon Butsuri Gakkaishi (Buturi) 21, Sept. 1966, pp-619-25, Tokyo Univ., Sendai.
5625. Takeuchi, S., Kobayashi, C., et al., Nucl. Instr. and Meth. 158, 1979, pp-333-8.
5626. Takeuchi, S., Kobayashi, C., et al., Nucl. Instr. and Meth. 158, 1979, p-333.
5627. Takigawa, N., Lee, S. Y., et al., Phys. Lett., B, 76(2), 22 May 1978, pp-187-191.
5628. Takimoto, K., Yamaya, T., et al., AIP Conf. Proc. ISSN 0094-243X, 47, 1978, pp-710-11.
5629. Takimoto, K., Inada, R., et al., Kyoto Univ., Kyoto (Japan), Dept. of Phys., Ann. Report 1977.
5630. Talbert, W. L., Jr., Wohn, K. F., et al., Nucl. Instr. and Meth. 139, 1 Dec. 1976, pp-257-266.

5631. Talutis, S., New England Nuclear Internal Report, 1980.
5632. Tamain, B., Plasil, F., et al., Phys. Rev. Lett. 36(1), 5 Jan. 1976, pp-18-21.
5633. Tamas, G., CEA-CR-7 (nd), 1975, pp-87-96.
5634. Tamas, G., Paris-11 Univ., 91 - Orsay (France), These (D. es S.), 1976, p-187.
5635. Tanaka, J., and Hagiwara, S., INS-TL-135, Report 1978, p-30.
5636. Taneichi, H., Yamagata Daigaku Kiyo, Shizenkagaku 9(2), Feb. 1977, pp-237-54.
5637. Tanimura, O., and Tazawa, T., Phys. Lett., B, ISSN 0370-2693, 83(1), 23 April 1979, pp-22-26.
5638. Tanis, A. J., and Shafroth, M. S., IEEE Trans. Nucl. Sci., NS-26(1), Feb. 1979.
5639. Tarara, R. W., Goss, J. D., et al., Phys. Rev., C, 13(1), Jan. 1976, pp-109-117.
5640. Taras, P., Schulz, N., et al., Europ. Phys. Soc., Geneva (Switz.), Europ. Conf. on Nucl. Phys. with Heavy Ions, 6-10 Sept. 1976, p-61.
5641. Tarina, E., and Timis, P., Rev. Roum. Phys. 12, 1967, pp-545-9.
5642. Tarutin, I. G., and Pilyavets, V. I., Instr. Exp. Tech. 14 (6), 1971, pp-1625-8.
5643. Tarvin, J. A., Sigler, R. D., et al., Appl. Opt. ISSN 0003-6935, 18(17), 1 Sept. 1979, pp-2971-74.
5644. Tatcher, M., Nucl. Instr. and Meth. 46, 1967, pp-171-2.
5645. Tatischeff, B., Bimbot, L., et al., Phys. Lett., B, 63(2), 19 July 1976, pp-158-160.
5646. Tatischeff, B., Bimbot, L., et al., AIP Conf. Proc. 33, 1976, pp-470-471.
5647. Tauc, J., et al., Proc. of Int. Mat. Conf. on Amorphous and Liquid Semiconductors, Cambridge (Eng.), 1969, pp-279-288.
5648. Tawara, H., and Hyakutake, M., et al., Genshikaku Kenkyu 13, June 1968, pp-13-20.
5649. Taylor, R. C., J. Vac. Sci. Technol. 11(6), Nov. 1974, pp-1148-50.
5650. Taylor, T., and Cameron, J. A., Nucl. Phys., A, 257(3), 9 Feb. 1976, pp-427-437.
5651. Taylor, T., Davidson, J. M., et al., Nucl. Phys., A, 274 (1-2), 7 Dec. 1976, pp-53-60.
5652. Taylor, T., and Summers-Gill, R. G., Nucl. Phys., A, 295 (1), 23 Jan. 1978, pp-77-85.
5653. Tebin, V. V., and Yudkevich, M. S., INIS-mf-5290, Voprosy Atomnoj Nauki i Tekhniki, 2, 1978, pp-2-7.
5654. Teer, D., and Salama, M., Proc. Conf. on Ion Plating and Allied Tech., Edinburg (CEP Consultants Ltd., Edinburgh, 1977), p-103.
5655. Tegart, W. J. McG., The Electrolytic and Chem. Polishing of Metals, Pergamon, NY, 1959.

5656. Tekou, A., CEA-CONF-3179, 1975, p-6.
5657. Templer, J. L., Chaudhri, M. A., et al., 7th AINSE Nucl. Phys. Conf., 1978, p-70.
5658. Tenhaken, R. K., and Quin, P. A., Nucl. Phys., A, 271(1), 26 Oct. 1976, pp-173-184.
5659. Terreault, B., et al., Can. J. Phys. 56, 1978, p-235.
5660. Terwilliger, K. M., Mich. Univ., Ann Arbor, CONF-660918, pp-389-408.
5661. Tessler, G., and Glickstein, S. S., NBS (U. S.), Spec. Publ., 425, Oct. 1975, pp-934-37.
5662. Thain, E., Nucl. Instr. and Meth. 108(3), 1973, pp-571-72.
5663. Thaxter, M. D., Proc. Health Phys. Soc. 1, 1956, pp-227-30.
5664. Thebado, E. A., IEEE Trans. Nucl. Sci., NS-16, June 1969, pp-648-9.
5665. Thibeau, H. L., Goodart, C. D., et al., Proc. Conf. Remote Syst. Technol. 27, 1979, pp-307-9.
5666. Thierens, H., De Frenne, D., et al., Phys. Rev., C, 14(3), Sept. 1976, pp-1058-1067.
5667. Thies, M., AIP Conf. Proc. 33, 1976, pp-71-73.
5668. Thies, M., Phys. Lett., B, 63(1), 5 July 1976, pp-43-46.
5669. Thiessen, H. A., LA-7892C, 22-24 Jan. 1979, pp-85-91.
5670. Thirion, J., Proc. of Int. School of Nucl. Phys., Erice (Italy), 22 Sept.-1 Oct. 1974, pp-227-243.
5671. Thomas, A. W., AIP Conf. Proc. 33, 1976, pp-375-383.
5672. Thomas, B. W., Proc. of 2nd Int. Symp. on Neutron Capture Gamma Ray Spectroscopy and Related Topics, Petten, the Netherlands, 2-6 Sept. 1974, pp-670-672.
5673. Thomas, E. W., and Bent, G. D., Georgia Inst. of Tech., Atlanta, Eng. Exp. Sta., Dec. 1, 1965, p-37.
5674. Thomas, G. E., INTDS Newsletter, Feb. 1980.
5675. Thomas, G. E., INTDS Newsletter, Jan. 1979.
5676. Thomas, G. E., INTDS Newsletter, July 1979.
5677. Thomas, G. E., and Lam, S. K., INTDS Proc., Boston, U. S. A., Oct. 1-3, 1979.
5678. Thomas, G. E., Den Hartog, P. K., et al., Nucl. Instr. and Meth. 167(1), 1979, pp-29-31.
5679. Thomas, G. E., Den Hartog, P. K., et al., INTDS Proc., Garching, FRG., Sept. 11-14, 1978.
5680. Thomas, G. E., Den Hartog, P. K., et al., INTDS Proc., Berkeley, U. S. A., Oct. 19-20, 1977.
5681. Thomas, G. E., and Dusza, P. J., INTDS Proc., LASL, U. S. A., Oct. 19-21, 1976.
5682. Thomas, G. E., and Karasek, F. J., INTDS Proc., ANL, U. S. A., Sept. 30-Oct. 2, 1975.
5683. Thomas, G. E., INTDS Proc., Argonne, U. S. A., Sept. 30-Oct. 2, 1975.
5684. Thomas, G. E., ANL/PHY/MSD-76-1, 1975, pp-255-267.
5685. Thompson, D. M., ANL/PHY/MSD-76-1, 1975, pp-44-51.

5686. Thompson, D. R., and Tang, Y. C., Phys. Rev., C, 12(5), Nov. 1975, pp-1432-1446.
5687. Thompson, D. R., and Tang, Y. C., Phys. Rev., C, 14(1), July 1976, pp-372-375.
5688. Thompson, M. N., Pywell, R. F., et al., Kakuriken Kenkyu Hokoku 8(2), Dec. 1975, pp-266-271.
5689. Thompson, M. D., INTDS Proc., ANL, U. S. A., Sept. 30-Oct. 2, 1975.
5690. Thompson, M. N., 7th AINSE Nucl. Phys. Conf., 1978, p-1.
5691. Thompson, P. A., LA-5348-MS, Contract W-7405-eng-36, Aug. 1973, p-15.
5692. Thompson, R. C., Huizenga, J. R., et al., Phys. Rev., C, 13(2), Feb. 1976, pp-638-647.
5693. Thompson, R. C., Wilcke, W., et al., Phys. Rev., C, 15(6), June 1977, pp-2019-2027.
5694. Thompson, W. J., Univ. of North Carolina, Chapel Hill, ORO-4856-26, 1975, pp-230-231.
5695. Thomson, J. J., Phys. Fluids 21(11), 1978, pp-2082-5.
5696. Thon, N., Proc. of Am. Electroplaters Soc., Jenkintown, PA, 1949, pp-241-49.
5697. Thornton, S. T., McKnight, R. H., et al., Nucl. Instr. and Meth. 101, 1972, p-607.
5698. Thornton, S. T., Schweizer, T. C., et al., Phys. Rev., C, 13(5), May 1976, pp-1936-1943.
5699. Thornton, S. T., Schweizer, T. C., et al., Nucl. Phys., A, 270(2), 19 Oct. 1976, pp-428-436.
5700. Throop, M. J., Cheng, Y. T., et al., Nucl. Phys., A, 283 (3), 20 June 1977, pp-475-492.
5701. Thurber, W. C., and Lotts, A. L., Proc. Hot Lab. Equip. Conf., 10th, Washington, D. C., 1962, pp-27-38.
5702. Thuriere, E., Paris Univ., Faculte des Sciences (France), 1970, p-117, Thesis.
5703. Tietsch, W., Bethge, K., et al., Nucl. Instr. and Meth. 158(1), 1979, pp-41-50.
5704. Tietsch, W., Feist, H., et al., INTDS Proc., Berkeley, U. S. A., Oct. 19-20, 1977.
5705. Tietsch, W., Feist, H., et al., Verh. Dtsch. Phys. Ges. 10(7), 1975, p-709.
5706. Tietsch, W., Bethge, K., et al., Gesellschaft fuer Schwerionenforschung m.b.H., Darmstadt (FRG.), LA-tr-76-18 (nd), p-37.
5707. Tietsch, W., Bethge, K., et al., GSI-79-11, Oct. 1979, p-160.
5708. Tietsch, W., Diss., Frankfurt, 1976.
5709. Tilbury. R. S., Gelbard, S. A., et al., IEEE Trans. Nucl. Sci., NS-26(1), Feb. 1979.
5710. Tilbury, R. S., Mamacos, J. P., et al., Int. Conf. on the Use of Cyc. in Chem., Metall., and Biol., Oxford (England), 1970, pp-117-24.

5711. Timofeev-Resovskii, N. V., Ivanov, V. I., et al., Moscow, Atomizdat, 1968, p-228.
5712. Timsit, R. S., and Daniels, J. M., Can. J. Phys. 49, 1 March 1971, pp-545-59.
5713. Timsit, R. S., Daniels, J. M., et al., Can. J. Phys. 49, 1 March 1971, pp-560-75.
5714. Timsit, R. S., Daniels, J. M., et al., Can. J. Phys. 49, 1 March 1971, pp-508-16.
5715. Timsit, R. S., Daniels, J. M., et al., 3rd Int. Symp. on Polar. Phenom. in Nucl. Reactions, Madison, WI, Aug.-Sept. 1970, pp-903-5.
5716. Tin, E. S., Manitoba Univ., Winnipeg (Canada), Dept. of Phys., Thesis (Ph. D), 1974, p-104.
5717. Tindall, W., Calif. Univ., Livermore (U. S. A.), Lawrence Livermore Lab., 6 Jan. 1977, p-7.
5718. Tishchenko, B. I., and Inopin, E. V., Yadern. Fiz. 7, May 1968, pp-1029-36.
5719. Tjoem, P. O., Espe, I., et al., Phys. Lett., B, 72(4), 16 Jan. 1978, pp-439-442.
5720. Toepfer, A. J., Sandia Lab., Albuquerque, NM, CONF-770611-8, Report 1977, p-6.
5721. Toevs, J. W., and Martin, J. L., Proc. of 4th Conf. on Sci. and Ind. Appl. of Small Accel., NY, 1976, pp-476-479.
5722. Toffer, H., and Amrein, R. L., Rev. Sci. Instr. 31, 1960, pp-348-9.
5723. Tokarevskij, V. V., and Shcherbin, V. N., Yad. Fiz. 25(1), Jan. 1977, pp-16-20.
5724. Toledo, A. Szanto de, Rao, M. N., et al., Phys. Rev., C, 15(1), Jan. 1977, pp-238-245.
5725. Tolfree, D. W. L., INTDS Newsletter, Feb. 1980.
5726. Tolfree, D. W. L., INTDS Newsletter, July 1979.
5727. Tolfree, D. W. L., INTDS Proc., Boston, U. S. A., Oct. 1-3, 1979.
5728. Tolfree, D. W. L., and N. R. S. Tait, Nucl. Instr. and Meth. 163, 1979, pp-1-14.
5729. Tolfree, D. W. L., INTDS Newsletter, Jan. 1979.
5730. Tolfree, D. W. L., Nucl. Instr. and Meth. 155, 1978, p-565.
5731. Tolstov, K. D., Nucl. Instr. and Meth. 95(1), 1971, pp-99-101.
5732. Tombrello, T. A., LA-7852-C, June 1978, pp-127-187.
5733. Tomikova, I., Prib. Tekh. Eksp. 6, Nov.-Dec. 1969, pp-202-4.
5734. Tomozov, A. P., Andreev, G. B., et al., Yad. Fiz. ISSN 0044-0027, 29(4), April 1979, pp-852-856.
5735. Tonapetyan, S. G., Dorofeev, M. T., et al., AN Ukrainskoj SSR, Kharkov, Fiziko-Tekhnicheskij Inst., Nucl. Sci. and Eng. Problems, KFTI-74-23, 3(12), 1975, pp-3-4.
5736. Tonn, J. F., Northwestern Univ., Thesis, Ph. D.
5737. Tonn, J. F., Segel, R. E., et al., Phys. Rev., C, 16(4), Oct. 1977, pp-1357-1362.

5738. Tornow, W., Lisowski, P. W., et al., Nucl. Phys., A, 296 (1), 13 Feb. 1978, pp-23-49.
5739. Toro, M. Di, and Russo, G., Nucl. Phys., A, 284(2), 4 July 1977, pp-177-188.
5740. Tostevin, J. A., and Johnson, R. C., Phys. Lett., B, ISSN 0370-2693, 85(1), 30 July 1979, pp-14-16.
5741. Toten, A., McDaniel, D. F., et al., IEEE Trans. Nucl. Sci., NS-26(1), Feb. 1979.
5742. Toth, K. S., Rainis, A. E., et al., ORNL-5025, May 1975, pp-51-54.
5743. Toth, K. S., Ford, J. L. C., Jr., et al., ORNL-5025, May 1975, pp-42-44.
5744. Toth, K. S., Ford, J. L. C., Jr., et al., Phys. Rev., C, 14(4), Oct. 1976, pp-1471-1483.
5745. Toulemonde, M., Chevallier, J., et al., Nucl. Phys., A, 262(2), 17 May 1976, pp-307-316.
5746. Toulemonde, M., and Haas, F., Phys. Rev., C, 15(1), Jan. 1977, pp-49-52.
5747. Toulemonde, M., Beck, F. A., et al., CRN-PN-79-4, 1979, p-14.
5748. Touloukian, Y. S. (ed.), Thermophysical Prop. of High Temp. Solid Mat., MacMillan, NY, 1967.
5749. Towsley, C. W., Bindal, P. K., et al., Phys. Rev., C, 15 (1), Jan. 1977, pp-281-286.
5750. Tracy, J. G., and Dagenhart, W. K., ORNL, CONF-750968-1, Report 1975, p-14.
5751. Tracy, J. G., and Dagenhart, W. K., INTDS Proc., ANL, U. S. A., Sept. 30-Oct. 2, 1975.
5752. Tracy, J. G., and Newman, E., INTDS Newsletter, Aug. 1980.
5753. Trainor, T. A., Back, N. L., et al., Wash. Univ., Seattle (U. S. A.), 1975, p-2.
5754. Trainor, T. A., Wash. Univ., Seattle (U. S. A.), Nucl. Phys. Lab., Annual Report 1977, pp-43-48.
5755. Trautmann, D., Roesel, F., et al., Helv. Phys. Acta. 50(2), 25 March 1977, p-195.
5756. Trautmann, W., Sharpey-Schafer, J. F., et al., Phys. Rev. Lett. ISSN 0031-9007, 43(14), 1 Oct. 1979, pp-991-4.
5757. Trautvetter, H. P. P., Toronto Univ., Ontario (Canada), Thesis (Ph. D), 1973, p-187.
5758. Treacy, P. B., and Bowkett, N. F., Nucl. Instr. 1, March 1957, pp-86-9.
5759. Treado, P. A., Lambert, J. M., et al., Proc. of 7th Int. Conf. on Few Body Prob. in Nucl. and Particle Phys., Delhi (India), 29 Dec. 1975-3 Jan. 1976, pp-232-235.
5760. Trefil, J. S., and Von Hippel, F., Phys. Rev., D, 7(7), 1 April 1973, pp-2000-12.
5761. Trehan, P. N., Nucl. Instr. and Meth. 24, Nov. 1963, pp-471-2.

5762. Treu, W., Dueck, P., et al., Erlangen-Nuernberg Univ., Erlangen (FRG.), Physikalisches Inst., Annual Report, 1976, p-25.
5763. Treu, W., Galster, W., et al., Phys. Lett., B, 72(3), 2 Jan. 1978, pp-315-318.
5764. Tribble, R. E., Kenefick, R. A., et al., Phys. Rev., C, 13(1), Jan. 1976, pp-50-54.
5765. Tribble, R. E., and Kubo, K. I., Nucl. Phys., A, 282(2), 23 May 1977, pp-269-290.
5766. Tribble, R. E., Cossairt, J. D., et al., Phys. Rev., C, 15(6), June 1977, pp-2028-2031.
5767. Tribble, R. E., Cossairt, J. D., et al., Phys. Rev., C, 16(2), Aug. 1977, pp-917-919.
5768. Tribble, R. E., Cossairt, J. D., et al., Phys. Rev. Lett. 40(1), 2 Jan. 1978, pp-13-16.
5769. Tripard, G. E., and White, B. L., Rev. Sci. Instr. 38, March 1967, pp-435-6.
5770. Trochon, J., Simon, G., et al., Acta Phys. Slovaca 26(1), 1976, pp-25-31.
5771. Troitskij, V. E., Yad. Fiz. ISSN 0044-0027, 29(2), Feb. 1979, pp-456-462.
5772. Truoel, P., AIP Conf. Proc. 33, 1976, pp-581-590.
5773. Trutia, E., and Tatiana, M., Studii Cercetari Fiz. 14, 1963, pp-723-4.
5774. Truttmann, P., Braendle, H., et al., Schweizerisches Inst. fuer Nuklearforschung, Villigen, 83(1), 1979, pp-48-50.
5775. Tryti, S., Holtebekk, T., et al., Nucl. Phys., A, 251(2), 20 Oct. 1975, pp-206-224.
5776. Tsai, S. F., and Bertsch, G. F., Phys. Lett., B, 59(5), 8 Dec. 1975, pp-425-426.
5777. Tsai, S., and Bertsch, G. F., Phys. Lett., B, 73(3), 27 Feb. 1978, pp-247-249.
5778. Tsang, F. Y., Missouri Univ., Columbia (U. S. A.), Thesis (Ph. D), 1978, p-127.
5779. Tsang, F. Y., and Brugger, R. M., Trans. Am. Nucl. Soc. 26, June 1977, p-163.
5780. Tschalaer, C., and Schwedtmann, K. A., IEEE Trans. Nucl. Sci. 18(3), 1971, pp-732-3.
5781. Tschalaer, C., Batty, C. J., et al., Nucl. Instr. and Meth. 78, 1970, pp-141-50.
5782. Tschalaer, C., and Bichsel, H., Nucl. Instr. and Meth. 62, June 15, 1968, pp-208-16.
5783. Tsin, T. L., Ohio State Univ., Columbus (U. S. A.), Thesis (Ph. D), 1976, p-207.
5784. Tsoupas, N., Hausman, H. J., et al., Phys. Rev., C, 13(2), Feb. 1976, pp-510-523.
5785. Tsoybun, V. I., JINR-R-16-11132, 1977, p-8.

5786. Tsubota, H., Oikawa, S., et al., Kakuriken Kenkyu Hokoku 8(1), June 1975, pp-71-81.
5787. Tucker, C. W., and Duke, C. B., Surf. Sci. 23, 1970, p-411.
5788. Tuerck, D., Technische Hochschule Darmstadt (FRG.), Fachbereich Physik, Diss. (D. Sc.), 23 June 1975, p-97.
5789. Tuerck, D., Clerc, H. G., et al., Phys. Lett., B, 2 August 1976.
5790. Turk, M., Fizika (Zagreb) 8(2/3), 1976, pp-173-180.
5791. Turkevich, A., Cadieux, J. R., et al., Phys. Rev. Lett. 38(20), 16 May 1977, pp-1129-1131.
5792. Tustanovskii, V. T., At. Energ. 32(2), Feb. 1972, pp-175-7.
5793. Tuttle, W. K., Tenn. Univ., Knoxville (U. S. A.), Thesis (Ph. D), 1976, p-139.
5794. Tzara, C., Nucl. Phys., A, 256(3), 19 Jan. 1976, pp-381-6.
5795. Tzeng, H. S., Liu, J. Y., et al., Nucl. Instr. and Meth. 150(2), 1 April 1978, pp-143-144.
5796. Udy, M. J., Chromium, Vol. I, Reinhold, NY, 1956, p-118.
5797. Uegaki, J., and Shoda, K., Kakuriken Kenkyu Hokoku 8(1), June 1975, pp-93-98.
5798. Uegaki, E., Abe, Y., et al., AIP Conf. Proc. ISSN 0094-243X, 47, 1978, pp-566-67.
5799. Uemura, Y., Kakigi, S., et al., Bull. Inst. Chem. Res., Kyoto Univ. 47, March 1969, pp-114-22.
5800. Ueno, K., Nucl. Instr. and Meth. 146(2), 15 Oct. 1977, pp-347-355.
5801. Ugglas, M. Af, Toth-Pal, E., et al., Annu. Rep. Res. Inst. Phys., 1978, pp-91-2. (Sweden).
5802. Ulbricht, J., Clausnitzer, G., et al., Nucl. Instr. and Meth. 102, 1972, p-93.
5803. Ullo, J. J., and Goldsmith, M., Nucl. Sci. and Eng. 60(3), July 1976, pp-239-250.
5804. Ulrickson, M., Hartwig, W., et al., Phys. Rev., C, 13(2), Feb. 1976, pp-536-539.
5805. Ulrickson, M., Benczer-Koller, N., et al., Phys. Rev., C, 15(1), Jan. 1977, pp-186-192.
5806. Unternaehrer, J., Lang, J., et al., Phys. Rev. Lett. 40 (16), 17 April 1978, pp-1077-1079.
5807. Unvala, B. A., and Booker, G. R., Phil. Mag. 9, 1964, p-691.
5808. Usova, N. I., Instr. and Experimental Techniques 6, Nov.-Dec. 1959, pp-876-9.
5809. Uttley, C. A., NBS (U. S.), Spec. Publ., 493, Oct. 1977, pp-47-53.
5810. Uzureau, J., Ardouin, D., et al., CEA-N-1798, June 1975, pp-73-81.
5811. Uzureau, J., Adam, A., et al., Nucl. Phys., A, 267(2), 23 Aug. 1976, pp-217-236.
5812. Val'dner, O. A. (ed.), Moscow, Atomizdat 1970, p-208.

5813. Valdre, V., Pashley, D. W., et al., Proc. 6th Int. Cong. Electron Microscopy, Kyoto, Maruzen-Tokyo, 1966.
5814. Valentin, M., and Champion, J., AERE-R-5097, Paper 15, p-2.
5815. Valentin, M., and Champion, J., PSPSITF, AERE, Oct. 20 & 21, 1965.
5816. Valenzuela, A., and Eckardt, J. C., Rev. Sci. Instr. 42 (1), 1971, pp-127-8.
5817. Valkonen, M., and Kantele, J., Nucl. Instr. and Meth. 103 (3), 1972, pp-549-53.
5818. Vance, D. W., 27th Annual Conf. Phys. Electronics, Cambridge, MA, March 20-22, 1967, pp-90-4.
5819. Vandenbosch, R., Webb, M. P., et al., Phys. Rev. Lett. 36(9), 1 March 1976, pp-459-462.
5820. Vandenbosch, R., Webb, M. P., et al., Phys. Rev., C, 13 (5), May 1976, pp-1893-1899.
5821. Vandenbosch, R., Webb, M. P., et al., Phys. Rev., C, 14 (1), July 1976, pp-143-151.
5822. Vandenbosch, R., Webb, M. P., et al., Nucl. Phys., A, 269 (1), 28 Sept. 1976, pp-210-222.
5823. Vandenbosch, R., Webb, M. P., et al., Phys. Rev., C, 17 (5), May 1978, pp-1672-1681.
5824. Vandenbosch, R., Wash. Univ., Seattle (U. S. A.), Int. Conf. on Dynamical Prop. of Heavy-Ion Reactions, Johannesburg, South Africa, 1-3 Aug. 1978, p-26.
5825. Vandenbulcke, L., and Vuillard, G., Chem. Vapor Deposition, Princeton, NJ, The Electrochem. Soc., Inc., 1975, pp-763-776.
5826. Varkonyi, L., and Krebsz, F., Magy. Tud. Akad. Kozp. Fiz. Kut. Int. Kozlemen. 12, 1964, pp-313-16.
5827. Varlamov, V. V., Kocharova, Zh. L., et al., Ukr. Fiz. Zh. 20(8), Aug. 1975, pp-1377-1380.
5828. Varlamov, V. V., Ishkhanov, B. S., et al., Izv. Akad. Nauk SSSR, Ser. Fiz. ISSN 0367-6765, 43(1), Jan. 1979, pp-186-193.
5829. Varma, G. K., Phys. Rev., C, 17(1), Jan. 1978, pp-267-271.
5830. Vasil'ev, L. M., Dmitrevskii, Yu. P., et al., Prib. Tekh. Ehksp. 6, Nov. 1974, pp-30-32.
5831. Vasil'ev, L. M., Dmitrevskii, Yu. P., et al., Instr. Exp. Tech. 19(3), Pt. 1, May-June 1976, pp-634-36.
5832. Vasil'ev, L. M., Dmitrevskii, Yu. P., et al., Cryogenics 13(3), March 1973, pp-180-1.
5833. Vasil'ev, L. M., Dmitrevskii, Yu. P., et al., IFVE-SEF-72-23, 1973, p-19.
5834. Vasil'ev, L. M., Dmitrevskii, Yu. P., et al., Prib. Tekh. Eksp. 15(6), 1972, pp-30-33.
5835. Vasil'ev, L. M., Dmitrevskii, Yu. P., et al., Prib. Tekh. Eksp. 15(4), 1972, pp-35-36.
5836. Vasil'ev, L. M., Radkevich, I. A., et al., Prib. Tekh. Eksp. 14(4), July-Aug. 1971, pp-56-7.

5837. Vasil'ev, S. S., Golovkov, M. S., et al., Leningrad Nauka 1978, p-168.
5838. Vasil'eva, I. A., Gerasimov, Ya. I., et al., Dokl. Akad. Nauk SSSR 232(2), 11 Jan. 1977, pp-355-358.
5839. Vater, P., Becker, H., et al., Phys. Rev. Lett. 39(10), 5 Sept. 1977, pp-594-98.
5840. Vatset, P. I., Vinokurov, E. A., et al., AN Ukrainskoj SSR, Kharkov, Fiziko-Tekhnicheskij Inst., Nucl. Sci. and Eng. Prob., KFTI-74-23, 3(12), 1974, pp-32-33.
5841. Vdovin, A. I., Golikov, I. G., et al., Leningrad Nauka 1978, p-139.
5842. Vdovin, A. I., Golikov, I. G., et al., Izv. Akad. Nauk SSSR, Ser. Fiz. ISSN 0367-6765 43(1), Jan. 1979, pp-148-150.
5843. Vecchio, R. M. Del, Freedman, S. J., et al., Phys. Rev., C, 13(5), May 1976, pp-2089-92.
5844. Veeser, L. R., Arthur, E. D., et al., Phys. Rev., C, 16(5), Nov. 1977, pp-1792-1802.
5845. Veeser, L. R., Phys. Rev., C, 17(1), Jan. 1978, pp-385-87.
5846. Velyukhov, E. G., and Prokof'ev, N. A., Instr. and Exp. Tech. 4, July-Aug. 1964, pp-750-3.
5847. Vennink, R., Nederlandse Natuurkundige Vereniging, Amsterdam, Spring Meet., 26-27 April 1976, p-63.
5848. Vennink, R., Ratynski, W., Nederlandse Natuurkundige Vereniging, Utrecht, Sectie Kernfysica, Autumn Meet., Oct. 1977, p-11.
5849. Vennink, R., Ratynski, W., Nucl. Phys., A, 299(3), 1 May 1978, pp-429-441.
5850. Verbitskij, S. S., Lapik, A. M., et al., Pis'ma Zh. Ehksp. Teor. Fiz. 27(5), 5 March 1978, pp-315-319.
5851. Verbruggen, R., and McMurray, W. R., Nucl. Instr. and Meth. 104(1), 1972, pp-197-203.
5852. Vasil'eva, O. I., Ismail, L. Z., et al., Leningrad Nauka 1978, p-198.
5853. Verdingh, V., Nucl. Instr. and Meth. 102, 1972a, b, pp-431-4, 497-500.
5854. Verdingh, V., TISRMNM, ORNL, Oct. 5-8, 1971.
5855. Verdingh, V., Nucl. Instr. and Meth. 102, 1972, pp-431-34.
5856. Verdingh, V., and Lauer, K. F., Z. Anal. Chem. 235, 1968, p-311.
5857. Verdingh, V., and Lauer, K. F., Nucl. Instr. and Meth. 60, 1968, p-125.
5858. Verdingh, V., and Lauer, K. F., Nucl. Instr. and Meth. 9, 1967, p-197.
5859. Verdingh, V., AERE-R-5097, Paper 12, p-8.
5860. Verdingh, V., PSPSITF, AERE, Oct. 20 & 21, 1965.
5861. Verdingh, V., and Lauer, K. F., NASA Access. 65, EUR-2242.e, 1965, p-14.

5862. Verdingh, V., and Lauer, K. F., Anal. Chim. Acta 33, 1965, p-469
5863. Verdingh, V., and Lauer, K. F., Nucl. Instr. and Meth. 31, 1964, p-355.
5864. Vergnes, M. N., Rotbard, G., et al., Phys. Rev., C, 14(1), July 1976, pp-58-63.
5865. Vermeulen, J. C., Rijksuniversiteit Groningen, Proefschrift (Dr.), INIS-mf-5473, 17 Sept. 1979, p-144.
5866. Vermeulen, J. C., Meijer, R. J. de, et al., Nederlandse Natuurkundige Vereniging, Amsterdam, Spring Meet., 26-27 April, 1976, p-58.
5867. Vermeulen, J., 2nd Int. Conf. on Polar. Targ., Berkeley, CA, 30 Aug. 1971, pp-69-71.
5868. Vermeulen, J., Int. Conf. on Instrumentation for High Energy Phys., Dubna, USSR, 8 Sept. 1970, pp-591-4.
5869. Vernotte, J., Fortier, S., et al., Phys. Rev., C, 13(2), Feb. 1976, pp-461-472.
5870. Verschuur, K. A., ECN (Rep.), ECN-64, 1979, p-17.
5871. Vershinin, G. A., Izv. Vyssh. Uchebn. Zaved., Fiz. ISSN 0021-3411, (2), 1979, pp-44-50.
5872. Vetter, J. E. (ed.), Kernforschungszentrum Karlsruhe (FRG.), Dec. 1977, p-72.
5873. Vialettes, H., CEA-CONF-1766, 1971, p-17.
5874. Vialettes, H., CEA-CONF-1765, 1971, p-19.
5875. Vialettes, H., Rocchesani, J., et al., CEA-R-4188, 1971, p-32.
5876. Vialettes, H., Int. Conf. Accel. Dosim. Exper., (Proc.), 2nd, 1970, pp-121-38.
5877. Viatte, P., Micek, S., et al., Helv. Phys. Acta 49(4), 13 Aug. 1976, pp-569-598.
5878. Viatte, P., Micek, S., et al., Helv. Phys. Acta 48(4), 25 Nov. 1975, p-558.
5879. Videbaek, F., Goldstein, R. B., et al., BNL-21269, CONF-760424-16, 1976, p-14.
5880. Videbaek, F., Goldstein, R. B., et al., ANL/PHY-76-2 (Vol. 2), May 1976, pp-819-824.
5881. Videbaek, F., Goldstein, R. B., et al., BNL-21268, CONF-760424-18, 1976, p-8.
5882. Videbaek, F., Goldstein, R. B., et al., ANL/PHY-76-2 (Vol. 2), May 1976, pp-825-831.
5883. Viefers, W., Witsch, W. von, et al., Nucl. Phys., A, 248 (3), 1 Sept. 1975, pp-518-23.
5884. Vieira Junior, N. D., Quacchia, J. C. A., et al., Cienc. Cult. Supl. 29(7), July 1977, p-272, Sao Paulo.
5885. Viggars, D. A., Brady, F. P., et al., Europ. Phys. Soc., Geneva (Switz.), Europ. Conf. on Nucl. Phys. with Heavy Ions, 6-10 Sept. 1976, p-57.
5886. Viktorov, D. V., Rozman, I. M., et al., At. Energ. 31(1), Jan. 1971, p-49.

5887. Vincent, E., Biennial Session of Nucl. Phys., La Toussuire, 28 Feb.-4 March 1977, pp-S4.1-S4.4.
5888. Vincent, J. S., and Smith, W. R., Nucl. Instr. and Meth. 116(3), 1974, pp-551-4.
5889. Vinogradov, V. N., Manokhin, V. N., et al., IAEA, Vienna (Austria), Int. Nucl. Data Comm., April 1976, p-6.
5890. Vinogradov, V. N., Davletshin, A. N., et al., IAEA, Vienna (Austria), Int. Nucl. Data Comm., April 1976, p-5.
5891. Vinogradov, V. N., Davletshin, A. N., IAEA, Vienna (Austria), Int. Nucl. Data Comm., Feb. 1977, pp-2-5.
5892. Viola, V. E., Jr., Wolf, K. L., et al., Maryland Univ., College Park (U. S. A.), Dept. of Phys. and Astron., 1975, pp-597-598.
5893. Visser, J., Brinkman, G. A., et al., Int. J. Appl. Radiat. Isot. ISSN 0020-708X, 30(12), Dec. 1979, pp-745-48.
5894. Viyogi, Y. P., Beiser, F., et al., AIP Conf. Proc. ISSN 0094-243X, 47, 1978, pp-676-677.
5895. Vizir', V. A., Kalinin, B. N., et al., Proc. of the All-Union Conf., 3-5 Sept., 1975, Tomsk, p-292.
5896. Vlasenko, V. G., Gol'dshtejn, V. A., et al., AN Ukrainskoj SSR, Kharkov, Fiziko-Tekhnicheskij Inst., Nucl. Sci. and Eng. Prob., KFTI-75-11, 2(14), 1975, pp-17-18.
5897. Vlasenko, V. G., Gol'dshtejn, V. A., et al., KFTI-77-9, Voprosy Atomnoj Nauki i Tekhniki 2(19), 1977, pp-53-4.
5898. Vlatskii, F. D., Krupnyi, G. I., et al., IFVE-ORZ-74-142, 1974, p-13.
5899. Vo Kim Tkhan', Vtyurin, V. A., et al., Leningrad Nauka, 1978, p-162.
5900. Vodennikov, B. D., Danilyan, G. V., et al., Leningrad Nauka, 1978, p-374.
5901. Vogel, U., IEEE Trans. Nucl. Sci., NS-16, June 1969, pp-905-8.
5902. Voit, H., Hamm, M., et al., Phys. Lett., B, 58(2), 1 Sept. 1975, pp-152-154.
5903. Volant, C., Conjeaud, M., et al., Europ. Phys. Soc., Geneva (Switz.), Europ. Conf. on Nucl. Phys. with Heavy Ions, 6-10 Sept. 1976, p-122.
5904. Volchek, Yu. A., At. Energ. 47(3), 1979, pp-198-200.
5905. Vold, P. B., Cline, D., et al., Phys. Rev. Lett., 39(6), 8 Aug. 1977, pp-325-328.
5906. Volkar, M. A., and Wolke, R. L., Rev. Sci. Instr. 40, June 1969, pp-849-51.
5907. Vonach, H., Glaessel, P., et al., Nucl. Phys., A, 278(2), 7 March 1977, pp-189-203.
5908. Vonach, H., Glässel, P., et al., Sitzungsbericht Österr. Akad. d. Wiss., Mathem.-naturw. Kl. Abt. II, 183. Bd, (1974), 4. - 7. Heft.
5909. Vonberg, D. D., Baker, L. C., et al., Uses Cyc. Chem., Met. Biol., Proc. Conf., 1970, pp-258-69.

5910. Vorob'eva, V. G., D'yachenko, N. P., et al., Yad. Fiz. 26(5), 1977, pp-962-965.
5911. Vorotnikov, P. E., Dubrovina, S. M., et al., IAEA, Vienna (Austria), Int. Nucl. Data Comm., Nucl. Phys. Rsch. in the USSR, Feb. 1976, pp-6-7.
5912. Vorotnikov, P. E., Gurtovenko, Yu. F., et al., Nucl. Phys., A, 281(2), 2 May 1977, pp-295-309.
5913. Vosniakos, F. K., Davison, N. E., et al., Nucl. Phys., A, ISSN 0375-9474, 332(1/2), Dec. 1979, pp-157-172.
5914. Voss, F., Cierjacks, S., et al., NBS (U. S.), Spec. Publ., 425, Oct. 1975, pp-916-919.
5915. Vouros, P., Masters, J. I., et al., Rev. Sci. Instr. 39, May 1968, pp-741-3.
5916. Vourvopoulos, G., and Paradellis, T., Phys. Rev., C, 17(5), May 1978, pp-1885-1887.
5917. Vries, R. M. De, AIP Conf. Proc. ISSN 0094-243X, 47, 1978, pp-344-351.
5918. Vtyurin, V. A., Popov, Yu. P., et al., JINR-R-3-10733, 1977, p-14.
5919. Vukanic, J., and Sigmund, P., Appl. Phys., 11(3), 1976, pp-265-72.
5920. Vylov, Ts., Gromov, K. Ya., et al., Yad. Fiz. ISSN 0044-0027, 28(5), 1978, pp-1137-1143.
5921. Vyunsh, R., Korotkikh, V. L., et al., Yad. Fiz. ISSN 0044-0027, 29(2), Feb. 1979, pp-318-331.
5922. Vyver, R. van de, Devos, J., et al., Z. Phys., A, 284(1), Nov. 1977, pp-91-93.
5923. Wada, R., Schimizu, J., et al., Phys. Rev. Lett. 38(23), June 6, 1977, pp-1341-1344.
5924. Waddell, C. N., Diener, E. M., et al., Nucl. Phys., A, 281 (3), 9 May 1977, pp-418-442.
5925. Wade, W. Z., and Wolf, T., J. Inorg. Nucl. Chem. 29, 1967, p-2577.
5926. Wagemans, C., D'Hondt, P., et al., Nucl. Phys., A, 259(3), 22 March 1976, pp-423-428.
5927. Wagner, P., Engelstein, P., et al., ANL/PHY-76-2 (Vol. 2), May 1976, pp-833-840.
5928. Wagner, S. R., and Debertin, K., IAEA, Vienna (Austria), 15-19 Nov. 1976, IAEA-208 (Vol. 2), pp-261-263.
5929. Wagner, W. T., Mich. State Univ., East Lansing (U. S. A.), Thesis (Ph. D), 1974, p-201.
5930. Wagschal, J. J., Ya'ari, A., et al., NBS (U. S.), Spec. Publ. 425(Vol. 1), Oct. 1975, pp-405-408.
5931. Waldron, J. C., AERE-R-6141, 1969, p-13.
5932. Wales, G. L., and Johnson, R. C., Nucl. Phys., A, 274(1-2), 7 Dec. 1976, pp-168-176.
5933. Walker, G. E., AIP Conf. Proc. ISSN 0094-243X, 54(1), 15 July 1979, pp-206-221.
5934. Walker, J. K., Burq, J. P., et al., Nucl. Instr. and Meth. 22, March 1963, pp-138-40.

5935. Walker, W. D., BNL-17018, Report, 1972, p-4.
5936. Wall, N. S., and Irvine, J. W., Jr., Rev. Sci. Instr. 24, 1953, pp-1146-7.
5937. Wall, N. S., Craig, J. N., et al., Nucl. Phys., A, 268(3), 21 Sept. 1976, pp-459-468.
5938. Wall, N. S., Cowley, A. A., et al., Maryland Univ., College Park (U. S. A.), Dept. of Phys. and Astron., ORO-5128-7, Jan. 1977, p-22.
5939. Wallace, W. J., Bisson, A. E., et al., Nucl. Instr. and Meth. 68, Feb. 15, 1969, pp-337-40.
5940. Walraven, R. L., Nucl. Instr. and Meth. 99(1), 1972, pp-73-5.
5941. Walter, R. L., and Lisowski, P. W., Duke Univ., Durham, NC, CONF-760715-P2, 1976, pp-1061-1082.
5942. Walton, D., J. Chem. Phys. 37, 1962, p-2182.
5943. Wang, S. T., Onesto, F., et al., IEEE Trans. Magn. Vol. Mag-13(1), 1977 Jan., pp-74-7.
5944. Ward, C. M., and Bergquist, L. E., UCRL-81416(Rev. 1), Topical Meet. on Fusion Reactor Mat., Miami Beach, FL, U. S. A., 29-31 Jan. 1979, p-14.
5945. Ward, C. M., Hendricks, C. D., et al., UCRL-80269(Rev. 1), Scanning Electron Microscopy Conf., Los Angeles, CA, U. S. A., 16-21 April 1978, p-17.
5946. Warke, C. S., and Nogami, Y., Proc. of Nucl. Phys. and Solid State Phys. Symp., Pune, 20B, Dec. 26-30, 1977, pp-1-4.
5947. Warneke, M. L., Houston Univ., TX (U. S. A.), Thesis (Ph. D), 1975, p-190.
5948. Warner, L. B., Nucl. Instr. and Meth. 14, Jan. 1962, pp-315-17.
5949. Warner, R. E., Ball, G. C., et al., Nucl. Phys., A, 269(2), 5 Oct. 1976, pp-286-300.
5950. Warner, R. E., Martin, D. C., et al., Nucl. Phys., A, ISSN 0375-9474, 326(1), 3 Sept. 1979, pp-209-224.
5951. Warszawski, J., and Eisenberg, J. M., Israel Physical Soc., Jerusalem, 1977 Annual Meet., pp-12-13.
5952. Warszawski, J., Eisenberg, J. M., et al., Nucl. Phys., A, 294(3), 16 Jan. 1978, pp-321-347.
5953. Waterbeemd, J. van de, Philips Res. Rep. 21, 1966, p-27.
5954. Waterman, F. M., Kuchnir, F. T., et al., Med. Phys. ISSN 0094-2405, 6(5), Sept. 1979, pp-432-435.
5955. Watson, D. D., Rev. Sci. Instr. 37(11), 1966, pp-1605-6.
5956. Watson, D. L., and Brown, G., Nucl. Phys., A, 296(1), 13 Feb. 1978, pp-1-22.
5957. Watson, I. A., Waters, S. L., et al., MRC Cyc. Unit, Hammersmith Hosp., Ducane Rd., London W12 OHS, Eng.
5958. Watson, J. M., Sevcik, V., et al., ANL, Argonne, IL.
5959. Watson, J. W., Nucl. Instr. and Meth. 103(2), 1972, p-421.

5960. Watt, F., Fifield, L. K., et al., Nucl. Instr. and Meth. 151(1-2), 1978, pp-163-73.
5961. Waugh, J. B. S., and Carver, J. H., Rev. Sci. Instr. 34, Feb. 1963, pp-192-3.
5962. Webb, D. V., Peterson, G. A., et al., Nucl. Instr. and Meth. 120(2), 1 Sept. 1974, pp-359-361.
5963. Webb, M. P., Wash. Univ., Seattle (U. S. A.), Thesis (Ph. D), 1977, p-224.
5964. Webb, M. P., Vandenbosch, R., et al., Phys. Rev. Lett. 36 (14), 5 April 1976, pp-779-782.
5965. Weber, E., Kneupfer, W., et al., Phys. Lett., B, 65(3), 22 Nov. 1976, pp-189-192.
5966. Weber, J. W., BNWL-CC-427, Battelle Pacific NW Labs., Richland, WA, Cont. E(4501)-1830, 11 Jan. 1966, p-12.
5967. Weber, J. W., BNWL-CC-426, Contract E(45-1)-1830, 4 Jan. 1966, p-24.
5968. Weckstroem, T., Forsblom, I., et al., Phys. Fenn. 10(4), Dec. 1975, pp-167-171.
5969. Wecksung, G. W., Walker, J. J., et al., Nucl. Instr. and Meth. 95(3), 1971, pp-605-9.
5970. Wehner, G. K., Appl. Phys. Lett. 30(4), 1977, pp-185-7.
5971. Wehner, G. K., and Anderson, G. S., "The Nature of Physical Sputtering", Handbook of Thin Film Tech., L. I. Maissel and R. Glang, eds. (McGraw-Hill, 1970).
5972. Wehner, G. K., J. Appl. Phys. 30, 1969, p-1762.
5973. Wei, Pax S. P., and Kuppermann, A., Rev. Sci. Instr. 40, June 1969, pp-783-5.
5974. Weidenmueller, H. A., Physikertagung, Frankfurt (M)-Hoechst, Oct. 4-9, 1965, pp-12-48.
5975. Weidinger, A., Eberhard, K. A., et al., Nucl. Phys., A, 257(1), 26 Jan. 1976, pp-144-164.
5976. Weidinger, A., Busch, F., et al., Nucl. Phys., A, 263(3), 14 June 1976, pp-511-532.
5977. Weil, A. S., Reid, A. F., et al., Rev. Sci. Instr. 18, 1947, p-501, 556-8.
5978. Weinberg, R. B., Mitchell, G. E., et al., Phys. Rev., B, 133, 1964, p-884.
5979. Weinreich, R., Bräutigam, W., et al., 7th Int. Conf. on Cyc. and Their Appl., Zurich, Aug. 19-22, 1975.
5980. Weinreich, R., Schult, O., et al., Int. J. Appl. Radiat. Isot. 25(11-12), 1974, pp-535-43.
5981. Weinstein, B. W., and Hendricks, C. D., Appl. Opt. 17(22), 1978, pp-3641-6.
5982. Weinstein, B. W., and Hendricks, C. D., UCRL-78477, Report, 1976, p-20.
5983. Weinstein, B. W., Hendricks, C. D., et al., Rev. Sci. Instr. 49(6), June 1978, pp-870-871.
5984. Weinstein, M. S., HASL-58, New York Oper. Off., Health and Safety Lab., pp-180-4.

5985. Weisehahn, W. J., Dautet, H., et al., Nucl. Instr. and Meth. 109(3), 1973, pp-613-14.

5986. Weisemiller, R. B., Wilcox, K. H., et al., LBL-5075, 1975, pp-82-83.

5987. Weisenmiller, R. B., LBL-5077, 3 Dec. 1976, p-132.

5988. Weisenmiller, R. B., Calif. Univ., Berkeley (U. S. A.), Thesis (Ph. D), 1977, p-131.

5989. Weishaar, M., and Meens, A., INTDS Proc., Boston, U. S. A., Oct. 1-3, 1979.

5990. Weishaar, M., INTDS Newsletter, Jan. 1979.

5991. Weisman, J., ed., Westinghouse Elect. Corp., Atomic Power Div., Pittsburgh, April 14, 1964, p-22.

5992. Weiss, H. V., and Chew, K., Anal. Chim. Acta 67(2), 1973, pp-444-7.

5993. Weiss, M. M., Health Phys. 15, Oct. 1968, p-372.

5994. Weiss, M. S., Proc. of Int. Symp. on Interaction Studies in Nuclei, Mainz, FRG., 17-20 Feb. 1975, pp-599-634.

5995. Weiss, M. S., Calif. Univ., Livermore (U. S. A.), Lawrence Livermore Lab, UCRL-80264, CONF-770987, 1977 Sept.

5996. Weiss, U., Fick, D., Nucl. Phys., A, 274(1-2), 7 Dec. 1976, pp-253-261.

5997. Weisser, D. C., and Hobbie, R. K., Rev. Sci. Instr. 40(5), 1969, pp-683-5.

5998. Weller, H. R., Blue, R. A., et al., Phys. Rev., C, 13(3), March 1976, pp-922-932.

5999. Weller, H. R., Szucs, J., et al., Phys. Rev., C, 13(3), March 1976, pp-1055-1060.

6000. Weller, H. R., Roberson, N. R., et al., Phys. Rev., C, 13 (5), May 1976, p-2062.

6001. Weller, H. R., Blue, R. A., et al., Phys. Rev., C, 17(3), March 1978, pp-1260-1262.

6002. Weller, H. R., Roberson, N. R., et al., Phys. Rev., C, 18 (1), July 1978, pp-65-70.

6003. Weller, H. R., Manglos, S., et al., Phys. Rev., C, ISSN 0556-2813, 20(4), Oct. 1979, p-1589.

6004. Wells, J. C., Jr., Robinson, R. L., et al., Phys. Rev., C, 12(5), Nov. 1975, pp-1529-1539.

6005. Wender, S. A., Oberley, L. W., et al., Nucl. Instr. and Meth. 89, 1970, pp-61-4.

6006. Wender, S. A., and Martin, D. J., Nucl. Phys., A, 259(2), 15 March 1976, pp-246-252.

6007. Wender, S. A., Gould, C. R., et al., Phys. Rev., C, 17(4), April 1978, pp-1365-1367.

6008. Wendling, R. D., Walther, V. H., et al., Nucl. Instr. and Meth. 105(1), 1972, pp-125-7.

6009. Wene, C. O., Nucl. Instr. and Meth. 91, 1971, pp-547-53.

6010. Werby, M. F., Phys. Rev., C, 16(5), Nov. 1977, pp-1882-89.

6011. Werf, S. Y. van der, Put, L. W., et al., Nederlandse Natuurkundige Vereniging, Amsterdam, Spring Meet., 26-27 April 1976, p-50.

6012. Werf, S. Y. van der, Fryszczyn, B., et al., Nucl. Phys., A, 273(1), 23 Nov. 1976, pp-15-28.
6013. Werner, Z. G., and Whitehouse, J. E., Rev. Sci. Instr. 41, Jan. 1970, pp-134-5.
6014. Wery, M., and Riehl, F., Nucl. Instr. and Meth. 96(3), 1971, pp-425-9.
6015. Wery, M., Seltz, R., et al., Rev. Phys. Appl. 4, June 1969, p-4.
6016. West, D., and Sherwood, A. C., AERE-R 9195, Report, 1978, p-5.
6017. West, L., Jr., and Fletcher, N. R., Phys. Rev., C, 15(6), June 1977, pp-2052-2058.
6018. West, L. A., Kozak, E. I., et al., J. Vac. Sci. Technol. 8, March-April 1971, pp-430-6.
6019. West, R. L., Notre Dame Univ., IN (U. S. A.), Thesis (Ph. D), 1977, p-150.
6020. Westfall, G. D., Texas Univ., Austin (U. S. A.), Thesis (Ph. D), 1975, p-115.
6021. Westfall, G. D., and Zaidi, S. A. A., Phys. Rev., C, 14(2), Aug. 1976, pp-610-18.
6022. Westfall, G. D., Sextro, R. G., et al., Phys. Rev., C, 17 (4), April 1978, pp-1368-1381.
6023. Westfall, G. D., LBL-9009, CONF-790140-5, Jan. 1979, p-5.
6024. Westgaard, L., and Bjoernholm, S., Nucl. Instr. and Meth. 42, June 1966, pp-77-80.
6025. Weston, L. W., and Todd, J. H., NBS (U. S.), Spec. Publ. 1, Oct. 1975, pp-229-231.
6026. Weston, L. W., and Todd, J. H., Nucl. Sci. and Eng. 61(3), Nov. 1976, pp-356-365.
6027. Weston, L. W., and Todd, J. H., Nucl. Sci. and Eng. 63(2), June 1977, pp-143-148.
6028. Westwood, W. D., Prog. Surf. Sci. 7, 1976, p-71.
6029. Whalen, J. F., and Smith, A. B., Nucl. Sci. and Eng. 67 (1), July 1978, pp-129-130.
6030. Whalin, A. E., Jr., Reitz, A. R., et al., Rev. Sci. Instr. 26(1), Jan. 1955, pp-59-65.
6031. Whaling, W., Handbuch der Physik 34, S. Flugge (ed.), Stringer-Verlag, Berlin, 1958, pp-193-217.
6032. Wheatley, J. C., 2nd Int. Conf. on Polar. Targ., Berkeley, CA, 30 Aug. 1971, pp-73-5.
6033. White, D. C. S., McDonald, W. J., et al., Nucl. Instr. and Meth. 121(3), 1 Nov. 1974, pp-439-443.
6034. White, P. H., AERE-R-5097, Paper 29, p-7.
6035. White, P. H., PSPSITF, AERE, Oct. 20 & 21, 1965.
6036. White, R. E., Barker, P. H., et al., 6th AINSE Nucl. Phys. Conf. 9-11 Feb. 1976, p-6.
6037. White, R. L., Charlton, L. A., et al., Phys. Rev., C, 12 (6), Dec. 1975, pp-1918-1926.
6038. White, R. M., Ohio Univ., Athens (U. S. A.), June 1977, p-197.

6039. Whitman, R. L., and Day, R. H., Appl. Opt. 19(10), 1980, pp-1718-22.

6040. Whitmell, D. S., Armitage, B. H., et al., Daresbury Lab. Report, 1978, DL/NSF/p-86.

6041. Whitmell, D. S., Armitage, B. H., et al., Rev. Phys. Appl. 12, 1977, p-1335.

6042. Whitmell, D. S., Armitage, B. H., et al., Proc. Int. Conf. on Technol. of Electrostatic Accel., Daresbury Lab. Report DNPL/NSF/R5, 1977.

6043. Whitmell, D. S., and Williamson, R., Thin Solid Films 35, 1976, pp-255-61.

6044. Whitmell, D. S., Armitage, B. H., et al., Proc. Int. Conf. on Technol. of Electrostatic Accel., Daresbury Lab. Report, DNPL/NSF/R5, 1973, p-265.

6045. Whitmell, D. S., J. Phys., E, 4, Feb. 1971, pp-151-2.

6046. Whitmell, D. S., Armitage, B. H., et al., Daresbury Lab. Report, DL/NSF/p-86.

6047. Whitton, J. L., Proc. Roy. Soc., Ser. A, 311(1504), 1969, pp-63-73.

6048. Wiborg, J. C., Cramer, J. G., et al., Wash. Univ., Seattle (U. S. A.), Nucl. Phys. Lab., Annual Report, 1977, June 1977, pp-83-86.

6049. Wick, O. C. (ed.), Plutonium Handbook, Vol. I, Gordon and Breach, 1967, p-36.

6050. Widmer, A. E., Rev. Sci. Instr. 36, July 1965, p-1054.

6051. Widner, M. M., Perry, F. C., et al., J. Appl. Phys. 48(3), March 1977, pp-1047-1053.

6052. Wieland, B. W., Highfill, R. R., et al., IEEE Trans. Nucl. Sci., NS-26(1), Pt. 2, Feb. 1979, pp-1713-17.

6053. Wieland, R. M., Stokstad, R. G., et al., Phys. Rev. Lett. 37(22), 29 Nov. 1976, pp-1458-1461.

6054. Wiele, J. van de, Gerlic, E., et al., Paris-11 Univ., 91 - Orsay (France), Inst. de Physique Nucleaire, 1977, p-64.

6055. Wieman, H. H., Wash. Univ., Seattle (U. S. A.), Thesis (Ph. D), 1975, p-219.

6056. Wieman, H. H., Anderson, R. E., et al., Colo. Univ., Boulder (U. S. A.), Dept. of Phys. and Astrophys., Technical Prog. Report, 1 Nov. 1975, pp-72-73.

6057. Wienke, B. R., and Seamon, R. E., LA-6537-MS, Sept. 1976, p-23.

6058. Wigle, G. L., and Fleischer, E. S., Proc. 10th Hot Lab. Equip. Conf., Washington, D. C., 1962, UCRL-10372, pp-269-73.

6059. Wigle, G. L., Spencer, N. C., UCRL-9663, Contract W-7405-eng-48, Sept 6, 1961, p-10.

6060. Wigmans, M. E. J., Heynis, R. J., et al., Phys. Rev., C, 14(1), July 1976, pp-229-242.

6061. Wilcke, W., Feix, W., et al., Nucl. Phys., A, 286(2), 15 Aug. 1977, pp-279-306.

6062. Wilde, N., Boland, T. J., et al., Phillips Petroleum Co., Idaho Falls, ID, Atomic Energy Div., Contract AT(10-1)-205, July 1967, p-99.
6063. Wilkin, C., Europ. Organiz. for Nucl. Rsch., Geneva (Switz.), CEA-CR-2 (nd), 8-10 May 1974, pp-171-78.
6064. Wilkin, C., AIP Conf. Proc. ISSN 0094-243X, 54(1), July 15, 1979, pp-537-549.
6065. Wilkinson, M. K., and Young, F. W., Jr., ORNL-5328, Oct. 1977, pp-168-175.
6066. Wilkinson, R., Switkowski, Z. E., et al., 7th AINSE Nucl. Phys. Conf., 6-8 Feb. 1978, p-10.
6067. Wilkniss, P. E., and Wynne, G. J., Int. J. Appl. Radiat. Isotop. 18, Jan. 1967, pp-77-81.
6068. Wilkniss, P. E., Hoover, J. I., et al., Nucl. Instr. and Meth. 56, Nov. 1967, pp-120-4.
6069. Williams, D. C., Knight, J. D., et al., Phys. Lett. 22, 1966, p-162.
6070. Williams, K. E., and Seaborg, G. T., LBL-8151, 1978, pp-33-35.
6071. Williams, W. G., Nicholas, D. J., et al., Nucl. Instr. and Meth. 88, 1970, pp-73-6.
6072. Williams-Norton, M. E., Petrovich, F., et al., Nucl. Phys., A, 275(2), 10 Jan. 1977, pp-509-518.
6073. Williams-Norton, M. E., Hudson, G. M., et al., Phys. Rev., C, 12(6), Dec. 1975, pp-1899-1906.
6074. Williamson, C. F., Boujat, J. P., et al., CEA-R-3042, 1966.
6075. Williamson, C., and Boujot, J. P., CEA-2189, 1962, p-460.
6076. Williamson, C. F., Rad, F. N., et al., Phys. Rev. Lett. 40 (26), 26 June 1978, pp-1702-1705.
6077. Williamson, F. S., Radiobiol. Rsch. Unit, Harwell, Berks., Eng., EUR-1815.e(Rev.), pp-258-62.
6078. Williamson, F. S., Proc. of 3rd Conf. on Accel. Targ. Des. for the Prod. of Neutrons, Liege (Belgium), Sept. 18-19, 1967, pp-165-79.
6079. Williamson, K. D., Jr., Simmons, J. E., et al., Adv. Cryog. Eng. 19, 1974, pp-241-7.
6080. Williamson, K. D., Jr., Simmons, J. E., et al., LA-UR-73-413, Contract W-7405-eng-36, 1973, p-21.
6081. Williamson, R., LA-UR-79-2057, CONF-790529-10, 1979, p-8.
6082. Willis, N., Brissaud, I., et al., Nucl. Phys., A, 261(1), 19 April 1976, pp-45-58.
6083. Willis, N., Paris-11 Univ., 91-Orsay (France), Inst. de Physique Nucleaire, These (D. es S.), 1975, p-100.
6084. Willis, R. D., Walter, R. L., et al., Nucl. Instr. and Meth. 142(1-2), 1-15 April 1977, pp-67-77.
6085. Willis, S. E., LA-8030-T, Sept. 1979, p-157.
6086. Wilmott, D. J., J. Appl. Phys. 43, 1972, p-4865.
6087. Wilson, M. T., Thorn, L. L., et al., LA-UR-77-644, Contract W-7405-eng-36, 1977, p-4.

6088. Wilson, M. T., Thorn, L. L., et al., IEEE Trans. Nucl. Sci., NS-24(3), June 1977, pp-1574-1576.
6089. Wilson, M. T., Nucl. News 12(3), March 1969, pp-35-7.
6090. Wilson, R. R., and Kamen, M. D., Phys. Rev. 54, 1938, pp-1031-6.
6091. Wilson, R. R., Rev. Sci. Instr. 29, Aug. 1958, p-732.
6092. Wilson, R., Proc. of Int. Symp. on Interaction Studies in Nuclei, Mainz, FRG., 17-20 Feb. 1975, pp-23-43.
6093. Wimpey, J. F., Mitchell, G. E., et al., Nucl. Phys., A, 269(1), 28 Sept. 1976, pp-46-60.
6094. Winsberg, L., Nucl. Instr. and Meth. 150(3), 1978, pp-465-77.
6095. Winsberg, L., Nucl. Instr. and Meth. 95(1), 1971, pp-19-22.
6096. Winsberg, L., Nucl. Instr. and Meth. 95(1), 1971, pp-23-7.
6097. Winsberg, L., Nucl. Instr. and Meth. 96(2), 1971, pp-301-7.
6098. Winter, G., Doering, J., et al., Contributed papers of Int. Symp. on High-Spin States and Nucl. Structure, Dresden, 19-24 Sept. 1977, pp-33-34.
6099. Winters, R. R., Denison Univ., Granville, OH (U. S. A.), 1975 Dec., p-5.
6100. Wise, K. D., Jackson, T. N., et al., J. Nucl. Mater. 85, 86(A), 1979, pp-103-6.
6101. Wisshak, K., Wickenhauser, J., et al., Kernforschungszentrum Karlsruhe G.m.b.H. (FRG.), Inst. fuer Angewandte Kernphysik, Annual Report, KFK-2868, Oct. 1979, pp-24-25.
6102. Wisshak, K., and Kaeppeler, F., KFK-2868, Oct. 1979, p-26.
6103. Witsch, W. von, and Willaschek, J. G., Nucl. Instr. and Meth. 138(1), 1976, pp-13-17.
6104. Wittchow, F., Ulbricht, J., et al., Phys. Lett., B, 59B (1), 1975, pp-29-31.
6105. Wittkopf, W. A., BNL-NCS-50446, April 1975, pp-192-206.
6106. Woerner, R. L., Bell, J. W., et al., UCRL-80877, CONF-781110-1, Report, 1978, p-15.
6107. Woerner, R. L., and Hendricks, C. D., UCRL-79442, Report, 1977, p-6.
6108. Wohlfarth, H., Lang, W., et al., Phys. Lett., B, 63(3), 2 Aug. 1976, pp-275-278.
6109. Wohlfahrt, H. D., Shera, E. B., et al., Int. Workshop on Gross Prop. of Nuclei and Nucl. Excitations, Hirschegg (Austria), 15-27 Jan. 1979, pp-1-17.
6110. Wojciechowski, H., Gustafson, D. E., et al., ANL/PHY-76-2 (Vol. 2), May 1976, pp-841-848.
6111. Wolber, G., Hartmann, G., et al., 7th Int. Conf. on Cyc. and Their Appl., Zurich (Switz.), Aug. 19-22, 1975.
6112. Wolf, K. L., Roche, C. T., et al., Europ. Phys. Soc., Geneva (Switz.), Europ. Conf. on Nucl. Phys. with Heavy Ions, 6-10 Sept. 1976, p-176.
6113. Wolf, K. L., Gosset, J., et al., LBL-8151, 1978, pp-134-36.

6114. Wolf, P., IAEA Symp. on Med. Radionuclide Imaging, Oct. 25-29, 1976, Los Angeles, CA (U. S. A.).
6115. Wolf, P. A., BNL Report 1976, BNL-22021, p-25.
6116. Wolfe, G. W., Calif. Univ., Davis (U. S. A.), Thesis (Ph. D), 1976, p-135.
6117. Wolke, R. L., Pitts. Univ., Wherrett Lab. of Nucl. Chem., Pitts., PA, Contract AT(30-1)-2771, Jan.-Dec. 1969, p-56.
6118. Wolke, R. L., Pitts. Univ., Contract AT(30-1)-2771, Jan. 1965-Jan. 1966, p-55.
6119. Wolke, R. L., and Sodd, V. J., Rev. Sci. Instr. 32, Dec. 1961, p-1415.
6120. Wollan, D. S., Nucl. Instr. and Meth. 119(2), 15 July 1974, pp-397-9.
6121. Wollan, D. S., 2nd Int. Conf. on Polar. Targ., Berkeley, CA, 30 Aug. 1971, pp-43-6.
6122. Wong, C. W., and Young, S. K., Nucl. Phys., A, 273(2), 30 Nov. 1976, pp-445-450.
6123. Wong, C., Anderson, J. D., et al., Phys. Rev., C, 12(6), Dec. 1975, pp-2115-2117.
6124. Wong, C. Y., ORNL, CONF-790743-5, 1979, p-20.
6125. Wong, E., Sheppard, D. M., et al., Nucl. Instr. and Meth. 129(2), 15 Nov. 1975, pp-537-542.
6126. Wong, V. K., Phys. Lett., A, ISSN 0375-9601, 73(5-6), 15 Oct. 1979, pp-398-400.
6127. Wong, W. H., and Quin, D. A., Nucl. Phys., A, 258(1), 16 Feb. 1976, pp-29-33.
6128. Woo, T. W., and Salaita, G. N., Trans. Am. Nucl. Soc. 28, June 1978, pp-91-92.
6129. Woodburn, E. H., Barrette, M., et al., Nucl. Instr. and Meth. 109(3), 1973, pp-561-63.
6130. Woodburn, E. H., Kundu, S., et al., Nucl. Instr. and Meth. 111(3), 1973, p-611.
6131. Woodruff, G. L., and Eccleston, G. W., Wash. Univ., Seattle (U. S. A.), Dept. of Nucl. Eng., Aug. 1976, p-18
6132. Woodworth, J. G., McNeill, K. G., et al., Can. J. Phys., 55 (19), 1 Oct. 1977, pp-1704-1715.
6133. Woodworth, J. G., McNeill, K. G., et al., Nucl. Phys., A, ISSN 0375-9474, 327(1), 17 Sept. 1979, pp-53-63.
6134. Woolfenden, J. M., Alberts, D. S., et al., Cancer ISSN 0008-543X, 43(5), May 1979, pp-1652-1657.
6135. Wormald, M. R., Underwood, B. Y., et al., Nucl. Instr. and Meth. 107(2), 1973, pp-233-5.
6136. Worth, J. H., Clark, P. A., et al., J. Brit. Nucl. Soc. 10(4), Oct. 1971, pp-329-33.
6137. Worth, J. H., Proc. of a Conf. held at St. Catherine's Coll., Oxford, 22-23 Sept. 1969.
6138. Worthington, J. N., Jedlowski, R. J., et al., INTDS Proc., Argonne (U. S. A.), Sept. 30-Oct. 2, 1975.

6139. Worthington, J. N., Jedlowski, R. J., et al., INTDS Proc., Argonne (U. S. A.), Sept. 30-Oct. 2, 1975.
6140. Wotke, H., Tortschanoff, T., et al., Acta Phys. Austr. 29, 1969, pp-183-9.
6141. Woude, A. van der, Meijer, R. J. de, et al., Rijksuniversiteit Groningen (Netherlands), Kernfysisch Versneller Inst., Annual Report, 1975, pp-20-21.
6142. Woude, A. van der, and Meijer, R. J. de, Nucl. Phys., A, 258(2), 23 Feb. 1976, pp-199-220.
6143. Wozniak, G. J., Glassel, P., et al., LBL-5075, 1975, pp-138-139.
6144. Wright, P. C., Storer, R. G., et al., Phys. Rev., C, 17 (2), Feb. 1978, pp-473-492.
6145. Wright, P. C., Storer, R. G., et al., Flinders Univ. of South Australia, Bedford Park, Inst. for Atomic Studies, June 1977, p-81.
6146. Wu, J. R., Chang, C. C., et al., Univ. of Maryland, College Park, ORO-4856-26, 1975, pp-360-361.
6147. Wu, J. R., and Chang, C. C., Maryland Univ., College Park (U. S. A.), Dept. of Phys. and Astron., ORO-5128-4, Nov. 1976, p-40.
6148. Wu, J. R., and Chang, C. C., Phys. Rev., C, 16(5), Nov. 1977, pp-1812-1824.
6149. Wu, J. R., and Chang, C. C., Phys. Rev., C, 17(5), May 1978, pp-1540-1549.
6150. Wu, J. R., Chang, C. C., et al., Phys. Rev., C, ISSN 0556-2813, 20(4), Oct. 1979, pp-1284-1300.
6151. Wullschleger, A., Phys. Rev. Lett. 40(10), 6 March 1978, pp-638-641.
6152. Wullschleger, A., and Scheck, F., Nucl. Phys., A, ISSN 0375-9474, 326(2-3), 10 Sept. 1979, pp-325-351.
6153. Wylie, W. R., Cybulska, E. W., et al., Cienc. Cult. 27(7), July 1975, p-40.
6154. Wylie, W. R., Bahnsen, R. M., et al., Nucl. Instr. and Meth. 79, 1970, pp-245-50.
6155. Wynchank, S., Nucl. Instr. and Meth. 100(2), 1972, pp-361-4.
6156. Wynchank, S., Nucl. Instr. and Meth. 93(1), 1971, pp-85-91.
6157. Yadav, H. L., and Geramb, H. V., Phys. Rev., C, 14(4), Oct. 1976, pp-1369-1380.
6158. Yaffe, L., Proc. of Conf. held at St. Catherine's Coll., Oxford, 22-23 Sept. 1969.
6159. Yaffe, L., Ann. Rev. Nucl. Sci. 12, 1962, pp-153-88.
6160. Yagi, K., Hendrie, D. L., et al., Phys. Rev., C, 14(1), July 1976, pp-351-353.
6161. Yagi, K., Hendrie, D. L., et al., Europ. Phys. Soc., Geneva (Switz.), Europ. Conf. on Nucl. Phys. with Heavy Ions, 6-10 Sept. 1976, p-37.
6162. Yamabayashi, H., and Shikata, E., Report JAERI-M-7972, 1978, p-75.

6163. Yamada, R., Proc. Int. Symp. High Energy Phys., 1973, pp-171-83.
6164. Yamaji, S., Scheid, W., et al., Z. Phys., A, 278(1), July 1976, pp-69-76.
6165. Yamamoto, S., Shimomura, Y., et al., J. Phys. Soc. Jpn. 48(3), 1980, pp-1053-4.
6166. Yamanaka, C., Yokoyama, M., et al., Osaka Univ., Inst. of Laser Eng., Jpn., 1976, p-15.
6167. Yamaya, T., Umeda, K., et al., AIP Conf. Proc. ISSN 0094-243X, 47, 1978, pp-712-13.
6168. Yamaya, T., Tohei, T., et al., Nucl. Instr. and Meth. 49, March 1967, pp-173-5.
6169. Yamazaki, Y., and Sheline, R. K., Phys. Rev., C, 14(2), Aug. 1976, pp-531-542.
6170. Yanabu, T., Bull. Inst. Chem. Res., Kyoto Univ. 47, March 1969, pp-154-61.
6171. Yang, K., Holstein, B. R., et al., Phys. Rev., C, 14(3), Sept. 1976, pp-1083-89.
6172. Yang, Yu-tung, Nucl. Instr. and Meth. 66, Dec. 15, 1968, pp-341-2.
6173. Yasue, M., Yokomizo, H., et al., J. Phys. Soc. Jpn. 42(2), Feb. 1977, pp-367-375.
6174. Yates, M. A., and Caretto, A. A., Jr., Carnegie-Mellon Univ., Pittsburgh, PA (U. S. A.), Dept. of Chem., Nucl. Chem. Rsch. of High Energy Nucl. Reactions at Carnegie-Mellon Univ., Prog. Report, 1974-1975, pp-41-46.
6175. Yavin, A. I., BNL-7534, pp-406-13.
6176. Yeh, T. R., and Lancman, H., Phys. Rev., C, 16(3), Sept. 1977, pp-1268-1270.
6177. Yissum Rsch. Dev. Co., Truman Bldg., Mt. Scopus, POB 20434, Jerusalem, Israel.
6178. Yntema, J. L., ANL, CONF-690301, pp-321-5.
6179. Yntema, J. L., Den Hartog, P. K., et al., INTDS Proc., Boston, U. S. A., Oct. 1-3, 1979.
6180. Yntema J., and Nickel, F., in Lecture notes in Phys. (ed. K. Bethge, Springer-Verlag, Berlin, Heidelberg, NY, 1978), 83, p-206.
6181. Yntema, J. L., IEEE Trans. Nucl. Sci., NS-23(2), 1976, p-1133.
6182. Yntema, J. L., INTDS Proc., ANL, U. S. A., Sept. 30-Oct. 2, 1975.
6183. Yntema, J. L., Nucl. Instr. and Meth. 122, 1974, p-45, ibid 113, 1973, p-605, ibid 98, 1972, p-379.
6184. Yntema, J. L., Nucl. Instr. and Meth. 113, 1973, p-605.
6185. Yntema, J. L., Nucl. Instr. and Meth. 98, 1972, p-379.
6186. Yntema, J. L., Conf. on Angular Correlations in Nucl. Disintegration, Delft (Netherlands), 17 Aug. 1970, pp-74-77.

6187. Yoh, J. W. A., Notre Dame Univ., IND (U. S. A.), Thesis (Ph. D), 1975, p-164.
6188. Yokosawa, A., ANL Report RL-73-88, 1973, pp-82-92.
6189. Yokosawa, A., Proc. of Meet. on Phys. with Polar. Targ. at High Energy, 1973, pp-82-92.
6190. Yokosawa, A., ANL/HEP-7208(Vol. 3), 1971, pp-1061-71.
6191. Yokota, H., and Suzuki, Y., AIP Conf. Proc. ISSN 0094-243X, 47, 1978, pp-574-575.
6192. Yonas, G., Nucl. Fusion 1978, (Pub. 1979), pp-125-33.
6193. Yoo, K., and Sternheim, M. M., Phys. Rev. Lett. 40(8), 20 Feb. 1978, pp-498-501.
6194. York, R. C., Iowa Univ., Iowa City (U. S. A.), Thesis (Ph. D), 1976, p-93.
6195. Yoshida, H., Nucl. Phys., A, 257(2), 2 Feb. 1976, pp-348-64.
6196. Yoshida, M., Kaneko, H., et al., Bull. Chem. Soc. Jpn. 49(6), June 1976, pp-1697-1700.
6197. Yoshida, T., and Maruyama, M., JAERI-Memo-4556, Aug. 1971, p-12.
6198. Young, G., Blum, E. B., et al., ANL/PHY-76-2(Vol. 2), May 1976, pp-849-855.
6199. Young, L., Anodic Oxide Films (London, Academic), 1961.
6200. Young, P. G., and Hale, G. M., BNL-NCS-50446, April 1975, pp-7-13.
6201. Young, P. G., and Stewart, L., LA-7932-MS, ENDF-283, July 1979, p-66.
6202. Young, S. K., and Gibbs, W. R., Phys. Rev., C, 17(2), Feb. 1978, pp-837-841.
6203. Youngblood, D. H., Moss, J. M., et al., Phys. Rev., C, 13 (3), March 1976, pp-994-1008.
6204. Youngblood, D. H., Rozsa, C. M., et al., Phys. Rev., C, 15(5), May 1977, pp-1644-1649.
6205. Youngblood, D. H., Rozsa, C. M., et al., Phys. Rev. Lett. 39(19), 7 Nov. 1977, pp-1188-1191.
6206. Yousef, M. I., and Reif, R., Zentralinstitut fuer Isotopen- und Strahlenforschung, Zentralinstitut fuer Kernforschung, Rossendorf bei Dresden (FRG.), Annual Report, 1976, pp-71-73.
6207. Yuasa, T., Prof. of the VII Int. Conf. on Few Body Prob. in Nucl. and Particle Phys, Delhi, India, 29 Dec. 1975-3 Jan. 1976, pp-181-185.
6208. Zaider, M., Ashery, D., et al., AIP Conf. Proc. 33, 1976, pp-307-309.
6209. Zaidins, C. S., Nucl. Instr. and Meth. 158(1), 1979, pp-237-41.
6210. Zajchenko, A. K., and Ol'khovskij, V. S., Ukr. Fiz. Zh. 21(7), July 1976, pp-1107-1111.
6211. Zapol'skij, N. A., Zelenko, K. P., et al., Erevanskij Fizicheskij Inst. (USSR), 1974, p-13.

6212. Zaritskii, A. R., Zakharov, S. D., et al., Zh. Eksp. Teor. Fiz., Pis'ma Red. 15(4), 20 Feb. 1972, pp-184-6.
6213. Zavattini, E., Muon Phys. Vol. II, NY, Academic Press, Inc., 1975, pp-219-261.
6214. Zegers, P., SORA Sci.-Exp. Programme, EUR 5080, 1974, pp-123-36.
6215. Zell, K. O., Dewald, A., et al., INTDS Newsletter, Feb. 1980.
6216. Zell, K. O., Gast, W., et al., Z. Phys., A, 292(2), 1979, pp-135-143.
6217. Zeller, A. F., Williams-Norton, M. E., et al., Phys. Rev., C., 14(6), Dec. 1976, pp-2162-2167.
6218. Zeller, A. F., Weisser, D. C., et al., 7th AINSE Nucl. Phys. Conf. 1978, p-64.
6219. Zerbst, H., Vakuum Tech. 12, 1963, p-173.
6220. Zheltov, K. A., Matakov, V. I., et al., Instr. Exp. Tech. 19(6) Pt. 1, Nov.-Dec. 1976, pp-1605-07.
6221. Zhivopistsev, F. A., and Rzhevskij, E. S., Izv. Akad. Nauk SSSR, Ser. Fiz. ISSN 0367-6765, 43(1), Jan. 1979, pp-183-185.
6222. Zhivopistsev, F. A., Vestn. Mosk. Univ. Fiz. Astron. (USSR), 16(1), Jan.-Feb. 1975, pp-16-21.
6223. Zhuchko, V. E., Tsipenyuk, Yu. M., et al., Yad. Fiz. ISSN 0044-0027, 28(5), 1978, pp-1170-1184.
6224. Zhuchko, V. E., Tsipenyuk, Yu. M., et al., Yad. Fiz. ISSN 0044-0027, 28(5), 1978, pp-1185-1194.
6225. Zhucko, V. E., Ostapenko, Yu. B., et al., Nucl. Instr. and Meth. 136(2), 15 July 1976, pp-373-378.
6226. Zhukova, O. A., Izv. Akad. Nauk Kaz. SSR, Ser. Fiz. Mat. 2, March-April 1971, pp-70-2.
6227. Zhusupov, M. A., Ibraeva, E. T., et al., Summ. of Reports of 26th Conf. on Nucl. Spectro. and Nucl. Structure, Baku, 3-6 Feb. 1976, p-225.
6228. Ziegler, J. F., Helium Stopping Powers and Ranges in All Elements, Pergamon Press, NY, 1977.
6229. Zimelev, A. G., and Kuz'min, R. N., Instr. Exp. Tech. 3, 1960, pp-498-9.
6230. Zimmerman, R. L., Boyd, R. M., et al., Georgia Inst. of Tech., Atlanta, Contract AT(40-1)-1346, April 1968, p-6.
6231. Zinamon, Z., Nardi, E., et al., Phys. Rev. Lett. 34(20), 1975, pp-1262-6.
6232. Zinsmeister, G., Vacuum 16, 1966, p-529; Thin Solid Films 2, 1968, p-497; 4, 1969, p-363; 7, 1971, p-51.
6233. Zint, P. G., and Mosel, U., Phys. Rev., C, 14(4), Oct. 1976, pp-1488-1498.
6234. Zisman, M. S., Cramer, J. G., et al., Symp. on Heavy Ion Phys. from 10-200 MeV/A, Upton, NY, U. S. A., LBL-9405, CONF-790743, 16-20 July 1979, p-12.

6235. Zisman, M. S., Cramer, J. G., et al., LBL-8151, 1978, pp-72-73.
6236. Zisman, M. S., Vandenbosch, R., et al., ANL/PHY-76-2(Vol. 2), May 1976, pp-857-866.
6237. Zlobin, V. G., Kolyada, V. M., et al., Instr. Exp. Tech. 17(4), 1974, pp-967-968.
6238. Znojil, M., J. Math. Phys. ISSN 0022-2488, 20(11), Nov. 1979, pp-2330-2333.
6239. Zolin, L. S., Nikitin, V. A., et al., BNL-tr-346, JINR-P13-3425, p-26.
6240. Zolin, L. S., Nikitin, V. A., et al., JINR-P-13-3425, 1967, p-20.
6241. Zoran, V., Berinde, A., et al., J. Phys., G, Nucl. Phys. ISSN 0305-4616, 6(1), Jan. 1980, pp-117-126.
6242. Zrelov, V. P., Kollarova, L., et al., Int. Conf. on Instrumentation for High Energy Phys., Dubna, USSR 8 Sept. 1970, pp-479-81.
6243. Zubarev, A. L., Irgaziev, B. F., et al., Yad. Fiz. ISSN 0044-0027, 29(4), April 1979, pp-912-918.
6244. Zuber, K., Helv. Phys. Acta 8, 1935, p-488.
6245. Zuber, K., Nature 136, 1935, p-796.
6246. Zuber, K., Helv. Phys. Acta 9, 1936, p-285.
6247. Zul'karneev, R. Ya., and Kutuev, R. Kh., JINR-R-1-9760, 1976, p-12.
6248. Zwieglinski, B., Saganek, A., et al., Nucl. Phys., A, 250 (1), 29 Sept. 1975, pp-93-105.

SUBJECT INDEX
HEADINGS

TARGET BACKING METALS
TARGET BACKING ORGANIC
TARGET BALL
TARGET BOX
TARGET BLOCK
TARGET CAPSULE
TARGET CARBIDES
TARGET CERAMIC (OR CERMET OR GLASS)
TARGET CHAMBER
TARGET CHARCOAL
TARGET CHEMICAL ANALYSIS
TARGET CLADDING
TARGET COMPLEX
TARGET CONCAVE
TARGET CONCEPT
TARGET CONICAL
TARGET CONSTRUCTION
TARGET CONTAINER
TARGET CONTAMINATION
TARGET CONTROL
TARGET CONVEX
TARGET COOLING
TARGET COOLING AIR
TARGET COOLING FLOW CONDITION
TARGET COOLING GAS
TARGET COOLING LIQUID NITROGEN
TARGET COOLING WATER
TARGET CORROSION
TARGET CRYOGENIC
TARGET CRYOGENIC EQUIPMENT
TARGET CYLINDRICAL
TARGET DAMAGE
TARGET DEFECT
TARGET DENSITY
TARGET DESIGN
TARGET DEUTERIUM
TARGET DEUTERATED
TARGET DIAMOND
TARGET DIMENSION
TARGET DISC
TARGET ECONOMY
TARGET ELEMENT-ACTINIUM (Ac)
-ACTINIDES
-ALUMINUM (Al)
-ALKALINE EARTH METALS
-AMERICIUM (Am)
-ANTIMONY (Sb)
-ARGON (Ar)
-ARSENIC (As)
-ASTATINE (At)

TARGET ELEMENT-NEON (Ne)
-NEPTUNTIUM (Np)
-NICKEL (Ni)
-NIOBIUM (Columbium-Nb)
-NITROGEN (N)
-NOBELIUM (No)
-OSMIUM (Os)
-OXYGEN (O)
-PALLADIUM (Pd)
-PHOSPHORUS (P)
-PLATINUM (Pt)
-PLUTONIUM (Pu)
-POLONIUM (Po)
-POTASSIUM (K)
-PRAESEODYMIUM (Pr)
-PROMETHIUM (Pm)
-PROTOACTINIUM (Pa)
-RADIUM (Ra)
-RADON (Rn)
-RARE EARTH
-RHENIUM (Re)
-RHODIUM (Rh)
-RUBIDIUM (Rb)
-RUTHENIUM (Ru)
-SAMARIUM (Sm)
-SCANDIUM (Sc)
-SELENIUM (Se)
-SILICON (Si)
-SILVER (Ag)
-SODIUM (Na)
-STRONTIUM (Sr)
-SULFUR (S)
-TANTALUM (Ta)
-TECHNETIUM (Tc)
-TELLURIUM (Te)
-TERBIUM (Tb)
-THALLIUM (Tl)
-THORIUM (Th)
-THULIUM (Tm)
-TIN (Sn)
-TITANIUM (Ti)
-TUNGSTEN (W)
-URANIUM (U)
-TRANSURANIUM ELEMENTS
-VANADIUM (V)
-XENON (Xe)
-YTTERBIUM (Yb)
-YTTRIUM (Y)
-ZINC (Z)
-ZIRCONIUM (Zr)

TARGET STORING
TARGET STRIPPER
TARGET STRUCTURE
TARGET SUBLIMATION
TARGET SUPPORTED
TARGET SURFACES
TARGET TEMPERATURE
TARGET TESTING
TARGET THEMRAL STRESSES
TARGET THICK
TARGET THICKNESS MEASUREMENT
TARGET THIN
TARGET TITANIUM DEUTERIDE
TARGET TITANIUM TRITIDE
TARGET TRANSPORTATION
TARGET TRITIUM
TARGET TRITIATED METALLIC
TARGET TRITIATED ORGANIC
TARGET UNUSUAL
TARGET VACUUM SYSTEM
TARGET VACUUM CHAMBER
TARGET VACUUM CONTAMINATION
TARGET VACUUM GETTER
TARGET VACUUM PUMPS
TARGET VACUUM ULTRA HIGH
TARGET VACUUM VESSEL
TARGET VAPOR
TARGET WATER
TARGET WEIGHING

SUBJECT INDEX

COUNTRY INDEX

PATENT INDEX

INDEX #	PATENT #	PATENT SOURCE
BELGIUM		
6449	665670	Guillaume, M.
6250	713308	Audenhove, J. Van
6251	771775	Nuken Gmbh
6252	777159	Union Carbide Corp.
6253	807540	Jongen, Y.
6254	810775	Com. L'Energie Atomique
6255	844248	Devienne, F.
6256	863415	General Electric Co.
6257	867986	Westinghouse Elec. Corp.
CANADA		
6258	553646	Ashley, R.W.
6259	739184	Curnow, E.
6260	758195	Watt, J.S.
6261	896622	Guillaume, M.
6262	957086	Ormrod, J.H.
6263	960377	France, D.W.
6264	968071	Mobil Oil Co.
6265	1003892	Schriber, S.O.
6266	3623130	Desmond, D.G.
CZECHOSLAVAKIA		
6267	131890	Svoboda, K.
EUROPEAN		
6269	3819	Kato, M.
FEDERAL REPUBLIC OF GERMANY (FRG)		
6270	1078245	Com. L'Energie Atomique
6271	1564024	Ges. Fuer. Kernfurschung MB
6272	1814888	Inst. Fuer Plasmaphysics
6273	1816130	Matsushita Elect. Ind. Co. L.
6274	1963957	Nukem Nuclear-Chemie U. Me.
6275	2009049	Nukem, Nuklear-Chemie Und.
6276	2029444	Nukem, Nuklear-Chemie Und.
6277	2053881	Nukem, Gmbh

PATENT INDEX

INDEX #	PATENT #	PATENT SOURCE
FEDERAL REPUBLIC OF GERMANY (FRG) (con't.)		
6278	2063261	Philips NV
6279	2102405	Commissariat A L'Energie
6280	2133671	Philips NV
6281	2262044	Nukem, Gmbh
6282	2331850	Nukem, Gmbh
6283	2353671	Wilke, W.
6284	2403349	Krimmel, E
6285	2533348	Atomic Energy Canada Ltd.
6286	2611314	Riedmuller, W.
6287	2655354	Seiler, G.
6288	2707390	Schulz, F.
6289	2726890	Langford, O.M.
6290	2734895	Dietrich, W.
6291	2749346	Fabian, H.
6292	2758975	Stock, E.
6293	2807374	Ebinger, H.
6294	2813964	Bensussan, A.
6295	2928663	Masnari, N.A.
6296	2941096	Colditz, J.K.E.
FRANCE		
6297	1502719	Latrous, H.
6298	1544089	Detaint, J.
6299	2070526	Debiesse, J.
6300	2076568	Roche, M.
6301	2081241	Pozwolski, A.E.
6302	2086724	Descleve, P.
6303	2156432	Commissariat A L'Energie
6304	2253255	Bargmann, H.W.
6305	2264581	Devienne, F.
6306	2297452	Comiss. Energie Atomique
6307	2309973	Comiss. Energie Atomique
6308	2335917	Comiss. Energie Atomiqie
6309	2344935	Air Industrie
6310	2422226	Romani, L.F.N.
GERMAN DEMOCRATIC REPUBLIC (GDR or DDR)		
6268	135951	Naehring, F.
JAPAN		
6311	1967-5315	Fujinaga, A.
6312	1965-7440	Kitagawa, Kazue
6313	1966-7755	Suganomata, S.
6314	1965-7960	Iwashita, F.
6315	1969-8480	Inoue, M.
6316	1966-9748	Ogawa, K.

PATENT INDEX

INDEX #	PATENT #	PATENT SOURCE
JAPAN (con't.)		
6317	1969-12797	Inoue, M.
6318	1968-13836	Toi, A.
6319	1966-15400	Adachi, T.
6320	1965-20831	Iwashita, F.
6321	1965-20832	Iwashita, F.
6322	1968-22792	Suganomata, S.
6323	1964-23459	Inou, Kazuo
6324	1975-26715	Yamaguchi, S.
6325	1964-27174	Suematsu, S.
6326	1964-30364	Iwashita, F.
6327	1974-36469	Fujibayashi, T.
6328	7011356	Fujitsu Co. Ltd.
6329	7018520	Tokyo Shibaura Electronic
6330	7021079	Tokyo Shibaura Elec. Co. Lt.
6331	50054796	Inst. Physical Ch. Re.
6332	1975-237	Shinohara, K.
6333	53135877	Tokyo Shibaura Elec. Ltd.
6334	54120399	Japan Atomic Energy Res.
6335	54129892	Hitachi KK
NETHERLANDS		
6336	6608385	European Atomic Energy Community
6337	6912098	Commissariat A L'Energie
6338	7013146	UK Atomic Energy Authorit
6339	7205034	Thomson-CSF SA
6340	7306027	Atomic Energy of Canada
6341	7707357	Philips Gloeilampen NV
6342	7800942	Yokogawa Electric WKS KK
SLOVAK		
6343	165865	Jesenak, V.
UNITED KINGDOM		
6344	736512	Holland, L.
6345	836348	Corning Glass Works
6346	974622	Redstone, R.
6347	978788	European Atomic Energy Community
6348	985286	Ault, D.J.
6349	1056938	Carpenter Steel Co
6350	1075411	NRA, Inc.
6351	1082940	Varian Associates
6352	1092768	Atomic Energy of Canada Ltd.
6353	1139209	Guillaume, M.
6354	1164780	Commissariat A L'Energie Atomique

PATENT INDEX

INDEX #	PATENT #	PATENT SOURCE
UNITED KINGDOM (con't.)		
6355	1203925	Auguin, B.
6356	1205359	Detaint, J.
6357	1231396	Perry, K.E.G.
6358	1243262	Stark, D.S.
6359	1259997	Commissariat A L'Energie Atomique
6360	2000634	Tronc, D.
6361	2006176	Whittaker, B.
6362	2010795	Chen, C.L.
6363	2024186	Shimada, T.
6364	2025683	Solomon, D.E.
UNITED STATES OF AMERICA (USA)		
6365	340863	Blue, J.W.
6366	2549596	Hamilton, G.J.
6367	2798178	Heard, G.H.
6368	2847331	Ashley
6369	2868987	Martin, D.M.
6370	2885584	Graaf, J.R. Van de
6371	2964710	Stone, F.K.
6372	3167655	Redstone, R.
6373	3179634	
6374	3183356	Cherubini, J.H.
6375	3219200	Ainsworth, A.
6376	3276969	Antal, E.F.
6377	3287592	Hirschfield, J.J.
6378	3320422	St. John, E.A.
6379	3324540	Lotts, A.L.
6380	3349001	Stanton, R.M.
6381	3360647	Avery, R.T.
6382	3388359	Lambertson, G.R.
6383	3409413	Burns, R.E.
6384	3431502	Stone, K.F.
6385	3435232	Sorenson, H.D.
6386	3440420	Attix, F.H.
6387	3448314	Bounden, J.E.
6388	3500098	Atomic Energy of Canada Ltd.
6389	3501377	Lindgren, J.R.
6390	3510270	United States Atomic Energy
6391	3525228	Anderson, R.L.
6392	3546575	Jeffries, C.D.
6393	3571594	Varian Associates
6394	3581093	Kaman Sciences Corp.
6395	3586744	Triggiani, L.V.

PATENT INDEX

INDEX #	PATENT #	PATENT SOURCE
UNITED STATES OF AMERICA (USA) (con't.)		
6396	3586745	Triggiani, L.V.
6397	3586746	Triggiani, L.V.
6398	3594275	Neutron Products Inc.
6399	3622830	Clay, H.S.
6400	3623130	Dalrymple, D.G.
6401	3540597	Varian Assoc.
6402	3694313	Blue, J.W.
6403	3698996	Cooperstein, R.
6404	3725663	Sanders Nuclear Corp.
6405	3746859	U.S. Atomic Energy Com.
6406	3766389	Fabian, H.
6407	3775216	Schlumberger Technology
6408	3800145	Mobil Oil Corp.
6409	3805081	Barthel, R.
6410	3860827	Cranberg, L.
6411	3864640	Bennett, W.H.
6412	3892970	U.S. Energy R & D Admin.
6413	3899681	U.S. Atomic Energy Com.
6414	3924137	Alger, L.D.
6415	3925676	Atomic Energy of Canada Ltd.
6416	3940318	Arino, H.
6417	3955093	Quinby, T.C.
6418	3957945	Grant, P.M.
6419	3963934	Ormrod, J.H.
6420	3966547	Blue, J.W.
6421	3968377	Radiation Dynamics Inc.
6422	3968378	U.S. Sec. of the Army
6423	3969631	U.S. Energy R & D Admin.
6424	3971697	Blue, J.W.
6425	3981769	Beaudry, S.B.
6426	3993538	Greene, M.W.
6427	3993910	Parkin, D.M.
6428	3999096	Funk, L.W.
6429	4017578	U.S. Energy R & D Admin.
6430	4021280	U.S. Energy R & D Admin.
6431	4034032	U.S. Energy R & D Admin.
6432	4038125	U.S. Energy R & D Admin.
6433	4063098	Haimson Res. Corp.
6434	4068495	Alger, D.L.
6435	4069457	Martin, R.L.
6436	4076990	Hendry, G.D.
6437	4088532	Blue, J.W.
6438	4092381	Halpern, G.M.
6439	4106982	Christman, D.R.

PATENT INDEX

INDEX #	PATENT #	PATENT SOURCE
UNITED STATES OF AMERICA (USA) (con't.)		
6440	4109612	Mayer, A.
6441	4118042	Booth, R.
6442	4118627	Porter, G.D.
6443	4119858	Cranberg, L.
6444	4154868	Woerner, R.L.
6445	4178196	Nonaka, Y.
6445a	4190016	Hendricks, Ch. D.
6446	829315	Pat-Appl-Alger, D.L.
6447	880677	Pat-Appl-Sahlin, H.L.
UNION SOVIET SOCIALIST REPUBLIC (USSR)		
6448	245214	Biophysics Inst. Min. of He.
6449	288178	United Nuclear Res. Inst.
6450	301882	Malyshev, I.F.
6451	315106	Artemev, Yu. M.
6452	335635	Polar Kola Br. Sm. Kirov. Ge.
6453	353282	Rodin Am. Surenyants Vv. Si.
6454	387541	Kalmykov, A.A.
6455	392829	Refractory Metal Re.
6456	430740	Khalin, N.F.
6457	456344	Gusev, O.A.
6458	459169	Komarov, U.L.
6459	468556	Khalin, N.F.
6460	470210	Barkov, B.P.
6461	486292	Kiselevich, I.L.
6262	486629	Komarov, V.L.
6463	488122	Bajramashvili, I.A.
6464	489438	Knyazev, V.A.
6465	496757	Levin, V.I.
6466	497824	Lebedev. N.A.
6467	512510	Tomsk Poly. Eltrn.
6468	534718	Nuclear Res. Inst.
6469	555786	Dyachkov, B.A.
6470	569196	Tikhonov, V.I.
6471	580000	Domanov, V.P.
6472	599619	Gorodkov, Yu. V.
6473	712873	Gerasimenko, N.N.